World Sustainability Series

Series Editor

Walter Leal Filho, European School of Sustainability Science and Research, Research and Transfer Centre "Sustainable Development and Climate Change Management", Hamburg University of Applied Sciences, Hamburg, Germany

Due to its scope and nature, sustainable development is a matter which is very interdisciplinary, and draws from knowledge and inputs from the social sciences and environmental sciences on the one hand, but also from physical sciences and arts on the other. As such, there is a perceived need to foster integrative approaches, whereby the combination of inputs from various fields may contribute to a better understanding of what sustainability is, and means to people. But despite the need for and the relevance of integrative approaches towards sustainable development, there is a paucity of literature which address matters related to sustainability in an integrated way.

Notes on the quality assurance and peer review of this publication

Prior to publication, the works published in this book are initially assessed and reviewed by an in-house editor. If suitable for publication, manuscripts are sent for further review, which includes a combined effort by the editorial board and appointed subject experts, who provide independent peer-review. The feedback obtained in this way was communicated to authors, and with manuscripts checked upon return before finally accepted. The peer-reviewed nature of the books in the "World Sustainability Series" means that contributions to them have, over many years, been officially accepted for tenure and promotion purposes.

Pardeep Singh · Shikha Daga · Kiran Yadav
Editors

Integrating Environmental Sustainability and Corporate Social Responsibility: A Move Beyond Profit

Corporate Social Responsibility for Environmental Sustainability

Springer

Editors
Pardeep Singh
Department of Enviornmental Studies
PGDAV College
University of Delhi
New Delhi, Delhi, India

Shikha Daga
Department of Commerce
PGDAV College
University of Delhi
New Delhi, Delhi, India

Kiran Yadav
Department of Commerce
PGDAV College
University of Delhi
New Delhi, Delhi, India

ISSN 2199-7373 ISSN 2199-7381 (electronic)
World Sustainability Series
ISBN 978-3-031-81468-6 ISBN 978-3-031-81469-3 (eBook)
https://doi.org/10.1007/978-3-031-81469-3

This Springer imprint is published by the registered company Springer Nature Switzerland AG
The registered company address is: Gewerbestrasse 11, 6330 Cham, Switzerland

Contents

"Harmonizing Business and Planet: Unveiling the Nexus of Corporate Social Responsibility, and Environmental Sustainability, for a Symbiotic Tomorrow"

Mahua Basu and Atanu Basu

Abstract In the times when environmental degradation threatens the very foundation of our existence, harmonizing business practices with planetary well-being becomes imperative. This chapter looks into the interconnectedness between Corporate Social Responsibility (CSR) and environmental sustainability, illuminating the possibility of a symbiotic affiliation in determining a sustainable tomorrow. CSR has shifted to a premeditated necessity for companies from just a humanitarian effort. CSR comprises a positive vow to ethical conduct, engagement of stakeholders and sustainable practices along with acquiescence to rules and regulations. With corporations identifying their role as universal citizens, CSR appears to be more of a persuasive instrument to foster social equity and environmental stewardship. The well-being of the current and future generations is guaranteed by environmental sustainability which can be achieved through actions aimed at preserving ecological integrity. An increase in the loss of biodiversity and exhaustion of resources along with unpredictable climate change and other issues makes sustainability a critical concern for both businesses and society in general. Such sustainable approaches and practices bestow economic benefits, boost the brand image and repute of the businesses, and endorse novelty and resilience besides mitigating harmful environmental impacts. The field of transformative action lies at the junction of CSR and environmental sustainability. Corporations can incorporate environmental issues in their CSR draft and catalyze positive environmental results while simultaneously dealing with social needs. This synergy can extend and move beyond boundaries in the form of collaborations that involve governments, communities and other stakeholders. Thus harmony between business practices and planetary well-being is the fundamental imperative for creating a sustainable future.

M. Basu (✉)
St. Xavier's College (Autonomous), Kolkata, India
e-mail: mbasu@sxccal.edu

A. Basu
Chief, Inland Waterways and Multimodal Logistics, TM International Logistics Ltd. (a subsidiary of Tata Steel Ltd.), CSR and Sustainability, Kolkata, India

P. Singh et al. (eds.), *Integrating Environmental Sustainability and Corporate Social Responsibility: A Move Beyond Profit*, World Sustainability Series,
https://doi.org/10.1007/978-3-031-81469-3_1

Keywords Corporate Social Responsibility (CSR) · Environmental sustainability · Climate change · Stakeholders · Collaboration

1 Introduction

Over the past few years, all over the world, businessmen, industrial practitioners, policymakers, academicians, and others have agreed that the environment is degrading owing to the goods they manufacture in their organizations. The influence of such activities was not acknowledged earlier. Manufacturing industries contribute considerably to the global problems of climate change, air and water pollution, rising water, resource depletion, and use of hazardous substances (Kraus et al. 2020). Stakeholders started exerting severe pressure on the organizations to reduce their influence and impact on the environment (Yu et al. 2018).

In the field of modern business, a transformative story is unfurling- a story that surpasses profit margins and shareholders' value. This narrative speaks of the responsibility of the corporations that realize their role as stewards of economic development along with the well-being of society and the environment. Hence this tale begins with an insightful perception of Corporate Social Responsibility (CSR) and environmental sustainability, which forms the two towering pillars underpinning a new paradigm of ethics and conduct. Individually social organization, economic development and environmental consciousness are not new in concept and practice, but combining these three issues and dealing with these in a unified manner is imperative (Hernández et al. 2020) and challenging. Though most of the studies focused on Corporate Social Responsibility (CSR) driven financial performance (Ali et al. 2020), little attention was given to CSR-driven environmental issues. Yet several studies revealed the positive impact of CSR on organizational performances (Javed et al. 2020).

The Berle-Dodd debate has caught public notice of the issue of Corporate Social Responsibility (CSR). While Berle opined that firms should focus on maximizing shareholders' value, Dodd on the other hand backs the opinion of firms held accountable to stakeholders and people in general. Currently, Corporate Social Responsibility (CSR) is held accountable to stakeholders (employees, creditors, consumers, environment, society, etc.) besides the shareholders (Hong et al. 2020). As CSR is gaining significance, its performance shows considerable national and regional differences that are further influenced by social, political, and economic factors (Baughn et al. 2007). Corporate Social Responsibility (CSR) can be explained as a combination of activities targeted to achieve several social good ahead of the unequivocal interests of the enterprises that are not necessary by law (McWilliams and Siegel 2000). Vogel (2007) asserts CSR to be authentic and substantive. It needs to be sincerely considered by everyone fascinated with political affairs, business, and others. According to him, the Corporate Social Responsibility (CSR) concept comprises "*activities associated with corporate virtue typically represent firms' efforts to do more to address a wide variety of social problems than they would have done in the course of their normal pursuit of profits"*. CSR according to the World Business for Sustainable Development (WBCSD), 2000 is *"the continuing commitment by business to behave ethically and contribute to economic development while improving the quality of life of the workforce and their families as well as local community and society at large"* (Najam 2013). CSR comprises a set of management practices to ensure that companies make the most of the positive impacts of their operations on society. The operations call for meeting and sometimes exceeding legal, ethical, commercial, and public expectations (Jamali and Mirshak 2007). CSR at its core reflects social essentials and social consequences of business triumph. It comprises clearly expressed and conversed policies and practices of the firms reflecting their business responsibility for broad societal good (Matten and Moon 2008). The CSR centres on the concept of business organizations transiting from a condition of simple compliance to a form of engagement; from minimizing impairment to creating value (Luetkenhorst 2004). CSR thus is built on a vigorous appreciation of the businesses' role as an active partner in the world of paucity and diminishing resources. The concept of CSR was not adopted uniformly; diverging opinions remain regarding its impending use and application.

1.1 Literature Review and Genesis of CSR

The beginning of social elements in corporate behaviour can be traced back to the early Roman laws. The state laws acknowledged discrete groups with separate identities with strong social facets organized for social purposes. Such components were recorded in asylums, hospitals, orphanages, old age homes, and others. Corporations or 'corpus' served religious, charitable, and social purposes rather than business purposes in ancient times (Chaffee 2017; Kaittani 2023). A similar opinion on some centuries-old business concerns on society was also opined by Carroll. The business managers started to presume the accountability of striking an equilibrium between profit maximization and client demands, labour groups, and community (Carroll 2008). Hence managers were regarded as trustees for the diverse set of external relations with the corporation, which in turn transformed into socio-economic responsibilities taken up by the companies (Carroll 2008; Heald 1991). Philanthropy appeared in the late 1800s when benefits were provided to immediate stakeholders rather than the general public (Wren 2005). A few early instances of corporate social responsibility were found in books like 'The Functions of the Executive' (Barnard 1968) and the Social Control of Business (Clark 1926). The roles of executives and the social performance of business organizations started to appear in literature only in the 1930s and 1940s (Carroll 2015a). The religious roots of the Victorian philanthropists inculcated a higher level of idealism and humanism. By the late 1800s, philanthropic endeavours were directed at the working group. Welfare schemes were thus created both in Europe and the USA (Harrison 1966). The late 1800s and early 1900s showed a paternalistic approach in the welfare schemes that aspire to protect and retain employees; a few companies even initiated to improve their quality of lives (Carroll 2008; Heald 1991). Heald referred to the social sensitivity of the businessmen. He stated about the R. H. Macy Company in USA that have contributed funds to orphan asylum in 1875 and labeled their donation as miscellaneous expenses in the company record by 1887. He also mentioned about Pullman Palace Car Company which established a model industrial community in 1893 to enhance the quality of life of its employees. The increase in industrialization and urbanization during this period was translated into large-scale productions that led to fresh apprehensions about the labour markets. The concerns can range from the new challenges the farmers and small firms were likely to confront to continue with the new interdependent economy, the creation of workers' unions for improved working conditions, or middle-class anxiety about losing religious and family values amidst the new industrial society. As a response to such challenges, and to achieve harmony between firms and employees, several business leaders formed organizations to promote values and improve working conditions. Take for example the Civic Federation of Chicago that was formed to improve working conditions and where religious ideals merge with economic goals along with a feeling of civic pride (Heald 1991).

It was only in the 1950s that the idea of specifically defining those responsibilities was addressed in literature for the first time marking the start of the contemporary construct of CSR, even though it was mostly referred to as social responsibility. The time after the second World War period as well as 1950 was thought of as a time of adjustment and altering approaches towards the dialogue of CSR (Carroll 2008). Bowen (2013) believed that big organizations accumulated great power and their measures had a substantial societal impact. The author felt the need for such firms to consider their impacts and include them in their decision-making, one of the noteworthy instances of changing attitudes towards corporate behaviour. Bowen forwarded a specific set of principles for business organizations towards the fulfilment of their social responsibilities. Concerns about corporate behaviour were seen in the works of authors after Bowen as well. Eells, in his book 'Corporation Giving in a Free Society', argued that big companies of that time were not bearing their responsibility in times of generalized inflation (Eells 1956). Selekman (1959), in his book, 'A moral philosophy for management' investigated the moral task of the firms concerning labour expectations of the time. Books like the "Silent Spring" by Rachel Carson in 1962 and the Population Bomb by Paul Ehrlich in 1968 raised questions about the limits of economic growth as well as the societal and company impact on the environment (Carson 2009; Carson and Darling 1962; Ehrlich 1968).

The 1960s witnessed a new social context that is noticeable by a culture of growing protest centering on civil rights and anti-war protests. CSR was still a new concept. These protests turned into a more radical political activism from students-led sit-ins, walk-outs, and rallies in the United States. They envisaged business houses as a fundamental part of the establishment that they aspire to transform. Their protests exerted enough pressure on establishments such as banks, financial institutions and large corporations and had a greater focus on those establishments having direct relations to war. Consider the instance of Dow Chemical Company which manufactured napalm employed in the Vietnam War (Waterhouse 2017). Keith Davis pointed out the pressure on businessmen to revisit their role in society in the context of significant social, economic, and political changes. He referred to the relevant societal obligation of the businessman in terms of economic and human values. He also emphasized the linkage of social responsibility to the economic returns of the corporations (Davis 1960). The initial half of the 20th Century was seen as a period of intellectual and institutional change that altered economic and social thinking and fetched increased economic supremacy to large business organizations. Frederick proposed business responsibility theory with five requirements to equilibrate the power of businessmen. The requisites are: 1. having a criterion of value, 2. to be founded on the recent concepts of management and administration, 3. to recognize historical and cultural customs behind the present social circumstance, 4. identifying the attitude of respective businessmen as a function of their task in the society, and 5. to recognize that responsible business attitude is not automatic but develops with purposeful and mindful efforts (Frederick 1960). A review of the development of business institutions showed visible change in the scale and types of organizations, public policies, regulatory controls, and social and economic settings. McGuire stated that a corporation's accountability reaches beyond legal and economic obligations and that they should take interest in politics, social welfare, education and happiness of its employees (Wood and Cochran 1992). Walton investigated the ideological changes that took place between the 1950s and 1960s and further reflected in public policies, corporations were considered as potential contributors to improve the social and economic conditions of the time (Walton 1982). In 1962, Milton Friedman, a Nobel Laureate in economics forwarded a specific viewpoint of the corporations' role in a free capitalist system where the corporations should be restricted to the quest for economic benefits. According to him, CSR activities involve improper use of the firm's resources that would result in untenable expenditure of currency for broad social interest (Friedman 2007).

The business sector confidence experienced a fall in 1969 owing to an oil spill on the coast of Santa Barbara, California. The massive protests across the USA ultimately resulted in the declaration of Earth Day. On this Earth Day in 1970, 20 million people joined protests. They demanded a clean and sustainable environment, and fight against corporate-led pollution like oil spills, toxic waste dumping, etc. This Earth day influenced the US political agenda significantly and hard-pressed for the establishment of the Environmental Protection Agency (EPA) by the end of 1970 that framed new regulations to influence the corporate attitude and render supplementary responsibilities for them (Day 2018). Along with it other organizations like the Consumer Product Safety Commission (CPSC), the Equal Employment Opportunity Commission (EEOC), and the Occupational Safety and Health Administration also formalized business responsibilities in the context of social concerns (Carroll 2015b). The pyramid of CSR for sustainable marketing developed by Archie Carroll in 1979 is worth mentioning as this framework proves to stand the test of time. The structure exhibits the levels of responsibility that a business can accomplish as it nurtures and expands its sustainable practices. It serves as a four-tier guide for self-regulation. The base level represents economic responsibility, the second level exhibits legal responsibility, the third level to ethical responsibility, and the topmost level implies philanthropic responsibility (Carroll 1991).

Ronald Reagan and Margaret Thatcher introduced a new line of thought into politics during the 1980's. To reduce soaring levels of inflation in both the USA and the UK, they emphasized reducing pressure on corporations (Feldstein 2013). According to them, their nations' growth and economic strength were dependent on their capacity to retain a free-trade environment with very little state interference (Pillay 2015). Reagan intended to reduce regulations on the private sector complemented with tax diminution (Feldstein 2013). In the 1980's Thomas M Jones, an important contributor to the CSR theory was the first author to regard CSR as a decision-making course that persuades corporate behaviour. According to him CSR is a process and should not be taken as a set of outcomes. His perception gave rise to a new era of the debate centering around CSR where more emphasis was given to operationalization over the concept, leading to the creation of new frameworks, models and methods to evaluate CSR from an operational angle (Strand 1983). Strand developed a model to exhibit the link between the firm and its social responsibility, responsiveness and responses (Strand 1983). Such concerns were observed in a sequence of events which the approachability of the global community towards sustainable development, the most crucial being the European Commission's Environment Directorate-General in 1981, the World Commission on Environment and Development (WCED) in 1983, the Chernobyl nuclear disaster in 1986, publication of "Our Common Feature", the Montreal Protocol in 1987, and the formation of the Intergovernmental Panel on Climate Change (IPCC) in 1988.

The concept of globalization and CSR during the 1990s influenced the global standpoint towards social responsibility and sustainable development. The creation of internationals bodies and events like the Rio Declaration and the Rio Earth Summit of 1992, the European Environment Agency (1994), and the United Nations Convention on Climate Change (UNFCCC) as well as the Kyoto Protocol (1997) represents the adoption of global agreements represent global efforts to set high standards with respect to climate change issues and thus corporate behaviour. Corporations and MNCs comprehended that being socially responsible offers a potential safe path and also equilibrates globalization challenges and opportunities. The establishment of Business for Social Responsibility (BSR) in 1992, including 51 companies initially became a force that would become a force for positive social change. Thereafter BSR was relaunched in 1994 and focused on working with businesses to create a sustainable world. (https://www.bsr.org/en/about/story). The European Commission (EC) played a significant role in encouraging the implementation of CSR that started in 1995 with twenty business leaders adopting the European Business Declaration against Social Exclusion to fight social segregation and unemployment (Latapí Agudelo et al. 2019). Proponents of CSR believe that such activities pay off both for the company as well as for the stakeholders and society. The CSR-led value creations of the firms were evaluated under five strategic dimensions centrality, specificity, pro-activity, voluntarism, and visibility (Burke and Logsdon 1996). See table 1 for a summary of the timeline of events leading to the evolution of CSR.

Table 1 A summary of the timeline of events leading to the evolution of CSR

Year	Significant incidents and changes
1953	Mentioned about social responsibility of businessmen (Bowen 2013)
1960	Commensuration of social accountabilities of businessmen with their social power (Davis 1960)
1962	" Silent Spring" published by Rachel Carson
1963	Firms to take interest in politics, social welfare of community; education and happiness of its employees (Wood and Cochran 1992; Wood and Cochran 1992; Wood and Cochran 1992; Wood and Cochran 1992; Wood and Cochran 1992; Wood and Cochran 1992; Wood and Cochran 1992; Wood and Cochran 1992; Wood and Cochran 1992; Wood and Cochran 1992; Wood and Cochran 1992)
1968	Population Bomb by Paul Ehrlich published (Ehrlich 1968)
1970s	Formation of EPA in the USA
1971	In the USA, the Committee for Economic Development (CED), established in 1942, published "Social responsibilities of business corporations"
1972	"The limits to growth" published (Meadows et al. 2013)
1979	A new definition and understanding of CSR was developed and published in 1991 (Carroll 1979, 1991)
1980	CSR as a decision making process (Jones 1980) Ronald Reagan and Margaret Thatcher emphasized reducing pressure on corporations
1983	WCED established
1986	The Chernobyl Nuclear disaster occurred
1987	Brundtland Commission presents "Our Common Future" Montreal Protocol adopted
1988	IPCC formed
1991	(Carroll 1991)
1992	Rio Declaration under UNCED, UNFCCC formed Business for Social Responsibility launched
1996	Evaluation of the benefits of the strategic implementation of CSR (Burke and Logsdon 1996)
1997	Kyoto Protocol adopted
2000	UNMDGs and UN Global Compact launched
2001	linking the term "strategic" to CSR (SCSR) (Lantos 2001)
2002	The European Commission launched the European Strategy on CSR
2005	The European Commission launched the European Roadmap for Businesses Identifying a social responsibility in which CSR passes from a minimal commitment to strategic necessity (Chandler and Werther 2014)
2006	Exploration of SCSR to create value (Roszkowska-Menkes and Strategic 2021)
2007	SCSR is to be implemented with the aim of creating value and generating new opportunities linked with social demands (Husted and Allen 2007)
2008	7 principles guiding SCSR defined that is directed towards creating shared value (Heslin and Ochoa 2008)
2010	Introduction of ISO26000

(continued)

Table 1 (continued)

Year	Significant incidents and changes
2011	The objective of a corporation must be redefined as creating shared value (Roszkowska-Menkes and Strategic 2021) The European Commission revised and published a new European Strategy for CSR (2011–2014)
2015	UNSDGs and Paris Agreement
2016	Shift of the early SCSR definition towards sustainable value creation (Chandler 2020)
2018	Mandatory disclosure of non-financial information by large corporations (Szabó and Sørensen 2015)
2020	Post-pandemic, CSR shifted focus from sustainability and climate action to how the corporations are treating their employees and community during crisis

The United Nations had initiated key proposals to woo the private sector. Rio + 5 in 1997 witnessed a round table meeting between the business, government, and NGOs organized by the UN General Assembly President and the World Business for Sustainable Development (WBCSD). To recall, the World Business for Sustainable Development (WBCSD) was launched in 1995 by an amalgamation of the Business Council for Sustainable Development (BCSD) formed out of the 1992 Earth Summit and the World Industry Council for the Environment (WICE) established as a project of the International Chamber of Commerce. Mr. Kofi Annan, the Secretary-General, in 1999 sought a '*global compact of shared values and principles*' amongst the business leaders and an international body at the World Economic Forum in matters of environment. This led to the formation of the Global Sustainable Development Facility-2B2M (2 billion people to Market by 2020 program). The World Conservation Union (WCU), World Resources Institute (WRI), and Institute of Environment and Development (IIED) have marched forward to look for corporate partnerships along with universities like MIT and Cambridge University persuading the World Business for Sustainable Development (WBCSD). The WBCSD became one of the prominent business voices on sustainable development (Najam 2013). Already, the developed nations possess adequate resources to adopt measures and implement them to achieve environmental homeostasis. The USA and EU developed extensive environmental standards. But for developing countries, environmental issues feature low in their priority list. But they are the sites where most of the industrial manufacturing and processes take place (Shahzad et al. 2020).

Besides the momentum from the United Nations, the lineage of Corporate Social Responsibility (CSR) goes back to the beginning of the twentieth century, a period when corporate companies were perceived as profit-driven entities hardly with any societal obligations. The CSR notions were born with the evolution of societal expectations. Businesses acknowledging their impact on the environment, communities, employees, and other wider aspects of society marked the dawn of CSR. The evolutionary path of CSR exhibits a transition from a mere philanthropic approach to a comprehensive strategy that integrates values and social and environmental contemplations into the heart of business operations.

Currently, Corporate Social Responsibility (CSR) encompasses a triple bottom line- profit, people, and planet. It entails a commitment to ethical business conduct, revering human rights, and taking an active part in addressing social and environmental issues. This chapter probes into the history and evolution of Corporate Social Responsibility (CSR), explores its philosophical underpinnings, and illuminates the key components of a robust Corporate Social Responsibility (CSR) stratagem.

1.2 CSR Drivers

The answer to why companies need CSR activities lies in the CSR **drivers**. (See Fig. 1) The Non-governmental Organizations (NGOs) and the common people have strong powers. Corporations often change their policies based on external demands from the NGOs. In this respect, the NGOs exerting visible impact are the Global Reporting Initiative (GRI) and the United Nations Global Compact. The Global Reporting Initiative (GRI) sets the standards for reporting. Apart from the NGOs, CSR is driven by other stakeholders. The corporate social contract is another driver under which everyone is obligated to respect the law, act conscientiously, and give up some of the self-interests that compromise the general good. The companies manufacture products that society needs and pay fair taxes. In this context, only tax cannot suffice for corporate social contracts. Companies must help solve social issues. Protecting the natural environment can often add to cost savings, such as it can reduce the amount of water usage. Philanthropy is another CSR driver like promoting decent work conditions, improving living standards, eradicating poverty and hunger, and others. Different States prioritize differently. While Italy focuses on a safe workplace, Norway on establishing core labour standards, and Switzerland emphasizes eliminating salary differences among genders (KsiężaK and FischBach 2017).

1.3 CSR as a Business Strategy

The weight of Corporate Social Responsibility (CSR) as a business strategy becomes prominent in its relations with stakeholders like customers, business partners, suppliers, employees, and NGOs. Building trust and dependability is the foundation of building good relations. The broad scope of the CSR domain allows the company to select any or many fields of activities in tune with the company's mission. As a result, many CSR consulting companies were emerging with an increase in demand for their services. Companies now hire such consultants to prioritize their social efforts suitably. Three types of social issues may be categorized depending on the company's operation. The first types are the social issues not much affected by the company's operations. The second type includes those social concerns that are extensively impinged on by the businesses' operation. The third category comprises those

Fig. 1 CSR drivers

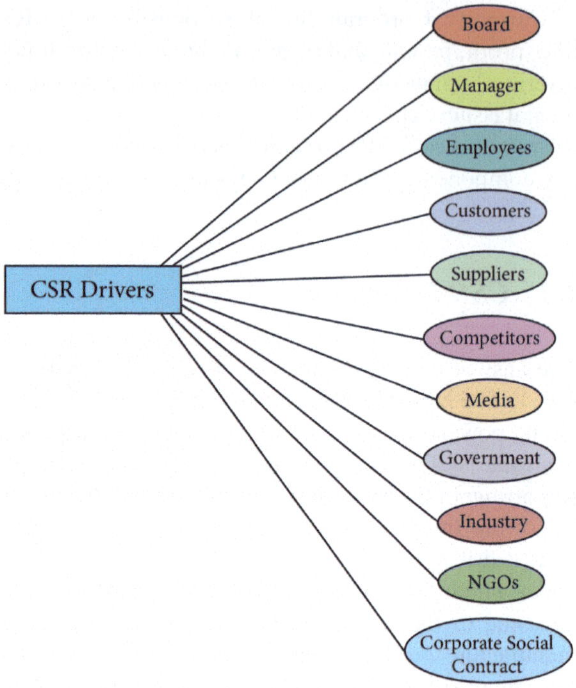

social issues in the external environment that can radically impact the drivers that underline the company's competitiveness in places where it operates (KsiężaK and FischBach 2017).

Many voiced for legitimizing CSR, particularly in the fields of environment such as CO_2 emissions and compliance with international human rights bills. In several countries, CSR is not legislated and any activities to the social problems seem voluntary. Since 2018 in Europe, the CSR regulations have become more stringent with the implementation of non-financial reporting complying with the 2014/95/EU directive. Companies hiring over 500 employees are obligated to this reporting and approximately 6000 companies including, banks, insurance, and others fall within this category (KsiężaK and FischBach 2017).

2 The Changing Perspectives of CSR: From Obligation to Opportunity with Sustainability Concept

Sustainability can be achieved from multiple dimensions. Here sustainability is discussed within the business context. All business operations exert a considerable impact on the environment. The environment comprises air, water, land, living creatures, our surroundings, and all conditions necessary for our existence. Sustainability in business is believed as a dynamic strategy for finding a greener choice in the corporation's environment. Corporate Social Responsibility (CSR) forms a practical beginning to establish sustainable goals that corporations can adopt and implement. As sustainability is understood as a necessary element in business, corporations' focus goes beyond consumer, socio-political, and ethical perspectives and addresses sustainable goals. These are beyond organizational goals. The CSR principles can be relevant to each and every type of business model and a responsible and sustainable move is not defined only by materials but the entire environment. A prerequisite for attaining sustainable goals lies in the provisioning of considerable information transfer to society. It seems quite difficult to sell an expensive product to customers without explaining to them the reason for extra payment. Similarly, implementing sustainable values in a business requires its employees to recognize these values as well. The public at large should also be acquainted with the values of sustainability. Therefore, it seems exceedingly important to boost business motivation (Demjanovičová and Varmus 2021).

In 1972, the United States enforced five legislations, the year the United Nations held the Stockholm Conference (UNCHE). Almost a year later, the Chipko movement against deforestation echoed. Other events that sparked in the later part cried for environmental responsibility. People demanded open transparent communication from industry and government about environmental risks. The Love Canal disaster, the Bhopal gas leakage, and Exxon Valdez oil spill, Chernobyl nuclear disaster all happened almost at the same time. Besides the environment, the Apartheid racial segregation policy of South Africa was also a growing concern. This was under constant attack from civil rights activists, and religious and student activists. The South African police firing on students at Soweto added to the fire. Investment from Europe and the USA began to drop in the firms that deferred to recognize human rights and equal opportunity in South Africa. With the sowing of the seeds of Apartheid's demise, socially responsible investing began to find a new meaning (Blackburn 2012).

All these events served as a foundation for the World Commission on Environment and Development (WCED) that published 'Our Common Future' and defined sustainable development. "..... *a development that meets the needs of the present without compromising the ability of the future generations to meet their own needs*". The 27 principles of 'Rio Declaration' not only linked environmental protection with economic development but also included social issues like peace, poverty, women empowerment, and others to attain sustainability.

In a world full of uncertainty, monumental challenges like global warming, ozone layer depletion, hunger, poverty, and inequality remain at the forefront. According to business leaders, sustainable business strategy is the need to address the planet's pressing challenges along with the company's success. To understand any sustainability efforts of a business, the triple bottom line seems to be one of the concepts.

2.1 Triple Bottom Line Philosophy

It was John Elkington, who named Triple Bottom Line (TBL) in 1994 as the theory of 3Ps for sustainability, stating the company's responsibility in profit, people, and planet- standing for economic, social, and environmental responsibility. (See Fig. 2) A company is equitable and fair when it cares for profit and people, but without environmental concerns, the planet is doomed. In other words, an enterprise should be committed to evaluating its environmental and social impact in addition to its fiscal performance.

A business's success depends much on its fiscal performance or turnover for shareholders. Premeditated planning and critical business conclusions are meticulously planned to make the most of costs whilst at the same time lowering costs and mitigating risks. Profit is mandatory for the company's development and it hopefully can ensure responsible behavior. The economic part is not only making a profit but to utilizing it properly. The indicators for the economic responsibility of a company can be the increase of gross Domestic Product (GDP) and Purchasing power parity which measures the corporation's connection in developing living standards. Transparency and open reporting enable the stakeholders to look into the company's work. Any positive changes in society define a corporation's accomplishment. The multiplier effect can be one of the aspects of economic responsibility, whose impact can be far-reaching when vast amounts of people in the locality work for that company. The high economic performance of the company yields higher salaries that are spent on products and taxes. The company can put more money from profits into socially responsible services. The second aspect comes in the way of more fair tax being paid to the government that comes from higher profit. The third aspect is keeping trust and escaping any activities that can bring mistrust and affect the company's license

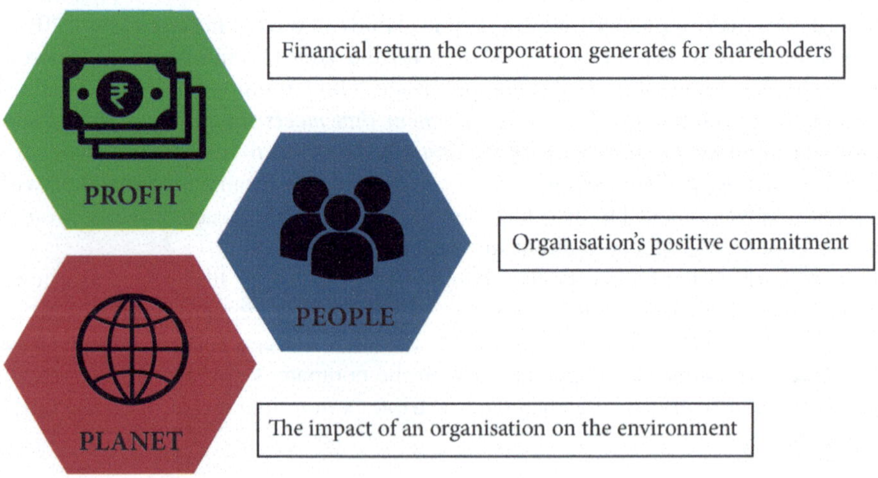

Fig. 2 Triple Bottom Line and the 3 P's (Planet, People, and Profit)

to function. Take for example the Nestle scandal in 1970 and it’s baby formula sold to developing nations (KsiężaK and FischBach 2017).

The next component of TBL illuminates the social impact of a business or precisely its commitment to people. The distinction between shareholders and stakeholders is important to understand. Conventionally shareholders are valued more as key success indicators but sustainable businesses have a focus on all stakeholders. The company’s social responsibility lies in improving living standards where CSR can serve as a tool to establish and maintain cordial relationships between society and enterprise. Small and medium enterprises often hire people from the locality where they operate. This doubles the responsibility of the enterprise. Social responsibility comprises everyone that is affected by the company and vice versa. A business that follows the triple bottom line will never exploit people or promote child labour rather, offer fair treatment and fair salary to its employees. This also extends to the subcontractors who are governed by the same set of rules. The US clothing company, Gap, Inc. was unaware of its products made by children in a subcontractor’s factory. Three aspects of social responsibility seem prominent. The first are the customers who put their confidence in the company to buy their products. There can also be new consumers who are well-informed and look for authenticity. They seek good quality products and after-sales service. They also act as critic having the power to boycott a company and destroy its reputation with negative reviews on the internet. A product with a value for money decides the success or failure of a product. The second aspect of social responsibility lies with employees. Though employment itself benefits people, CSR should guarantee optimum use of people’s skills, ensure safety measures, and provide impartial treatment. The third aspect of social responsibility is the corporation’s image to the local community. The image can be built in ways of sponsorship, training, donations, and recruiting and is crucial for the company’s reputation against its competitors (KsiężaK and FischBach 2017). One of Unilever’s programs is “Self-Esteem” by Dove which aims to abolish any type of discrimination based on skin color and physical features. The handwashing Education project by Lifebuoy helped spread awareness during the COVID-19 pandemic.

Ever since the beginning of the Industrial Revolution, the activities of big companies have polluted the environment driving the planet toward destruction. Corporations contribute the maximum amount of GHGs contributing to climate change along with NO_x and SO_x. We are all accountable for our natural environment; the responsibility is even profound for the enterprises that cause enormous planetary damage. Overexploitation of resources, dumping of untreated wastes, and emitting toxic gases all contribute to negative impacts of the enterprises on the planet’s environment. Hence, corporations should endeavour all efforts to mitigate or prevent such deleterious impacts. Companies can make eco-friendly goods; recycle materials, take measures to reduce the levels of toxic wastes, use methods that are less energy and water-intensive, and so on. In a nutshell, responsible behaviour with regard to the environment can add extra returns to the company eventually. The first aspect of environmental responsibility is the environmental impact that counts for all the detrimental effects of the enterprise on the environment. An environmentally responsible business measures all such impacts such as ecological footprint, water footprint, and carbon footprint; and estimates the resource quantity used by the company every year against its supply still available on this planet. Life cycle assessment (LCA) is another way to appraise the environmental performance of goods and services right from the raw material procurement stage. The second aspect of environmental responsibility is the win–win situation where both sides win. Obvious cost cutting increases the viability of the business through more savings. Moreover, any abnormalities in production can be identified and possibly eliminated. This can enhance the business reputation and contribute to competitive advantage (KsiężaK and FischBach 2017). DHL's initiative (Logistics Company) to make eco-friendly supply chains reduces transport-related waste, emissions, and other environmental hazards.

2.2 Benefits and Challenges of Triple Bottom Line (TBL)

Most companies neglect the Triple Bottom Line (TBL) concept as it is a barrier to their financial goals. Recycling on the other hand may raise the cost. Implementing the TBL seems impractical where the enterprises care hardly about society and the environment. Measuring social and environmental impact is rigorous and difficult in contrast to quantifiable financial accounting. The three elements are contradictory; equitable allocation of material, human, and financial resources is difficult. The companies use the TBL internally. The environmental, social, and governance (ESG) measurement are a third-party evaluation of these actions thus publicly holding the business accountable. Many corporations want to appear 'green' and use the Triple Bottom Line (TBL) as a marketing approach resulting in fake efforts.

Corporations are increasingly under scrutiny. They are often held accountable for social inequities and environmental incidents for their activities. This raises concerns about the conventional way of management, complicates decision-making, and requires a broadening of expertise which was quite a difficult task. The World Business for Sustainable Development (WBCSD) represents many of the global multinational corporations seeking ways by which enterprises can achieve economic vitality along with social vigour and environmental protection; making sustainable development the strategic corporate objective (Charles Jr et al. 2017).

3 The Essence of Environmental Sustainability

Environmental sustainability has become one of the most commonly used phrases and indicates a balanced relationship between the available natural resources and human consumption. The maintenance of the qualities appreciated in the physical environment can be referred to as environmental sustainability. Company influences our natural environment at diverse levels in a variety of ways like mining, damming the rivers, construction activities, emissions, and effluents. Corporation has an unrequited accountability toward living beings impacted by their operations. Businesses must guarantee natural living conditions, and unproblematic existence for all living creatures by optimizing the use of natural ecosystems not compromising the ecosystem's health (Zsolnai 2011). An increasing public and government concern in the past 10–15 years has put the firms under monumental pressure to reduce their environmental impacts. For instance, the goods distribution affects atmospheric quality, produces noise, and vibration, and even causes accidents sometimes. It is reported that freight transport contributes around 8% of CO_2 emissions globally with warehousing and product handling adding another 2–3% to the total (McKinnon 2010).

3.1 Business and Environmental Sustainability

The environment was paid the least attention by industrial practitioners and businessmen even a few decades back as they denied any harmful impact of their actions on the environment. However, perceptions have changed with an increase in environmental degradation and increased pressure on the stakeholders on organizations. Globally the trends in business have changed owing to the competitive environment. Paying attention to 'green' has become the most crucial motivation for corporations now. Manufacturing industries exert one of the greatest impacts on subjects like climate change, resource depletion, pollution through emissions and effluents, waste generation and disposal, and others. It is not enough to make a profit and achieve a competitive image but the companies are now answerable for their environmental impacts (Kraus et al. 2020). The term eco-entrepreneur is used to describe entrepreneurs who aim to protect the environment and prevent environmental problems in industries together with introducing eco-friendly products and services in the market (Melay and Kraus 2012).

Three business models can be discussed where investing in sustainability results in instant and expected returns. These models are potent ways of making money through sustainable acts. The first model is based on preventing risk. Several industries undergo major changes in the context of competition with increasing environmental pressures. In many cases investing in sustainability is a cause of competitive advantage but often requires effective leadership as such investments lead the business to confront several possible futures. Such type of uncertainty challenges the usual tactics and requires the capacity to grasp the business tensions that arise out of investing in exploratory ideas. Some opine that the environmental crisis creates notable opportunities for the private sector. Nearly fifty per cent of the business openings for reducing greenhouse gas (GHG) emissions are Net Present Value (NPV) positive and economically viable. The widespread social media and increasing global media along with well-aware customers lead many corporations to invest in sustainable business practices to avert impending brand damage. Nike factories were reported to pollute the local waterways and this swayed Nike to invest in sustainability. Greenpeace accused Mc McDonald's of contributing to deforestation. In reaction, McDonald engaged in preserving the Amazon forest. Coca-Cola was accused of depleting and polluting the local water supply in Plachimada, India and in response, they announced their commitment to become 'water neutral' (Henderson 2015).

The second model is based on increasing the operational efficiency of a business. Many companies are of the idea of earning a great amount of money by increasing resource efficiency. Studies showed that the cost of initial construction increases by 0–8% when complying with LEED or equivalent standards. Normally the construction involves 5–10% of the total life cycle costs. The total cost of making a green building is increased by less than 1% but saves energy expenditure by 5–30%, and water consumption between 8 and 11%. Many companies reported to make considerable savings in energy expenditure. IBM reported to lower electricity consumption of 6.1kWh between 1990 and 2012. One of the largest shipping companies, Maersk committed to energy conservation through the introduction of 'slow steaming' by its ships and cutting fuel consumption by 7% between 2009 and 2012. An Indian company, Jain Irrigation specializes in 'micro-irrigation' experienced a revenue growth of 41% between 2005 and 2010. Walmart reported cutting waste production by over 80%. The aluminium industry uses recycled materials and saves 95% of the energy needed to make new aluminium from bauxite ore; it is predicted that from around 100 million tons of garbage collected annually, USD15 billion/annum could be generated by segregation and reselling alone (Henderson 2015).

The third model rests on selling to the environmental niche whereby many enterprises have developed specialized offerings for consumers having high value for environmental performance. 10% of Clorox's sales of $5 billion are 'green products' (Henderson 2015). Toyota's Prius, a hybrid car, was launched in 1997. Around 15 million units of this hybrid electric vehicle (HEV) were sold as of 2020. Renewable energy comprises a small proportion of the total energy supply. Walmart announced its complete shift towards renewable by 2020. Verizon announced to spend $100 million on solar and fuel cell technologies (Henderson 2015).

Environmentally responsible business initiatives are an element of CSR that have started for reasons not only to make a profit for the enterprise, not always required by law, but they benefit society at large. Their survey on sports businesses indicates that these companies are increasingly considering corporate environmental responsibility in their policies. By doing so they may prevent legal battles, save money, and bond strongly with the stakeholders This type of engagement in environmental CSR activities allows the companies to improve their reputation, address the expectations and demands of the customers, lower pessimistic media reports, and build a strong system of partners (Babiak and Trendafilova 2011). According to Campbell, several mitigating factors in the relationship between business and society assist in explaining why companies should conduct themselves in a socially responsible way. He assumed to pay more attention to institutional mechanisms in adopting and implementing CSR behaviours (Campbell 2007). The environment is one of the pillars of CSR (Babiak and Trendafilova 2011). Companies are now considering their impact on the surroundings while adopting management strategies (Sarkar 2008). An enhanced environmental performance led to improved financial performance (Klassen and McLaughlin 1996). Environmental responsibility can be evaluated using a number of variables starting from the use of raw materials to waste management. In management, Both CSR and corporate sustainability (CS) refer to social and environmental issues with no clear difference between these two terms. Legal compliance and social sanctions drive to address the environmental impacts of business and are components of corporate environmental management.(Montiel 2008).

3.2 *Business as Environmental Stewards: Embracing Eco-Friendly Practices*

The role of businesses has transcended conventional profit-making objectives and has realized environmental responsibility. This digs into the evolution of business as environmental stewards undergoing a motivational shift. Although several firms comply with environmental regulations on paper and hence provide the least protection; they see matters of environment, health, and safety as a center of cost. They fail to look into the real value.

Recently enterprises have become far more aware of their activities on the environment. Environmental stewardship involves the responsible use, management, and protection of the environment and its entities. Environmental stewardship for businesses comprises taking accountability for their activities and making conscious efforts to protect and conserve the environment and contribute towards a sustainable future (Hartman 2009). Conservation and use of sustainable practices is the route towards environmental stewardship. A vigorous corporate environmental policy can reduce the negative impact of the business on the environment and even reduce the existing damages. Stewardship includes a list of elements and specific practices. A manufacturing house shifts towards procuring renewable raw materials or using safe chemicals in production processes. In general, businesses can reduce paper consumption; alter transport use, and others. Using cutting-edge technology like hybrid vehicles and solar or wind power requires huge initial investments and hence comes with a cost. Subsidies and tax credits can help to offset these expenses. Hence stewardship management as a part of a budget poses a challenge to every business.

The imperative for firms to behave as environmental stewards primarily owes to heightened awareness about environmental degradation, resource depletion, climate change, and others. Regulatory bodies, investors, and consumers are now prioritizing sustainability. Companies hence align to keep with market relevance and legitimacy. The companies recognize the linkage between environmental health, social well-being, and economic prosperity. Increasing demand from consumers, B2B vendors, and shareholders can increase revenue collection as people are willing to spend more on products made in an eco-friendly manner. Environmental stewardship improves a firm's reputation as it opens access to environmentally oriented suppliers and vendors to create a sustainable supply chain. Sustainable supply chain practices include ethical sourcing, fair labour standards, reducing ecological and carbon footprint, etc. Innovations and efficient processes can contribute to savings for the companies. Stewardship helps in protecting public health by using safe chemicals, and safe processes, and reducing wastes and emissions. Companies may earn certification from environmental organizations. This can help in marketing by making their activities known to customers. Apple has a Gold rating from the Electronic Product Environmental Assessment Tool. Such a company acts as a stand-out model for other companies and leads others towards achieving sustainability (Hartman 2009). The company's transparent communication with the stakeholders helps to build trust and credibility as well as a positive brand image. The company's compliance with environmental policies and laws mitigates potential risks, avoids penalties and legal issues, and keeps its goodwill and reputation.

Businesses as environmental stewards can exert positive impacts like innovations, efficient operations, use less virgin materials, reduce operational costs while adding value, potentially lowering regulatory obligations, enhancing environmental performance, reducing waste generation, emitting fewer pollutants, reducing liability, and providing a long-term competitive advantage. Many companies take proactive steps to integrate ethical practices in business operations. Apart from encouraging and implementing sustainable practices, enterprises engage actively in CSR activities like supporting local organizations to address environmental concerns, participating in community-led clean-up operations, afforestation and reforestation programs, granting financial aid for research, invest in green technologies (energy efficient lighting, water conservation, efficient solid waste management, and others. Companies align their business objectives with the environmental goals.

Despite growing momentum towards stewardship, economic constraints like raised expenses of sustainable initiatives, pressure to deliver short-term profits, legal uncertainties, inconsistent enforcement, logistic hurdles, and conflicting stakeholder interests pose serious challenges to businesses and deter the achievement of sustainability.

3.3 Innovations in Sustainability: Green Technologies and Beyond

The world has become a rapidly changing place and so are the business and corporate landscape. Production and consumption have rapidly changed in recent times owing to the rising population that leads to over-exploitation and other environmental issues compromising our life on this planet. These issues put pressure on the nature as well as on the overall performance of the corporations. With newer and newer regulations, and stricter legislations all over the world, both countries and organizations are facing challenges in resolving these environmental issues; consumers started reshaping their lifestyles, and organizations are taking increasing interest in increasing resource efficiency and reducing all types of harmful emissions and discharges (Cisneros et al. 2023).

As the ecological environment was undergoing destruction, the environmental impact of human activities became a global and growing concern. This prompted the stakeholders to exercise tremendous pressure on the firms and corporations to persuade towards environmental sustainability thereby driving more and more companies to actively embrace environment-friendly management actions to enhance environmental performance (Yuan and Cao 2022). Many academicians believe innovations to be the crucial step to be taken as they determine competitiveness and business performance (Cisneros et al. 2023). Innovation offers a high chance of competitive advantage to the corporation (Kneipp et al. 2019; Soewarno et al. 2019; Szutowski 2021).

Many researchers proposed a category of innovation to strike a balance by reducing environmental damage but enhancing business efficiency, the notion that encompasses the concept of green innovation (Portney 2020). Because organizations are facing growth limits due to environmental constraints, immediate attention should be drawn towards green innovation (Handfield et al. 2005). In the face of environmental challenges, numerous innovations spurred forcing green techs and holistic approaches to develop. Renewable technologies experienced the booms most in innovations. Solar, wind, hydel power, and geothermal options have experienced exponential growth. Technological breakthroughs, economies of scale, and supportive policies are the main drivers in this case. Highly efficient solar panels, floating wind turbines, and grid-scale energy storage options have revolutionized the present world by presenting much more reliable alternatives to fossil fuels. Moreover, the decentralization approach empowers the communities to produce, store, and distribute their energy which fosters independence and resilience. Earlier, the need for safeguarding the environment was overridden by the need for economic development (Fosu et al. 2024).

With present companies facing environmental challenges, the management has recognized green innovation as a factor towards sustainable development. Green innovations are ecological in conception, construction, use and disposal; comprising novel ideas that aim to afford products or services without harming nature but revering the environment (Tseng et al. 2013). It refers to the development of eco-friendly products and processes involving the utilization of green raw materials creating ecological products designed to trim down material use, pollutant emissions and discharges, energy expenditure, water consumption, and others. Presently, green innovation spurts out as a logical response to the customers' requirements, they are willing to spend more for eco-friendly, expecting responsible products and services from the organizations (Chen et al. 2012; Tang et al. 2018) Thus green innovation becomes one of the most important strategic tools for effective sustainable development.

Green consumers are growing rapidly. Business organizations need to think about green competitors to gain sustainable competitive advantage (Soewarno et al. 2019). Several factors affecting green innovation have been explored. Greening of the suppliers can act as an external contextual factor affecting the capability of the business organizations' green innovation (Song and Yu 2018). Apart from green suppliers, market demand, and environmental regulations are also key external factors. Environmental ethics, environmental commitment, green human capital, and green adaptability are the key internal factors driving green innovation. Although both external and internal factors drive green innovations, it is the internal factors that can drive proactive green innovation (Fosu et al. 2024).

Green innovation strategies help to improve business models, alter management approaches, and strive to develop a green organizational identity. Organizational identity refers to the system where members together create and add meaning to their actions and choices. It influences the way people behave in the organization. And green organizational identity is defined as *'an interpretive scheme about environmental management and protection that members collectively construct to provide meaning to their behaviors'*. Green innovation necessitates green creativity that comprises developing novel and useful ideas about green products and services when a company staunchly believing in green organizational identity faces external pressures (Song and Yu 2018). Innovating green strategy paves the way to woo green innovative performance (Dai et al. 2019). Such a strategy erects the company's environmental responsiveness for preventing pollution, creating green technology and product stewardship (Fosu et al. 2024). Escalating the waste generation crisis and finite resource constraints led the circular economy to gain popularity as a viable alternative. Innovations in material science, product design, and waste management facilitate a shift toward a closed-loop system where resources are subjected to reuse, recycle, or regenerate. Green innovation brings changes in manufacturing processes. Modular construction techniques and biodegradable packaging are now reshaping industries and unlocking novel economic opportunities. Nature from time immemorial serves as a source of inspiration to solve complex challenges. Biomimicry has opened a diverse array of sustainable innovations across various disciplines. (Dhaliwal et al. 2012) opined the possibility of CSR to attract institutional investors and their investment by drawing analysts' attention and minimizing the analysis forecast errors. (Ren et al. 2020) reported that CSR information disclosure may lessen organization profits, but minimize sulphur emissions, and industrial wastewater thereby advancing the environmental performance of the organization. Many countries of the European Union, China, and the Unites States of America have embraced mandatory CSR disclosure. In countries like China, the Shanghai Stock Exchange was notified of a mandatory CSR information disclosure in 2008 (Hong et al. 2020). With the implementation of the 2014/95/EU directive, big European organizations need to reveal their environmental and social impacts; all the EU member nations were required to adopt this Directive into their national legislation by 6 December 2016 (Ren et al. 2023).

The question arises of the mandatory implementation of CSR information disclosure. Investigation reveals that such disclosure can indeed lower financing costs, ease symmetry of information (Xu et al. 2021) and promote green innovation though large firms, state-owned organizations and pollution-intensive industries are likely to be affected more by such disclosure (Hong et al. 2020). CSR along with green dynamic capabilities boosts both green product innovation and green process innovation, with green dynamic capability serving as an intermediary between Corporate

Social Responsibility (CSR) and green innovation (Yuan and Cao 2022). Dynamic capabilities can be defined as *"the firm's ability to integrate, build, and reconfigure internal and external resources/competences to address and shape rapidly changing business environments"* (Teece 2010).

Owing to globalization, the firms' operations and related disclosures changed rapidly. Both the shareholders and regulators seek economic, social, and environmental information while making decisions. Corporations consider Corporate Social Responsibility (CSR) as a management tool to attain their financial and non-financial goals. Green innovation can be used as one of the mediators to explain the connection between CSR and corporate financial performance (CFP). Green innovation serves as an advanced form of innovation including technological expressions that safeguard the environment and produce green products. Green innovation is a proactive method to contribute to environmental development (Homayoun et al. 2023). Green innovation comprises *"the innovation in technology applied to minimize wastage, global warming, use of water, air pollution, use of coal, oil, electricity, and conserving energy"*. Green innovation enhances environmental performances along with social and financial performances (Kraus et al. 2020). The environmental aspect of Corporate Social Responsibility (CSR) can reinforce corporate financial performance (CFP) by cutting the expenses involved in waste recycling, pollution abatement, water, and energy conservation, etc., and boost the company's image and reputation nationally and internationally. The concept of green innovation includes both virtual and physical novelty in software and hardware (Homayoun et al. 2023). Green innovation assists corporations in learning and exploring novel Corporate Social Responsibility (CSR) tactics to provide improved services that meet customer demands and suit community interests along with the protection of the environment. Corporate Social Responsibility (CSR) initiatives are now considered a remedy for achieving sustainable development.

3.4 Inspirational Stories of Companies Leading the Sustainability Charge

The New Leaf paper assisted in conducting one of their first LCAs for the paper industry. They were supported by the Environmental Defense Fund and the Paper Task Force. The assessment showed that New Leaf Paper customers saved 2.1 million trees, 650 million gallons of water, and 65 million pounds of solid waste between 1998 and 2010 by using their papers manufactured from 50% recycled waste on average. Patagonia offers repair services under the Worn Wear Program. Customers are encouraged to repair and reuse their clothing instead of purchasing new items. Patagonia commits to reducing waste, inculcates sustainable consumption, and adds longevity to its products.

From early times, the philanthropy of business people in India was firmly rooted in religious beliefs. In the 1900s business practices were more socially responsible like donating to charities and educational institutions, community services, and improving employee benefits. These actions were also to buy goodwill. JRD Tata believed in going beyond and behaving like honest citizens and contributing to public welfare activities besides normal activities. Apart from the donation of funds, their own financial, managerial, and human resources could supply task forces directly for providing relief and reconstruction measures. Gradually it was accepted that businesses have to undertake a proportion of social overhead costs (Gautam and Singh 2010).

The automobile industry is one of the greatest polluters. Ford Motor Company vehicles use sustainable fabrics. Eighty per cent of Focus and Escape vehicles can be recycled. Special emphasis is given to fuel efficiency; Ford uses geothermal cooling systems in its factories. They possess the largest green roof in the world. Their Crown Victoria Interceptor vehicle can run on both ethanol and gas. They are the winners of EPA Energy Star twice. Nike uses recycled polyester to make sustainable environmentally preferred products and uses renewable energy sources for production. It compelled its 650 suppliers to frame and implement environmental policies worldwide. The Starbucks Company purchases Certified organic and Fair Trade Certified coffee. It emphasizes LED certifications for all its outlets. The company pursues a green building strategy that involves temperature adjustments for its air-conditioned stores.

Google powers its facilities with renewable energy. It hosts farmer's markets and practices green supply chain management and sustainability. (https://www.smartcitiesdive.com/ex/sustainablecitiescollective/9-companies-great-environmental-initiatives/1193165/) **Hewlett Packard Enterprise** is improving the energy efficiency of its products by using sustainable materials, controlling supply chain emissions, reducing toxic gas emissions, stopping the use of toxic substances used in cartridge manufacture, and reducing the environmental cost of the customers. Their goal is to achieve net-zero emissions by 2040 and an absolute reduction of all scope emissions by 90%. **Microsoft** promised to go beyond net zero. They committed to becoming carbon-negative and water-positive by 2030. They are also spending in circularity and have set the target of zero waste by 2030. **Procter and Gamble** announced a 'Climate Transition Action Plan' for achieving net zero by 2040. The company aimed at reducing deforestation and reducing waste by promoting the use of recyclable and reusable packaging for all consumer products by 2030 and also creating a water-positive future. While IBM announced twenty-one climate-related goals and achieving net-zero by 2030, Accenture disclosed to achieve net-zero target by 2025. (https://justcapital.com/news/earth-day-2023-top-companies-for-the-environment/?gad_source=1&gclid=Cj0KCQiA5rGuBhCnARIsAN11vgRoopMVCf0B_HrcoZQcRat9TSR-fl1uAOEqDdYH7gcYPJhQhhz5choaAgy0EALw_wcB) The CSR activities of a few notable companies are summarized in Table 2.

Table 2 A few Companies and their CSR initiatives

S. no	Company	Focus areas	Programs
1	Infosys Limited- a flagship company ITES (Information Technology Enabled Services) sector in India	– Combat hunger, poverty, and malnutrition while advancing preventive healthcare, sanitation, and ensuring safe drinking water access – Promote inclusive education and vocational skills, particularly targeting children, women, the elderly, and people with disabilities; implement projects to enhance livelihoods – Strengthen rural areas by improving access, shelters, potable water, sanitation, electricity, and livelihood opportunities, building sustainable villages – Promote gender equality and empower women through the establishment of homes, hostels, daycare centers, and old age homes – Ensure environmental sustainability; protect flora and fauna, natural resources ensuring animal welfare and ecological balance – Safeguard national heritage, art, and culture to preserve and encourage cultural diversity and historical legacy	Infosys Foundation joined with the CITTA Education Foundation India to enhance educational opportunities for students residing in rural Rajasthan, especially in Sam, Salkha. Foster holistic development of the girls beyond traditional educations and nurture innate abilities Infosys Foundation is dedicated to safeguarding natural resources, mitigating climate change, and addressing human-induced damage The Infosys Foundation plays a crucial role in preserving and restoring heritage sites and water bodies; notably the Lingampet stepwell, also known as Naganna Baavi The Foundation, a key partner of Namma Metro, participated in the construction of the Infosys Foundation Electronic City Station of Phase 2 of the metro rail project of Bengaluru Infosys collaborated with the Museum of Art and Photography (MAP) in Bangalore which was formed by the Art and Photography Foundation With an emphasis on infrastructure and technological innovation, Infosys played an active role in constructing the digital museum building in Bangalore together with an encyclopedia, interactive website experiences, software visualizations, and exhibitions. The Infosys Foundation Gallery takes art ahead of the confinement of museums and galleries

(continued)

Table 2 (continued)

S. no	Company	Focus areas	Programs
2	Mahindra and Mahindra Ltd A farm solutions provider	– Focus on girls, youth, and farmers in the fields of education, health and environment	Some of the important programs on health include Swaraj Sehat—Mahindra Mobile Primary Health and Cancer Care Project, and AD Chakan Sehat—Integrated Health Project Some of the important programs on education include Project Nanhi Kali—Education of the Girl Child, Mahindra All India Talent Scholarship, FD Jaipur Hunnar—Livelihood through farm mechanization skilling (tractor and agri equipment), Powerol Hunnar—skill development, Swaraj Gyandeep—Promoting Quality Education through Critical Infrastructure and Scholarship Support in Government Schools Important programs on women empowerment include Women Empowerment with Regenerative Agriculture—UP and Punjab, Maharashtra, WEE—Apparel Sector, KCO Prerna—women empowerment, Swaraj Krishi Mitr—Prerna—Empowering women for livelihood enhancement Important programs on Environment and Natural Resource Management include Araku and Northern India Hariyali, Antim Prasthan, AD Nasik 1 Green Guardian—Natural Resource Management, Swaraj Green Guardian—Renovation of Ponds and Roof Rainwater Tanks in Govt Schools of Rajasthan for water conservation Important programs on Rural development include FD KND Gram Vikas—Integrated Community Development, FD Nagpur Krishi Mitr for youth and women, KCO Gram Vikas—Integrated Village Development through soil and water conservation, KCO Gram Vikas—Watershed Development Fund (WDF) and many others (https://www.mahindrafinance.com/wp-content/uploads/2023/08/CSR-Report_FY-2022-23.pdf)

(continued)

Table 2 (continued)

S. no	Company	Focus areas	Programs
3	ONGC	– Promotes healthcare and sanitation, water management with recharge of groundwater; – Promotes Education and vocational courses, girl child development, infrastructure for schools and colleges, and distribution of IT devices – Promotes gender parity, and women empowerment, establishes houses and hostels, daycare centers, and helps physically and mentally challenged people programs – Promotes environmental sustainability through protection, and conservation, plantation programs – Promotes heritage site protection, and promotes art, crafts, and culture – Programs for armed forces veterans, and war widows – Training for rural sports, para-Olympic, and Olympics – Fund the Prime Minister's Relief Fund related to welfare schemes for backward classes and the marginalized – Take part in disaster management initiatives – Allot incubator funds to the central and state governments and public-funded universities to promote the SDGs - Allocate sponsorship programs for workshops and conferences - Entrepreneurship schemes including self-help and livelihood programs – Undertake rural development and slum development projects. (https://ongcindia.com/web/eng/csr/indicative-list)	Promoted preventive health care along with Helpage India under 'Varisthajana Swasthya Sewa Abhiyan'. Opened 20 mobile medical units, treatment provided for 1.9 lakh people Partnered with Bharatiya Vidya Bhawan to operate 5 computer centers known as 'ONGC-GICEIT Computer Centre' and trained 1400 students Ventured with Mokshda Paryavaran Evam Van Suraksha Samiti to develop energy-efficient environment-friendly 'Moksha Green cremation System (MGCS)' in Delhi, Cambay, Ahmedabad, and other places Installation of 2309 tube wells and hand pumps, 121 RO water facilities, 26 water ATMs, several hundreds of school toilets and community toilets, cleaning initiatives of the high peaks of the Himalayas, revival of River Saraswati, cleaning River Sabarmati bank, beautification of places like Varanasi Ghats, installation of 36,597 solar street lamps, installing sanitary vending machine under Project Sakhi with the UNICEF, Bio-CNG cum Fertilizer and Bottling Plant at Hardwar, etc. promotes informal school with single teacher under Ekal Vidyalaya program Undertook the Eastern Swamp Deer Conservation Project in Kaziranga National Park Established multi-speciality hospitals in many places including, Assam, Gujarat, UP, etc.(Mishra) 330 beds for the underprivileged at Assam Cancer Institute in Nagpur MRPL Lady Goschen Hospital in Mangalore to name a few

(continued)

Table 2 (continued)

S. no	Company	Focus areas	Programs
4	State Bank of India (SBI), India's largest commercial bank pledges to deliver "Service Beyond Banking"	Envisioned making an evocative and quantifiable impact on the socially, economically, and physically challenged communities and ensuring environmental sustainability	Spearheaded Solar ATMs, Rooftop solar, energy-efficient lighting, and air conditioning system projects. Additionally, motion-activated lighting is installed to optimize energy usage Implemented a Branch Server Consolidation (BSC) project, removing Branch servers, and securing data in a virtual environment Introduced Power Management Utility on all desktops for energy conservation Indian Green Building Council (IGBC) rated seven SBI premises are rated as "Green Premises/Projects." SBI offers Green Reward Points to digital channel customers, redeemable for credit to SBI GREEN FUND Released additional Green Bonds of US$ 100 million in 2019–20, adding up to a total of US$ 800 million in Green Bonds SBI Green Marathon targets to raise awareness regarding environmental conservation Introduced E-Newsletter "SUSTAIN ON" to sensitize employees on sustainability-related issues (https://sbi.co.in/corporate/AR1920/download_center/english/11-4.10-Corporate%20Social%20Responsibility.pdf) SBI Foundation (SBIF)—promotes socio-economic well-being through the Gram Seva program; SBIF launched the India Health Alliance program to fight present and future healthcare issues. SBIF provides financial help for cochlear implants under the ADIP scheme, GOI SBI staff members donated 108 crores to the PM Cares Fund in 2020 to fight COVID-19 SBI partnered with Extension for Community Healthcare Outcomes (ECHO) India to train healthcare providers during the coronavirus outbreak Undertook 'Project Prana' in designing an electro-mechanical ventilator with locally available components SBI's Project SAMEIP encourages career options for 500 underserved youth with disabilities. SBI offers financial support to the National Sports Development Fund and Armed Forces Flag Day Fund. Along with UNFPA, its 'Project Samriddhi' empowers adolescent girls. SBI collaborated with 'Mauli Seva Prathisthan' to work for destitute women and children and their mental health SBI's "Sustainability and Business Responsibility (BR) Policy" ensures environmental sustainability. Approving the carbon–neutral policy by 2030, some of the key initiatives are the carbon neutrality project, SBI Green Fund, and Green Bonds. An online tutorial 'Astitva' is dedicated to the employees to ensure internal sustainability (Thacker 2023). (https://thecsrjournal.in/state-bank-of-india-sbi-csr-corporate-social-responsibility-report/)

3.5 *The CSR Case of TM International Logistics Limited (TMILL)*

TMILL is an international logistics provider. This joint venture company was formed by Tata Steel Limited, NYK Holding (Europe) BV, and IQ Martrade, which has the core purpose of improving the quality of life through the creation of long-term stakeholder value. They emphasize on living standards of the underprivileged, promote education that includes job-oriented vocational skills, promote sustainable healthcare and sanitation practices, make the availability of potable water, and ensure environmental sustainability by protecting plants, animals, and natural resources. The TMILL identifies the target beneficiaries. Based on their need they prepare an annual plan. The company serves alone or can partner with other organizations, train their CSR personnel, and thereafter deliver and implement the chosen CSR initiatives (Picture 1).

Currently, they have undertaken a mangrove sapling plantation project in the Gosaba Block of Sunderbans. The mangroves act as a natural barrier and shock absorber that protects us from the wrath of nature. The calamities first strike these coastal lands where its intensity is considerably lessened by the mangrove ecosystem thus reducing the extent of destructions and devastations than it would have otherwise.

TMILL thus aims at

- Building and strengthening the defence system
- Raise general awareness about such hazards and make them resilient
- Impart knowledge of healthcare and hygiene, as well as nature education
- To foster awareness regarding the importance of plastic recycling

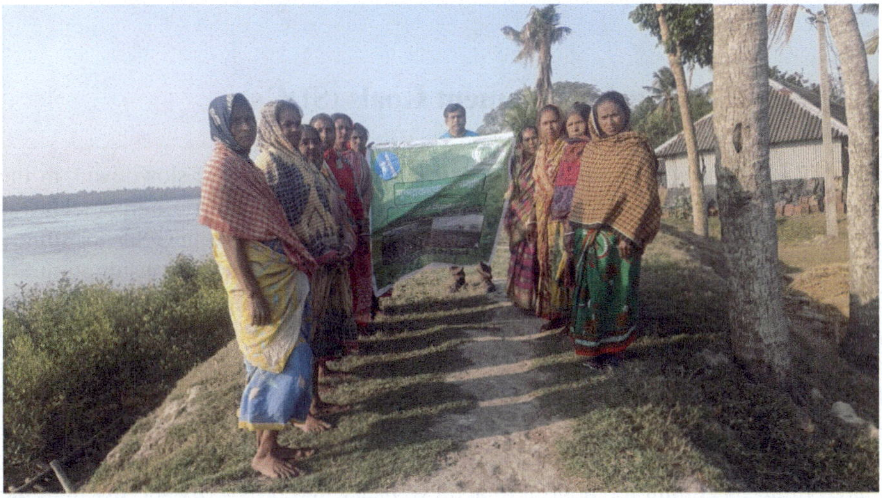

Picture 1 TMIL adopts mangrove sapling plantation project

- To promote women's empowerment by creating avenues for their income for women of those localities

Sunderbans provide livelihoods to millions of people living there as well as in its vicinity. It acts as a shelter bed. It protects people from storms and cyclones, tidal surges, seawater seepage, and saltwater intrusion. People here live in poverty and earn their livelihood as wood-cutters, fishermen, honey gatherers, leaves, and grass and twig gatherers. They are worst hit by Amphan 2020. After the water subsided only ruins of fallen and worn trees on the dilapidating shorelines. Turning over of the soil and saline encroachments were also seen. Not only did the mangroves get bulldozed but the cyclone also slowed down the recovery rate of the trees. Hence human intervention seems necessary to re-grow the mangroves, protect and make them sustainable and resilient before the next hit.

Based on the aims the first approach was to awareness camps in Gosaba Block to address the women about the uniqueness and significance of the mangrove ecosystem, pollution undergoing and ways of prevention, restoration practice with sapling plantation, plastic pollution, and its recycling, sustainable health and hygiene matters and others. The village women were made the stewards of the sapling plantation program where they took care of sowing, watering, and other aspects of nurturing. 17,000 sapling plantations were planned in the FY2024 in Kumirmari and Bali villages of Gosaba (Pictures 2, 3 and 4). The program comprises subsequent monitoring, videography, reporting, and review analysis. As a part of income generation, the saplings were bought from the women sellers. Women taking part in the plantation program received 5 months' worth of hygiene kits in turn to be bought from the village women through Self-Help Groups (SHG). These efforts cater to the upliftment of the underprivileged and poor as well as empower women towards gender parity.

4 The Sustainable Development Goals (SDGs)

Environmental deterioration, social degradations, and economic slowdowns in the last few decades have exerted so many pressures that a historic agreement was concluded in 2015 as the United Nations Sustainable Development Goals, or simply the SDGs. The burgeoning population and increase in per capita consumption are the key drivers (Pedersen 2018). This is an extension or replacement of the fifteen-year time-bound program known as Millennium Development Goals adopted in 2000 where eight goals were selected to fight poverty in all dimensions. A text outline with 17 goals and 169 indicators was laid in front of the United Nations General Assembly in 2014 as the beginning of intergovernmental negotiations. The 2030 Agenda for Sustainable Development was adopted by all the members of the UN. It provides a shared future, a common blueprint to achieve peace and prosperity serving both the people and the planet. Both MDGs and SDGs are proclamations of aspirations. The declarations are voluntary which serves both as a disadvantage and an advantage.

Picture 2 TMIL Chief—Inland Waterways and Multimodal Logistics, CSR and Sustainability initiates the plantation drive

Picture 3 Women in mangrove sapling plantation project

The absence of any legal compliance pressure may put us on the back foot. At the same time, countries might be more ambitious in broadening the scope and vision without any pressure of obligations (Pogge and Sengupta 2015). Nevertheless, the SDG document failed to inspire and guide a global concerted effort towards global development. (see Table 3) The table below shows the 17 goals of the SDGs.

It aspires to a prosperous, socially inclusive, and environmentally sustainable world. Though prosperity is uneven with nearly a billion people trapped in poverty, social inclusion is not very obvious in a society divided by class, race, ethnicity, gender, and further discrimination. The environment is endangered amidst innumerable disasters and catastrophes. Both global and national economy continues to generate huge income inequalities, hardly attend to the needs of the poor and marginalized, and head towards unprecedented global crisis. Traditionally markets accept the environment as a global commons as well as a universal disposal ground. Amidst all, the SDGs stand out as a hope for taking a new course in global politics and economies with the help of business sectors. The business sectors need to align their activities in tune with the SDGs and evaluate their environmental, social, and governance practices (Sachs and Sachs 2021).

Picture 4 Women in mangrove sapling plantation project under supervision

4.1 Aligning Corporate Strategies with SDGs

Despite drawbacks, in the present era where we confront unprecedented global challenges, the SDGs become a ray of hope and the roadmap for collective action transcending the geographical boundaries. With increased recognition of businesses as key stakeholders, aligning the corporate strategies with the SDGs becomes a strategic imperative. It is essential to comprehend the essence of the SDGs before tuning corporate approaches with the SDGs. All 17 SDGs include specific targets and indicators offering a holistic framework exhibiting the interconnectedness of social, economic, and environmental aspects of sustainable development. These goals also serve as a pressing call for corporations worldwide to enforce sustainable business approaches to guarantee value addition and certainty to the stakeholders and community. Corporations can identify prospects to create financial returns and also leave extended and enduring social and environmental impacts on society (Jimenez et al. 2021).

Businesses play a key role in the progress of the SDGs. They exert tremendous influence on economies, societies, and the environment. They serve as engines of innovation, investment, and employment to bring meaningful changes. They possess the resources and expertise to drive such changes. An outstanding attribute of the 2030 agenda is the accent on the role of the private sector in achieving the goals, and partnerships with governments, civil society, and other stakeholders. Effective

Table 3 The Sustainable Development Goals (SDGs) and their trends (Nations 2023)

Sustainable Development Goals (SDGs)		Trends of progress
Goal 1	No Poverty	Slow progress; predicted that 575 million people will live in extreme poverty by 2030
Goal 2	Zero hunger	Slow progress; as of 2022, nearly 9.2% of the global population (735 million people) faces chronic hunger
Goal 3	Good health and well-being	In 2020, nearly 8 women died /per day from pregnancy and childbirth-related issues; 146 countries are on track to fulfill the SDG target of under-5 mortality
Goal 4	Quality education	Six countries are predicted to achieve the target of secondary school completion by 2030; 84 million will be out of school
Goal 5	Gender equality	Only 15.4% of the indicators are on track, and 61.5% progressing moderately
Goal 6	Clean water and sanitation	9% rise in efficient water use; in 2020 2.4 billion people lived in water-stressed nations
Goal 7	Affordable and clean energy	Predicted 660 million people will have no access to electricity; nearly 2 billion people will depend on polluting fuels by 2030
Goal 8	Decent work and economic growth	Global GDP per capita growth is expected to slow down
Goal 9	Industry, innovation, and infrastructure	Globally the manufacturing growth fell from 7.4% in 2021 to 3.3% in 2022; CO_2 emissions rose by 0.9%
Goal 10	Reduced inequalities	34.6 million Refugees documented in 2022; globally one in six people experience discrimination in some way
Goal 11	Sustainable cities and communities	Presently, 50% of the global population living in urban areas will reach 70% by 2050. 1.1 billion People live in slums or similar conditions
Goal 12	Responsible consumption and production	High-income nations have 10 times more material footprint
Goal 13	Climate action	The action plans failed to tackle climate change leading to increasingly unprecedented disasters. CO_2 emissions rose by 0.9%
Goal 14	Life below water	An increase in eutrophication, acidification, ocean warming, plastic pollution, and fishing practices pose great threats to aquatic life
Goal 15	Life on land	Extinction, biodiversity loss, and deforestation continue
Goal 16	Peace, justice, and strong institutions	More than 50% increase in civilian deaths due to conflicts; 108.4 million people forcibly displaced as of 2022
Goal 17	Partnerships for the goals	Geopolitical tensions impede global cooperation and partnerships. Developing countries are immersed in debt along with several challenges

business models can aid the diffusion and integration of these goals into business policies and practices (Pirzada et al. 2023).

Conventionally business approaches focus on maximizing profits instead of value addition and prevent the shift of local and national economies towards a more sustainable model. Corporate purpose alternatively emphasizes purposeful profits, points out the contradiction between profit maximization and value addition, and finds synergy between people, the planet, and profits. Corporate Purpose serves as a budding model to bring sustainable development goals into action, permitting the business to be the vehicle of change. Meaningful impact on the stakeholders results by integrating social, economic, and environmental values into the core business vision and goals. It removes any incentives for unsustainable business practices. Sustainable business models not only benefit the stakeholders by engaging the customers and employees but also reduce costs, lower risks, value creation, and innovation, grab the competitive advantage, undergo brand differentiation, and leave a mark (Jimenez et al. 2021).

The corporations need to realize the current situation and understand the urgency to adapt and undergo deep transformations—a basic need to decarbonize the energy sector, practice sustainable land-use, bridge the gaps of inequalities and income, gender inequalities, ensure the right to healthcare and education, and others (Sachs and Sachs 2021).

Corporations can adopt several practices to support SDG 1 like creating decent work opportunities, supporting micro, small, and medium enterprises, developing inclusive business models, and undertaking social welfare programs. In support of SDG 2 businesses can promote food security, sponsor smallholder farmers, endorse fair trade practices, bring innovative agricultural practices like precision agriculture, vertical farming, and aquaponics, use quality seeds resilient to climatic shifts, and minimize post-harvest losses. Business organizations can be proactive in reducing food waste in their organizations, take initiatives in donating surplus food to the needy, provide training to the people in the food sectors, and advocate policies for food security, nutrition, and agriculture. SDG 3 is well connected with the previous 2 goals. Corporations bear the responsibility to provide a safe and secure workplace based on sound occupational health and safety policies, supply personal protective equipment, and organize wellness, mental health awareness, and tobacco cessation programs. Affordable healthcare services can include health insurance coverage, on-site clinic facilities, and telemedicine services, regular health check-up programs, all of which can boost productivity, and reduce absenteeism. SDG 4 ensures quality education. Corporate houses can invest in training and skill development programs, and sponsor employees to participate in external training schedules to upgrade and advance career options. They can even back up access to quality education by teaming up with educational institutions, government organizations, and non-profit organizations, sponsoring grants, and scholarships to deserving candidates, developing educational apps, creating e-learning platforms, and others to empower the next generation of learners. The business houses can support the construction of schools and provide resources to educate underprivileged children.

Business houses can contribute to SDG 5 through several initiatives by framing gender-inclusive workplace policies that ensure equal opportunities and fair treatment in recruitment, hiring, promotion, and compensation and take all necessary measures to prevent harassment. Women should be encouraged to leadership positions by providing opportunities. The organizations can prioritize sourcing goods and service supply chains from women-owned enterprises to empower women more. Enterprises can raise awareness of the importance of clean water and sanitation (SDF 6) through campaigns, training, and community engagement programs. They can support water conservation programs like watershed management, ecosystem restoration, etc. They can invest in infrastructural development like sanitation facilities, and wastewater treatment plants, both on the premises and as public facilities, promote rainwater and other natural water harvesting methods as well as comply with the regulatory standards. Industries are energy-intensive and hence have a lot to contribute towards SDG 7. Besides enhancing energy efficiency, they can deploy renewable energy sources like solar, wind, hydel, biomass/biogas, etc. They can pursue onsite installations or purchase renewable energy agreements, invest in such projects, and engage in its infrastructural development. They can reduce energy consumption by fitting energy-saving devices and lowering the emission of GHGs in their operational sites and supply chains. Enterprises should invest in off-grid decentralized energy solutions by developing microgrid systems and conducting regular energy audits. The role of businesses in creating decent work conditions and economic growth (SDG 8) is undoubted in fostering an environment of respect, diversity, and inclusion. Business bears the responsibility of decent jobs with fair wages, benefits, and opportunities for career advancement. They can uphold labour rights, prohibit forced labour, and child labour and promote ethical operations. They can lead by helping small and medium enterprises through supply chain partnerships. Investments in education, training, and skill development can further promote productivity, innovation, and brand image.

Businesses play a crucial role in advancing resilient infrastructural development, innovation, and sustainable industrial activities (SDG 9). Businesses can augment our access to essential services like energy, water, sanitation, transport, health, and settlements through themselves or by partnering with others. They drive innovation and technological advancements across their operations. They can tie up with startups, invest in research and development, and promote a circular economy. Business houses can alter their policies and policies to initiate the reduction of inequalities (SDG 10). As mentioned earlier, they can foster a culture of equality by affording fair wages to all irrespective of their social differences. In contribution to SDG 11, businesses can promote sustainable urban development, well connected and inclusive to better the quality of life of the inhabitants. Their area of investment can be in green spaces, roads, pavements, transportation, energy, water, sanitation, and resilient housing projects against social nuisance, hazards, and disasters.

Responsible consumption and production (SDG 12) call for promoting resource efficiency reducing wastes; minimize environmental impacts which are key to sustainable business operations. Using eco-friendly raw materials, minimizing packaging wastes, waste segregation at source, targeted waste reduction, reducing water and energy use, optimizing production processes, and designing eco-efficient products and services can trim down the ecological footprint of the business. Business activities and climate impacts are closely knotted. Addressing climate change (SDG 13) needs adaptation measures to reduce GHG emissions and use renewable energy resources. The important initiatives are discussed under SDG 7. Conservation and sustainable use of oceans, seas, and other water resources (SDG14) can be achieved if enterprises implement marine pollution reduction practices protect marine and freshwater ecosystems, and practice sustainable fisheries and aquacultures. Selecting responsible fishing methods and fishing gear, averting overfishing and bycatch, coastal region protection, eco-friendly tourism, preventing plastic disposals, cleanup programs, and awareness campaigns, can be promoted. Businesses can also participate in conservation initiatives for fragile ecosystems. Businesses can likewise protect, restore, and boost sustainable use of terrestrial ecosystems (SDG 15), combat desertification, stop deforestation, prevent land degradation and topsoil loss, and conserve biodiversity. Businesses dependent on forest resources can source forest materials and products certified by the Forest Stewardship Council. Agro-based businesses can implement crop rotation, organic farming, and integrated pest management, and reduce our dependence on artificial inputs. Business houses can take part in afforestation and reforestation programs to help restore degraded ecosystems. Under SDG 16, businesses can facilitate ethical governance, information disclosure to stakeholders, transparency in decision-making, framing codes of conduct and anti-corruption policies, upholding human rights, addressing stakeholders' concerns, and providing support thus contributing to peace, justice, and strong institutions. Both MDGs and SDGs emphasize partnerships. SDG 17 revitalizes the collaboration with governments, private sectors, NGOs, civil societies, and other stakeholders within and between the countries. Businesses can foster such associations to drive collective actions to address all other goals in terms of knowledge sharing, resource mobilization, capacity building, technology transfer, and allocation of funds. Healthy collaborations leverage strengths and eliminate weaknesses.

4.2 Spotlight on Corporations Making Strides in SDG Implementation

Business approaches should take different strides like choosing responsibility over profits and seeking responsible investing and long-term profits to mitigate harmful impacts on people and the planet. In this context, the Carbon Disclosure Project (Matisoff et al. 2013), the Sustainable Accounting Standards Board (SASB) (Hales 2021), the World Benchmarking Alliance (WBA) (Urlings 2020), and the Global Reporting Initiative (GRI) (Hedberg and Von Malmborg 2003) are worth mentioning.

Grameen Bank founded the model of microfinance by granting small loans to the impoverished (Strasser 2014). Founded in 1866 in Switzerland, Nestle promised a commitment to source raw materials responsibly (milk is the most important raw material), assist smallholder farmers, and promote healthy eating habits (Cambaza 2022). Johnson and Johnson Consumer Health launched the Healthy Lives Mission investing 8 billion dollars through 2030 which targets improving the health and well-being of 1 billion people through partnering with governments, NGOs, and healthcare service providers. They shall use 100% recyclable, reusable, and compostable plastic packaging and certified recycled paper and pulp for packaging (https://www.jnj.com/latest-news/johnson-johnson-commits-800-million-to-making-more-sustainable-products). Likewise, Novartis launched the Novartis Access program to supply necessary medicines for non-communicable diseases at reasonable prices in developing nations like Kenya (Manji and Pastakia 2019). Pharmaceutical companies like AstraZeneca, GSK, Takeda, Novo Nordisk, and Sanofi are leading toward sustainability with patient-centric innovative drugs, and specialty medicines. IBM SkillsBuild platforms provide free online training to learners to develop digital skills promoting employability (Cabardo, Gardon, and Rabago). The LEGO Foundation, the LEGO Group collaborated with UNICEF since 2015 to empower children into physical, creative, cognitive, social, and emotional learners through play, offer psychosocial help for susceptible children in classrooms, and offer playful means to alleviate trauma in forty-eight nations. Their play-to-learn program supplies play-based learning materials along with teacher training for quality education (Ferguson 2019).

Coca-Cola Company embarked on 5 $\times$ 20 programs and pledged to five million women entrepreneurs across its value chain by 2020 (Kaplan and VanderBrug 2014). In the context of SDG 6, Coca-Cola Company has begun its water stewardship programs covering three themes: reducing water consumption, recycling, and replenishment. They work on improving water quality through wastewater treatment compliant with legal requirements and standards in almost all the countries they operate. The company also supports the UNDP program on 'Every Drop Matters' by undertaking nearly 100 watershed restoration projects, and sustainable agriculture practices in more than 20 countries (Jones et al. 2015). The Replenish Africa Initiative (RAIN) by The Coca-Cola Africa Foundation (TCCAF) aimed to promote accessibility to clean water and sanitation for 6 million African people by 2020 under Water, Sanitation, and Hygiene (WASH), watershed protection, and productive use of water category (Rostorfer et al. 2015). Procter and Gamble started the Children's Safe Drinking Water Program to supply clean water through the use of water purification technologies to nearly 1 billion people (Jones et al. 2015). Google pledged to achieve net-zero emissions by 2030 and operate on carbon-free energy like wind, solar, etc. More than 2.4 metric tons of carbon emissions were offset through the use of Google Maps (Pichai 2023) (https://sustainability.google/). Tesla pioneered the development of electric vehicles to decarbonize the transportation sectors (Masoud et al. 2023). Danone Manifesto Ventures program, based on the shareholder value concept, helps entrepreneurship and social inclusion by investing in sustainable food and beverages (Stephens 2021). Tesla's Model S, Model 3, Model X, and Model Y have revolutionized the automobile industry with innovation offering high-performance electric vehicles. Their energy storage products like Powerwall and

Powerpack enable renewable energy integration (Masoud et al. 2023; Wu et al. 2021). Siemens helps businesses to simulate and optimize processes through the application of digital twins towards lowering environmental impact (Annanth et al. 2021). Google products and services like Google Maps and Google Earth Engine serve as valuable tools for urban planning, disaster response, and environmental monitoring (Amani et al. 2020). Unilever's 'Sustainable Living Plan' prioritized the promotion of supplier diversity and supporting small minority-owned businesses to reduce inequalities in its value chain (Lawrence et al. 2018). Cisco provides smart city solutions. The Cisco Kinetic for Cities platform, a cloud-based stage, collects data from sensors using a secure network connection and helps to optimize transportation, water, energy, and waste management systems (Campus 2019; Palka, Pasichnyk, Matsiuk, and Matsiuk). Schneider Electric helps communities optimize energy and reduce carbon emissions through its products and services (Aoun 2013). The Boklok concept of the IKEA Group emphasizes low and middle income households and offers sustainable housing solutions (Bougrain et al. 2010).

The apparel company Patagonia ensured sustainability across its supply chain by total compliance with environmental and labour standards, transparency, monitoring, and regular audits (Shourkaei et al. 2023). Interface's Mission Zero (Van Koert et al. 2019), Unilever's Sustainable Living Plan (Lawrence et al. 2018), and Coca-Cola's World Without Waste (Bista 2019) programs are steps towards achieving SDG 12. As mentioned earlier, Microsoft's commitment to become carbon-negative by 2030 is a step towards SDG 7 and SDG 13. It also plans to start an internal carbon fee to promote emission reductions across its operations (Joppa et al. 2021). As a step towards SDG 14, a leading seafood company, Bumble Bee Foods ensures sustainability of the seafood supply chain. Their operations are third-party standards certified by the Marine Stewardship Council and Aquaculture Stewardship Council (Coca 2023; Crawford 2022). Hilton Group of Hospitality implemented sustainable tourism practices to reduce the ecological footprint and conserve coastal destinations. In tune with SDG 15, Nestle works with farmers to promote sustainable farming and supports rural livelihood (Cambaza 2022). A global carpet tiles manufacturer, Interface was disclosed to become a climate-positive company through its Climate Take Back mission by 2040 (Van Koert et al. 2019). In the direction of SDG 16, Siemens AG enforced robust compliance and integrity programs to promote ethical behavior and prevent corruption within the company (Annanth et al. 2021). The Bill and Melinda Gates Foundation takes SDG 17 ahead and invests in innovation and technology transfer initiatives to improve global health, reduce poverty, and prevent the spread of infectious diseases (Kalfagianni 2022). The International Finance Corporation, an affiliate of the World Bank Group supports sustainable business practices and mobilizes capital through its Green Bond program (Lee 2001). The UN Global Compact with thousands of participating corporations commits to tune their business operations with ten doctrines on human rights, labour, environment, and anti-corruption (Compact 2006).

5 Conclusion

Though CSR has evolved significantly over time, it is driven by shifting societal expectations, environmental concerns, and sustainable practices in business. Not only does CSR need to consider societal impacts but also connects business processes to macro-ecological processes within the planetary boundary limits (Schönherr et al. 2017).

5.1 Emerging Trends in Corporate Responsibility

With continued recognition of the significance of sustainability and ethical business practices, various CSR trends are emerging. Companies now place great emphasis on climate change and its impacts, offset carbon emissions, shift to renewable energy resources, conserve water, treat wastes, and practice circular economy. Businesses are increasingly embracing greener technologies, cutting their carbon footprint, and enforcing eco-friendly practices throughout their business practice. In the future CSR may leverage technologies based on AI such as machine learning, blockchain, IoT, and big data analytics to address and tackle social challenges, more effectively and result in a win–win condition for society and the environment (Schönherr et al. 2017). Companies are increasingly creating inclusive workplaces, upholding human rights, and addressing various inequalities to deal with diversity in various aspects through sustainable supply chain management and fair labour practices. Companies are growing in recognition of the stakeholders' engagement in CSR practices. Companies are scaling impacts and streamlining processes through digital innovations and building capacity, fostering community resilience, and meeting regulatory requirements and legal frameworks (Sudirman et al. 2021).

5.2 Challenges in Corporate Responsibility

It is a fact that businesses generate jobs and create wealth. It becomes a threat to society when it acts irresponsibly. Here comes the CSR which can mitigate such damages inflicted by irresponsible business. Though businesses have invested in CSR activities, many such plans are not systematically designed (Fallah Shayan et al. 2022). Prioritizing stakeholders and transparency in CSR can be challenging but assist the company in building trust and credibility. More cross-sectoral collaboration will enable enterprises to pool resources, share expertise and enhance collective actions to tackle poverty, education, and healthcare (Schönherr et al. 2017). Striking equilibrium between the financial goals and social impacts at the same time ensuring sustainability is challenging. Companies need to balance profitability with purpose-driven initiatives. Companies operating in more than one jurisdiction face regulatory complexities and compliance-related CSR obligations. Smaller companies with limited resources and budget constraints face difficulties in the implementation of CSR programs. Meeting the diverse expectations of stakeholders though challenging can be worked out through proper communication and proactive engagement of the stakeholders. Geopolitical uncertainties and upcoming social issues can pose new challenges for CSR practitioners.

The road is long and tough but not impossible to walk. Corporations empower vulnerable people by reducing hunger, poverty, inequality, and other discrimination (Singh and Rahman 2021). It goes beyond ethical and philanthropic activities and adds value to the company itself. Such endeavours generate shared value for businesses and societies together, making this world an improved place to reside. SDGs are extensive and forward-looking, a basket of solutions to key global challenges; serving as a consistent program in the endlessly changing and random business world. Hence, the corporations' CSR should align with SDG goals to make the business sustainable (Fallah Shayan et al. 2022).

Acknowledgements The authors are grateful and acknowledges the contribution of Dejus as an illustrator.

References

Ali HY, Danish RQ, Asrar-ul-Haq M (2020) How corporate social responsibility boosts firm financial performance: the mediating role of corporate image and customer satisfaction. Corp Soc Responsib Environ Manag 27(1):166–177

Amani M, Ghorbanian A, Ahmadi SA, Kakooei M, Moghimi A, Mirmazloumi SM, ... Parsian S (2020) Google earth engine cloud computing platform for remote sensing big data applications: a comprehensive review. IEEE J Sel Topics Appl Earth Observ Remote Sens 13:5326–5350

Annanth VK, Abinash M, Rao LB (2021) Intelligent manufacturing in the context of industry 4.0: a case study of siemens industry. Paper presented at the journal of physics: conference series

Aoun C (2013) The smart city cornerstone: urban efficiency. Schneider Electric White Paper 1:1–13

Babiak K, Trendafilova S (2011) CSR and environmental responsibility: motives and pressures to adopt green management practices. Corp Soc Responsib Environ Manag 18(1):11–24

Barnard CI (1968) The functions of the executive (Vol. 11): Harvard University Press

Baughn CC, Bodie NL, McIntosh JC (2007) Corporate social and environmental responsibility in Asian countries and other geographical regions. Corp Soc Responsib Environ Manag 14(4):189–205

Bista S (2019) Sustainability in business (A critique of environmental sustainability practices in Coca-Cola and Unilever)

Blackburn WR (2012) The sustainability handbook: the complete management guide to achieving social, economic and environmental responsibility. Routledge.

Bougrain F, Forman M, Haugbølle K (2010) Industrialisation in construction: Multiple actors, multiple collaborative strategies

Bowen HR (2013) Social responsibilities of the businessman. University of Iowa Press

Burke L, Logsdon JM (1996) How corporate social responsibility pays off. Long Range Plan 29(4):495–502
Cabardo J, Gardon RW, Rabago L, SkillsBuild Innovation Camp: Learning Paradigm Innovation Spearheaded by Asia Pacific College. Issues Philippine Higher Educ, 53
Cambaza E (2022) A commentary on nestlé's corporate social responsibility. Academia Lett
Campbell JL (2007) Why would corporations behave in socially responsible ways? An institutional theory of corporate social responsibility. Acad Manag Rev 32(3):946–967
Campus PP (2019. Managing Information in the Smart City: a proposed framework
Carroll AB (1979) A three-dimensional conceptual model of corporate performance. Acad Manag Rev 4(4):497–505
Carroll AB (1991) The pyramid of corporate social responsibility: Toward the moral management of organizational stakeholders. Bus Horiz 34(4):39–48
Carroll AB (2008) A history of corporate social responsibility: concepts and practices
Carroll AB (2015a) Corporate social responsibility. Organ Dyn 44(2):87–96
Carroll AB (2015b) Corporate social responsibility: the centerpiece of competing and complementary frameworks. Organizational dynamics
Carson R (2009) Silent spring. 1962: Getty Publications Los Angeles, CA, USA
Carson R, Darling L (1962) Silent spring. Houghton Mifflin/Riverside Press, Boston et Cambridge
Chaffee EC (2017) The origins of corporate social responsibility. U. Cin. l. Rev. 85:353
Chandler D (2020) Sustainable value creation. Routledge
Chandler D, Werther WB (2014) Strategic corporate social responsibility: Stakeholders, globalization, and sustainable value creation
Charles Jr OH, Schmidheiny S, WattsP.(2017) Walking the talk: The business case for sustainable development. Routledge
Chen YS, Chang CH, Wu FS (2012) Origins of green innovations: the differences between proactive and reactive green innovations. Manag Decis 50(3):368–398
Cisneros P, Shamsuzzoha A, Kuusniemi H, Jovanovski B (2023) Defining green innovation, its impact, and cycle–a literature analysis. Cleaner Eng Technol, 100693
Clark JM (1926) Social control of business. University of Chicago Press, Chicago, c1926, 1923 printing
Coca N (2023) Why, for bumble bee, certification matters for the future of sustainable seafood. Sustainable Brands
Compact UG (2006) UN global compact. New York.[Online.] Available from http://www.unglobalcompact.org [8 January 2006]
Crawford E (2022) The Bumble Bee Seafood Co's sustainability success opens unexpected marketing opportunities. FoodNavigator
Dai J, Lu C, Qi J (2019) Corporate social responsibility disclosure and stock price crash risk: evidence from China. Sustainability 11(2):448
Davis K (1960) Can business afford to ignore social responsibilities? Calif Manage Rev 2(3):70–76
Day E (2018) The history of earth day: Diakses dalam https://www.earthday.org/about/the-history-of-earth-day/.(5 ….
Demjanovičová M, Varmus M (2021) Changing the perception of business values in the perspective of environmental sustainability. Sustainability 13(9):5226
Dhaliwal DS, Radhakrishnan S, Tsang A, Yang YG (2012) Nonfinancial disclosure and analyst forecast accuracy: International evidence on corporate social responsibility disclosure. Account Rev 87(3):723–759
Eells RSF (1956) Corporation giving in a free society
Ehrlich PR (1968) Population bomb
Fallah Shayan N, Mohabbati-Kalejahi N, Alavi S, Zahed MA (2022) Sustainable development goals (SDGs) as a framework for corporate social responsibility (CSR). Sustainability 14(3):1222
Feldstein M (2013) The reagan-thatcher revolution. Retrieved from Juillet 22:2014
Ferguson S (2019) UNICEF and The LEGO foundation team up to build better futures for children. Forbes

Fosu E, Fosu F, Akyina N, Asiedu D (2024) Do environmental CSR practices promote corporate social performance? The mediating role of green innovation and corporate image. Cleaner Responsible Consum 12:100155
Frederick WC (1960) The growing concern over business responsibility. Calif Manage Rev 2(4):54–61
Friedman M (2007) The social responsibility of business is to increase its profits Corporate ethics and corporate governance, pp 173–178. Springer.
Gautam R, Singh A (2010) Corporate social responsibility practices in India: a study of top 500 companies. Glob Bus Manag Res 2(1):41–56
Hales J (2021) Sustainability accounting standards board (SASB) world scientific encyclopedia of climate change: case studies of climate risk, action, and opportunity Volume 3, pp 37–41. World Scientific
Handfield R, Sroufe R, Walton S (2005) Integrating environmental management and supply chain strategies. Bus Strateg Environ 14(1):1–19
Harrison B (1966) Philanthropy and the Victorians. Victorian Studies 9(4):353–374
Hartman D (2009) What is stewardship in business? small business
Heald M (1991) The social responsibilities of business: company and community, 1900–1960. J Bus Ethics 10(1)
Hedberg CJ, Von Malmborg F (2003) The global reporting initiative and corporate sustainability reporting in Swedish companies. Corp Soc Responsib Environ Manag 10(3):153–164
Henderson RM (2015) Making the business case for environmental sustainability. Harvard business school working paper series# 15–068
Hernández JPS-I, Yañez-Araque B, Moreno-García J (2020) Moderating effect of firm size on the influence of corporate social responsibility in the economic performance of micro-, small-and medium-sized enterprises. Technol Forecast Soc Chang 151:119774
Heslin PA, Ochoa JD (2008) Understanding and developing strategic corporate social responsibility. Organ Dyn 37:125–144
Homayoun S, Mashayekhi B, Jahangard A, Samavat M, Rezaee Z (2023) The controversial link between CSR and financial performance: the mediating role of green innovation. Sustainability 15(13):10650
Hong M, Drakeford B, Zhang K (2020) The impact of mandatory CSR disclosure on green innovation: evidence from China. Green Finance 2(3):302–322
Husted BW, Allen DB (2007) Strategic corporate social responsibility and value creation among large firms: lessons from the Spanish experience. Long Range Plan 40(6):594–610
Jamali D, Mirshak R (2007) Corporate social responsibility (CSR): theory and practice in a developing country context. J Bus Ethics 72:243–262
Javed M, Rashid MA, Hussain G, Ali HY (2020) The effects of corporate social responsibility on corporate reputation and firm financial performance: Moderating role of responsible leadership. Corp Soc Responsib Environ Manag 27(3):1395–1409
Jimenez D, Franco IB, Smith T (2021) A review of corporate purpose: an approach to actioning the sustainable development goals (SDGs). Sustainability 13(7):3899
Jones P, Hillier D, Comfort D (2015) Water stewardship and corporate sustainability: a case study of reputation management in the food and drinks industry. J Public Aff 15(1):116–126
Jones TM (1980) Corporate social responsibility revisited, redefined. Calif Manage Rev 22(3):59–67
Joppa L, Luers A, Willmott E, Friedmann SJ, Hamburg SP, Broze R (2021) Microsoft's million-tonne CO_2-removal purchase—lessons for net zero: Nature Publishing Group
Kaittani N (2023) Corporate social responsibility: where did it originate from, how has it evolved and why is it important?: a literature review of the history and evolution of CSR
Kalfagianni A (2022) Philanthropic foundations as agents of justice in global sustainability governance. Global Studies Quart 2(3):ksac033
Kaplan S, VanderBrug J (2014) The rise of gender capitalism. Stanf Soc Innov Rev 12(4):36–41
Klassen RD, McLaughlin CP (1996) The impact of environmental management on firm performance. Manage Sci 42(8):1199–1214

Kneipp JM, Gomes CM, Bichueti RS, Frizzo K, Perlin AP (2019) Sustainable innovation practices and their relationship with the performance of industrial companies. Revista De Gestão 26(2):94–111

Kraus S, Rehman SU, García FJS (2020) Corporate social responsibility and environmental performance: the mediating role of environmental strategy and green innovation. Technol Forecast Soc Chang 160:120262

KsiężaK P, FischBach B (2017) Triple bottom line: the pillars of CSR. J Corporate Responsib Leadership 4(3):95–110

Lantos GP (2001) The boundaries of strategic corporate social responsibility. J Consum Mark 18(7):595–632

Latapí Agudelo MA, Jóhannsdóttir L, Davídsdóttir B (2019) A literature review of the history and evolution of corporate social responsibility. Int J Corp Soc Responsib 4(1):1–23

Lawrence J, Rasche A, Kenny K (2018) Sustainability as opportunity: Unilever's sustainable living plan managing sustainable business: an executive education case and textbook, pp 435–455. Springer

Lee CF (2001) International finance corporation: financing environmentally and socially sustainable private investment Liber Amicorum Ibrahim FI Shihata, pp 469–481. Brill Nijhoff

Luetkenhorst W (2004) Corporate social responsibility and the development agenda: the case for actively involving small and medium enterprises. Intereconomics 39(3):157–166

Manji I, Pastakia SD (2019) Novartis access: a small step towards increased access for non-communicable disease care. Lancet Glob Health 7(4):e398–e399

Masoud RM, Mostafa RM, Khattab AA, Ahmed AMHA, Wahib R (2023) Assessing the transformative impact of tesla's strategic change interventions and technology implementation on the electric vehicle and clean energy industries

Matisoff DC, Noonan DS, O'Brien JJ (2013) Convergence in environmental reporting: assessing the Carbon Disclosure Project. Bus Strateg Environ 22(5):285–305

Matten D, Moon J (2008) "Implicit" and "explicit" CSR: a conceptual framework for a comparative understanding of corporate social responsibility. Acad Manag Rev 33(2):404–424

McKinnon A (2010) Environmental sustainability. Green logistics: improving the environmental sustainability of logistics. London

McWilliams A, Siegel D (2000) Corporate social responsibility and financial performance: correlation or misspecification? Strateg Manag J 21(5):603–609

Meadows DH, Randers J, Meadows DL (2013) The limits to growth (1972) The future of nature (pp 101–116): Yale University Press

Melay I, Kraus S (2012) Green entrepreneurship: Definitions of related concepts. Int J Strategic Manag 12(2):1–12

Mishra MS, Corporate social responsibility activities at ONGC: impact of regulations. GENERAL MANAGEMENT, 224

Montiel I (2008) Corporate social responsibility and corporate sustainability: Separate pasts, common futures. Organ Environ 21(3):245–269

Najam A (2013) World business council for sustainable development: the greening of business or a greenwash? Yearbook of International Cooperation on Environment and Development 1999–2000, pp 69–81. Routledge

Nations U (2023) The sustainable development goals report. Special edition

Palka NK, Pasichnyk V, Matsiuk O, Matsiuk S, Comparative analysis of smart city platforms

Pedersen CS (2018) The UN sustainable development goals (SDGs) are a great gift to business! Procedia Cirp 69:21–24

Pichai S (2023) Our global progress toward a sustainable future. Google

Pillay R (2015) The changing nature of corporate social responsibility: CSR and development–the case of Mauritius: Routledge

Pirzada K, Ahmed K, Moens G (2023) Aligning corporate strategies with the sustainable development goals. Foresight and STI Governance 17(2):5–7

Pogge T, Sengupta M (2015) The sustainable development goals (SDGS) as drafted: nice idea, poor execution. Wash. Int'l LJ 24:571
Portney KE (2020) Sustainability and public policy Handbook of US Environmental Policy, pp 463–476. Edward Elgar Publishing
Ren S, Huang M, Liu D, Yan J (2023) Understanding the impact of mandatory CSR disclosure on green innovation: evidence from Chinese listed firms. Br J Manag 34(2):576–594
Ren S, Wei W, Sun H, Xu Q, Hu Y, Chen X (2020) Can mandatory environmental information disclosure achieve a win-win for a firm's environmental and economic performance? J Clean Prod 250:119530
Rostorfer D, Mitler D, Ledbetter K (2015) Creating shared value through corporate social investment: managing water-related risk and opportunity through Coca-Cola's Replenish Africa Initiative
Roszkowska-Menkes M, Strategic C (2021) Porter and Kramer's (2006)"Shared Value" encyclopedia of sustainable management. Springer Cham, Switzerland
Sachs JD, Sachs LE (2021) Business alignment for the "Decade of Action." J Int Bus Policy 4:22–27
Sarkar R (2008) Public policy and corporate environmental behaviour: a broader view. Corp Soc Responsib Environ Manag 15(5):281–297
Schönherr N, Findler F, Martinuzzi A (2017) Exploring the interface of CSR and the sustainable development goals. Transnational Corporations 24(3):33–47
Selekman BM (1959) A moral philosophy for management
Shahzad M, Qu Y, Javed SA, Zafar AU, Rehman SU (2020) Relation of environment sustainability to CSR and green innovation: a case of Pakistani manufacturing industry. J Clean Prod 253:119938
Shourkaei MM, Taylor KM, Dyck B (2023) Examining sustainable supply chain management via a social-symbolic work lens: Lessons from Patagonia. Business strategy and the environment
Singh AP, Rahman Z (2021) Integrating corporate sustainability and sustainable development goals: towards a multi-stakeholder framework. Cogent Bus Manag 8(1):1985686
Soewarno N, Tjahjadi B, Fithrianti F (2019) Green innovation strategy and green innovation: the roles of green organizational identity and environmental organizational legitimacy. Manag Decis 57(11):3061–3078
Song W, Yu H (2018) Green innovation strategy and green innovation: the roles of green creativity and green organizational identity. Corp Soc Responsib Environ Manag 25(2):135–150
Stephens P (2021) Social finance investing for a resilient food future. Sustainability 13(12):6512
Strand R (1983) A systems paradigm of organizational adaptations to the social environment. Acad Manag Rev 8(1):90–96
Strasser G (2014) The grameen bank and beyond the millennium development goals: microfinance and poverty alleviation
Sudirman FA, Upe A, La Ode Herman FTS (2021) Corporate social responsibility (CSR) contribution to achieve sustainable development goals (SDGs) in southeast Sulawesi. Paper presented at the Proceedings of the 11th Annual international conference on industrial engineering and operations management Singapore
Szabó DG, Sørensen KE (2015) New EU directive on the disclosure of non-financial information (CSR). Europ Company Financ Law Rev 12(3):307–340
Szutowski D (2021) The impact of eco-innovation on the stock returns of technology-based KIBS: role of eco-innovation type and its degree of novelty. Sustain Account Manag Policy J 12(1):130–155
Tang M, Walsh G, Lerner D, Fitza MA, Li Q (2018) Green innovation, managerial concern and firm performance: an empirical study. Bus Strateg Environ 27(1):39–51
Teece DJ (2010) Technological innovation and the theory of the firm: the role of enterprise-level knowledge, complementarities, and (dynamic) capabilities Handbook of the Economics of Innovation (Vol 1, pp 679–730): Elsevier
Thacker H (2023) State bank of India CSR report: providing service beyond banking. The CSR J
Tseng M-L, Tan RR, Siriban-Manalang AB (2013) Sustainable consumption and production for Asia: sustainability through green design and practice. J Clean Prod 40:1–5

Urlings L (2020) Benchmarking for a better world: assessing corporate performance on the SDGs Assessment of responsible innovation, pp 316–318. Routledge
Van Koert M, Sherwood D, Kennedy S (2019) Interface: creating a climate fit for life through carpet tiles
Vogel D (2007) The market for virtue: the potential and limits of corporate social responsibility: Brookings Institution Press
Walton CC (1982) Corporate social responsibility: the debate revisited. J Econ Bus 34(2):173–187
Waterhouse BC (2017) The personal, the political and, the profitable: business and protest culture, 1960s–1980s. Financ History(121):14
Wood DJ, Cochran PL (1992) Business and society in transition. Bus Soc 31(1):1–7
Wren DA (2005) The history of management thought
Wu Z-Q, Nguyen P-Q, Phankasemsan I, Wolf E (2021) Driving sustainability and electric vehicles evolution: Tesla Company's Success Strategies
Xu H, Xu X, Yu J (2021) The impact of mandatory CSR disclosure on the cost of debt financing: evidence from China. Emerg Mark Financ Trade 57(8):2191–2205
Yu W, Ramanathan R, Wang X, Yang J (2018) Operations capability, productivity and business performance: the moderating effect of environmental dynamism. Ind Manag Data Syst 118(1):126–143
Yuan B, Cao X (2022) Do corporate social responsibility practices contribute to green innovation? The mediating role of green dynamic capability. Technol Soc 68:101868
Zsolnai L (2011) Environmental ethics for business sustainability. Int J Soc Econ 38(11):892–899

What Influences Companies to Go Beyond Mandatory Corporate Social Responsibility Rule? Empirical Evidence from India

Kishore Kumar, Rubee Singh, Pooja Mishra, Bikram Paul Singh Lehri, and Amit Kumar

Abstract This study aims to empirically investigate how corporate governance (CG) characteristics influence firms to go beyond the mandatory minimum corporate social responsibility (CSR) expenditure rule to contribute towards sustainable development. It employs the panel regression technique for analysis of top seventy-five listed companies of the NIFTY100 index at Indian National Stock Exchange (NSE) for the years from 2014–15 to 2020–21. The empirical results revealed that CG attributes like large board size, large independent directors, and female directors significantly influence the CSR performance of the companies. However, no significant evidence was found in case of the impact of board meeting frequency and CSR practices of the companies. This chapter enables a better understanding of self-induced CSR practices to policymakers, regulators, practitioners, and other stakeholders. The findings suggest that various stakeholders should concentrate on specific CG attributes to focus on CSR performance. It is one of the first studies that determines what influence the adoption of self-induced CSR practices especially against the backdrop of major CG mechanism and CSR reforms in India. It provides additional empirical evidence

K. Kumar (✉)
School of Management, Bennett University, Greater Noida, India
e-mail: akishore001@gmail.com

R. Singh · P. Mishra
GLA University, Mathura, India
e-mail: dr.rubeerajput@gmail.com

P. Mishra
e-mail: pujumishra@gmail.com

B. P. S. Lehri
University School of Business (USB), Chandigarh University, Mohali, India
e-mail: lehribikram@gmail.com

A. Kumar
Christ University, Bengaluru, India
e-mail: amitgem@gmail.com

R. Singh
Department of Management Studies, Kumaun University, Nainital, India

P. Singh et al. (eds.), *Integrating Environmental Sustainability and Corporate Social Responsibility: A Move Beyond Profit*, World Sustainability Series,
https://doi.org/10.1007/978-3-031-81469-3_2

to the extant body of literature on the CG and CSR practices from the perspective of emerging economies.

Keywords Board attributes · Corporate sustainability · Mandatory CSR · Sustainable development · Developing countries · India

1 Introduction

In the present era, increased pressure on companies from various stakeholders to improve the sustainability performance of the business has become a common norm. Organizations are expected to be financially accountable and adopt socially responsible behavior. Recently, CSR has been developed as a very important tool by companies to contribute towards sustainable development (Wang et al. 2018; Muller and Kolk 2009; Kah and Akenroye 2020). Kumar and Prakash (2021) emphasized that practicing CSR has become a new norm for companies and a public license to exist. Cho et al. (2019) described CSR as the practices undertaken by firms to assume responsibility for the negative consequences of their operations on society and environment. It is investing a certain portion of the profit earned by the companies back into society (Ng 2022). Integrating CSR into the fundamental corporate strategy allows companies to gain a competitive edge that is sustainable over time, while also positively influencing their financial performance (FP). Adoption of CSR also helps corporate to build a better reputation and brand value of the business (Li et al. 2016; Seifi and Crowther 2018). Kumar and Prakash (2020) highlighted that disagreement on practicing CSR persists among the corporate as high CSR expenditure may contradict the primary objective of profit maximization. CSR has garnered considerable interest from policymakers, political leaders, practitioners, civil society, and researchers in recent years. Kumar and Prakash (2020) emphasized that CSR refers to serving communities, society, and the environment by going above and beyond what is required by the law. However, many countries from both the developed and developing worlds have introduced mandatory and voluntary guidelines and policies for companies to adopt CSR practices. Usman (2020) argued that there is a consistent debate about whether CSR should be limited to a voluntary approach or enforced through a legislated rule of mandatory minimum CSR expenditure.

There is an increasing body of literature on CSR and this field is receiving evergrowing interest from scholars across the globe. Considerable research has been conducted to explore the CSR practices and their association with the firms FP. However, the majority of the empirical research revolves around the role of CSR performance in influencing customer satisfaction, employee engagement, and firms' brand value (Elmaghrabi 2021; Prakash et al. 2023). Bacha et al. (2021) noted the direct association among the firm's FP and CSR practices. Partalidou et al. (2020) concluded that CSR positively influences firm value in cases of high customer awareness and this relationship was found to be insignificant in cases of low customer awareness. However, it is also pertinent to identify what influence the uptake of

CSR practices in business (Uyar et al. 2020; Eiselein and Dentchev 2020). Many prior studies have also explored the role of firm-specific factors like profitability, age, turnover leverage, etc. in influencing CSR adoption and implementation. The corporate governance mechanism is critical in policy formulation and strategy implementation in the company (Kolk and Pinkse 2010; Fahad and Rahman 2020; Dwekat et al. 2022). Welford (2007) emphasized that sound corporate governance mechanisms considerably improve accountability toward the various stakeholders and the firm's CSR practices. Although CG plays a crucial role in influencing CSR initiatives, there is a scarcity of empirical studies that have investigated this connection specifically from the perspective of emerging economies such as India. Furthermore, previous papers on the CG and the CSR practices of firms has yielded inconsistent outcomes. It is noteworthy to examine how CG influences companies to go beyond legislated minimum CSR expenditure and spend more than what is required to be spent as CSR by the companies. Therefore, this study aims to conduct an empirical analysis to inspect the CG mechanism and the CSR activities linkage.

India possesses several significant regulatory and institutional characteristics that make it an intriguing and distinctive context for empirically assessing the influence of CG features on the CSR performance of organizations. In 2013, India became the pioneer in enacting legislation that mandated CSR expenditure, making it the first country to do so (Aggarwal and Jha 2019; Prakash et al. 2023). The Indian Companies Act, 2013 (Sec. 135) mandates companies to spend a certain portion of income on CSR. This law imposes a statutory obligation on every company in India, having a turnover of Indian national rupee (₹) 10 bn (USD 140 mn approx.) or more, ₹ 50 mn (USD 699,000 approx.) or more profits, or ₹ 500 mn (USD 70 mn approx.) or more net worth in a year to spend 2% of the average profits of the past 3 years as CSR expenditure (Aswani et al. 2021; Kapoor and Dhamija 2017; Guha et al. 2019). Moreover, the Schedule VII provision of this act also provides 12 broad activities for undertaking CSR practices. Kumar and Prakash (2020) found that the firm's CSR performance varies significantly despite the mandatory CSR rule in India. Some companies consistently spend way above the minimum amount required and several companies do not spend any amount on CSR activities to fulfill their CSR obligation (Hickman et al. 2021; Balon et al. 2022; Suman and Singh 2021). Consequently, in the last decade, several amendments have been undertaken by the government of India concerning the mandatory minimum CSR rule (Kumar 2022a, b; Kapoor and Dhamija 2017). Therefore, it is worthwhile to assess the factors that influence the CSR performance of Indian firms. Second, past ten years has witnessed several major reforms in the field of CG regulatory framework. The Securities Exchange Board of India (SEBI) introduced Clause 49 in the Companies Act 2013, for enhancing fairness and accountability in the CG framework (Almaqtari et al. 2020). Major reforms in the CG mechanism were related to diversity, independent directors (ID), the role of board members in CSR policy, size of board (BS) etc., to promote effective CG in the companies (Prakash et al. 2023). The contribution and implications of these CG reforms in improving the firm's CSR performance are still unknown. In light of this, the goal of this study is to undertake an empirical analysis to determine how certain CG elements influence Indian firms' CSR practices.

The findings of this chapter offer several crucial implications. Essentially, CSR is giving back to society. Interventions like mandatory CSR expenditure-related rules might not be enough to promote the CSR practices of companies. The results indicate that several CG attributes (i.e. BS, presence of female board members, and ID) has a critical role in promoting the adoption and execution of CSR practices within companies. Policymakers, practitioners and other stakeholders should benefit from the practical implications of these empirical findings because they shed light on the effects of key reforms and how they relate to the CSR performance of businesses. This chapter employs a multi-theory approach to assess the CG and CSR practices linkage. Furthermore, by exploring the dichotomy of voluntary versus mandatory CSR, this chapter substantially extends the existing literature in this domain.

The subsequent sections of this chapter are as follows: Sect. 2 is the conceptual foundations and hypotheses. Research method employed in this chapter is provided in Sect. 3. Findings is in Sect. 4. Finally, Sect. 5 provides concluding remarks, discusses its implications, and limitations.

2 Theoretical Background

Previous research in the domain of corporate social responsibility (CSR) has employed diverse theoretical frameworks to explore the concept and factors that influence the CSR performance of companies. Kumar (2022a, b) highlighted that CSR is a complex and evolving notion and therefore, one theory may not be sufficient to explain this phenomenon. The agency and the legitimacy theories have been used in the present study to understand the linkages between CG attributes and CSR practices. The legitimacy theory postulates that the survival of the business depends upon society. Olateju et al. (2021) emphasized that firms have a kind of tacit social contract with society that compel companies to contribute to a certain portion of their earnings in society. From the legitimacy theory perspective, companies should consider the expectations of not only their shareholders but also the expectations of all the members of society (Arena et al. 2013; Dare 2016). Patten (2020) emphasized that these social expectations are the social norms for companies. Failure to comply with these social norms may adversely impact the firm image and its performance (Orazalin et al. 2019). Companies adopt better CSR to comply with the social standards and expectations of society (Silva 2021). Another widely used theory is the agency theory perspective which advocates firms should adopt CSR activities to diminish the agency cost between the agent and the principal (Li 2016). Pérez (2015) highlighted that managers undertake socially responsible business activities to limit information asymmetry and convey a positive image to all the stakeholders. As noted by Jo and Harjoto (2012) firms voluntarily adopt better CSR practices to mitigate perceived financial risk. Although, the agency theory perspective provides a strong explanation of the adoption of CSR practices, very often finding the accountable principal of the concerned agent is not easy (Patten 2020). Whereas in the case of legitimacy theory, the expectation of all the stakeholders varies and keeps on evolving (Ahmad et al. 2017). Hence, the present study applies both theories to present a complementary perspective on the linkages between CSR practices and CG attributes.

2.1 Hypotheses

2.1.1 BS and Firms CSR Practices

The existing research examining the affiliation between BS and CSR practices has presented contradictory perspectives and views. Kumar (2022) found that a large BS represents an effective CG mechanism in the firm to ensure socially responsible business conduct. Numerous researchers have found that larger board size has a better postion to control firm's socially responsible behavior (Rouf and Hossan 2022; Cuadrado-Ballesteros et al. 2017; Endrikat et al. 2021; Velte 2022; Al Fadli et al. 2020; Kumar and Prakash 2020). A board with more members tend to take rational decision taking into consideration the expectations of all the participants of the society and there are less chances that it may be impacted by the top management (Abdullah et al. 2011; Alabdullah et al. 2019; Kathy Rao et al. 2012). In contrast, Jensen (1993) argued that a large board also implies poor coordination, and decisions are most likely taken by the top management. Large board size is often considered inefficient because it results in weaker control of management (Laksmana 2008; Giannarakis 2014). Relatively small boards work more efficiently and mitigate the agency problem. If board is large, they are less chance that they carry out their responsibilities and duties effectively (Ahmed et al. 2006). Small boards have more control over management and tend to impact the decisions related to CSR practices (Shamil et al 2014). Kumar (2022a, b) and Eberhardt-Toth et al. (2017) reported a negative influence of the BS on the firm CSR. Whereas some studies also reported an insignificant influence of BS on the firm's CSR practices (Sufian and Zahan 2013). Thus, as per the agency theory, this chapter assumed that more members of the board lead to better CSR practices.

H1. Board size significantly impacts CSR performance.

2.1.2 FBM and Firms CSR Practices

The association between the FBM and the organizations' CSR practices is relatively less explored in the previous literature. Alshbili (2020) reported that very frequent board meetings facilitate proper monitoring and consequently result in better organizational performance. Kumar (2022) argued that frequent board meetings reflect that board members are active and dedicated to addressing business concerns. Frequent board meetings enhance board members' consideration to the social responsibility of the business (Kent and Monem 2008; Haji 2013). It implies that active board members implement more CSR activities to build a positive image for the business.

The legitimacy theory posits that frequent board meetings indicate a well-functioning board proactively taking different CSR initiatives to legitimize the business in society (Liu and Zhang 2017; Jizi et al. 2016). Based on the above-discussed argument and the outcome of past studies, this research assumed that FBM positively impact CSR practices.

H2. Board meeting frequency significantly impacts CSR performance.

2.1.3 Board Independence (BI) and Firms CSR Practices

As per the agency theory, the presence of an ID is instrumental in promoting socially responsible business activities. Ahmad and Rashid (2017) emphasized that more independent directors are crucial for improving CG mechanisms in the firm. Independent directors present the concerns of all the stakeholders of the firm and mitigate the agency problem (Al Farooque and Ahulu 2017; Barako 2007; Brooks et al. 2009). They offer better inputs on the stakeholder's view of the business (Chau and Gray 2010; García-Sánchez et al. 2019). They not only take part in important strategic decisions but also facilitate the monitoring of those decisions (Post et al. 2015; Patelli and Prencipe 2007; Kumar et al. (2021). Webb (2004) conducted a study and concluded that publically responsible firms have more number of IDs as against non-socially responsible companies. Kumar et al. (2021) highlighted that IDs are an important instrument to monitor management behavior toward the stakeholders. However, previous literature presents conflicting results on association of board independence and CSR performance. Brooks et al. (2009) reported that a higher proportion of IDs positively impact firm CSR. In contrast, some studies reported an insignificant association between ID and the CSR practices of the companies (Orazalin and Mahmood 2020; Amran et al. 2014). Hence, the following hypothesis is developed to revisit this issue:

H3. Board independence significantly impacts CSR performance.

2.1.4 Gender Diversity (GD) and Firms CSR Practices

GD in the CG mechanism has garnered considerable attention among scholars (Dienes and Velte 2016; Nielsen and Huse 2010). Many countries in recent years have introduced guidelines for the obligatory addition of female directors on the board (i.e., Spain, Italy, Kenya, Germany France, Israel, Finland, India, etc.). The Indian Companies Act, of 2013 made it compulsory to include atleast 01 woman director on the board for all listed companies operating in India (Reddy and Jadhav 2019; Kumar 2022). Dienes and Velte (2016) highlighted that women directors significantly improve the effectiveness of the CG mechanism in the firm. Halliday et al. (2021) emphasized that women's presence brings valuable perspectives to the boardroom. Several studies have explored gender diversity and financial performance perspectives (Kakabadse et al. 2015). Elsayih et al. (2019) concluded women board members help to ensure moral values and consider the interest of all stakeholders. Most of the existing literature reported that women board member postivly impact firm CSR practices (Elsayih et al. 2019; Bear et al. 2010; Kirsch 2018). As per the theoretical argument and prior findings, this study assumed that the more female directors positively impacts the firm's CSR.

H4. Gender diversity significantly impacts CSR performance.

2.1.5 Control Variables

This study has also taken into consideration certain firm-specific factors (i.e., age, level of leverage, return on assets, and size) as the control variable in the model.

Multiple research have reported positive correlation among CSR practices and the age of firms (Bhagat and Bolton 2008; Arora and Sharma 2016; Buallay et al. 2017). Kumar (2022) emphasized that relatively old firm has more experience and generally tend to contribute a certain portion of profits for the development of society. Established companies implement CSR to authenticate their societal existence (Withisuphakorn and Jiraporn 2016). The profitability variable has been consistently used by researchers as an important determinant of the CSR performance of companies (Crifo and Forget 2015). Several prior studies have found a positive influence of earnings on the firms CSR (Goering 2010). Similarly, the leverage variable is also extensively studied by prior researchers. However, the finding reported present conflicting views. Orazalin and Mahmood (2020) findings indicated a significant association between firm leverage and CSR, although a few studies have also observed a inverse impact of these variables. In the case of firm size, the most studies concluded a positive impact of FS and CSR practices.

3 Methods

3.1 Data

The sample of the present research comprised seventy-five companies (excluding banking companies) of the NIFTY100 Index. The NIFTY100 Index includes the top hundred listed corporations based on their market capitalization at the NSE, India. According to NSE, companies included in the NIFTY100 index represent 79.8 percent of the total turnover of NSE (NSE, 2019). A total of 523 observations spanning a period of seven years, from fiscal year 2015 to 2022 is the final sample used in this study. This study has taken FY 2015 because India implemented the new Companies Act, 2013 from the year 2014–15. This act ushered in a major reform related to CG and CSR for corporate in India. Table 1 provides the sample classification of industries. The data about firm-specific factors and CSR expenditure has been accessed using the Centre for Monitoring Indian Economy (CMIE) database

Table 1 Presents sample classification based on industry type

Industry type	No. of Companies
IT, telecom, and other consumer services	12
Energy	12
FMCG and others	11
Industrial manufacturing and realty	9
Metals and mining	8
Automobile	8
Cement and construction material	7
Chemicals, fertilizer and Pesticides	5
Miscellaneous	3

The information about CG attributes have been gathered from annual reports (AR), and the official websites of the companies.

3.2 Measurement

The independent variables (IV) in this chapter were BS, BI, GD, and board meetings frequency (FBM). BS was determined by the total number of directors. GD was determined by the proportion of women directors. BI was identified using the proportion of IDs and FBM was measured as the number of board meetings conducted in a year. Control variables (CV) used were age (AG), level of leverage (LV), size (SZ), and return on asset (ROA). The firm CSR performance is the dependent variable (DV) in this chapter. CSR performance was assessed as the percentage of profits spend by the company as CSR expenditure in the financial year. Operationalization and variable sources are presented in Table 2.

3.3 Empirical Model

The panel data regression model was employed to examine the effects of CG characteristics on the CSR performance of companies in India. The following regression model has been used in the present study:

$$\begin{aligned} CSRP_{it} = & \beta_0 + \beta_1 * BS_{it} + \beta_2 * BI_{it} * + \beta_3 * FBM_{it} + \beta_4 * GD_{it} \\ & + \beta_5 * AG + \beta_6 * LV_{it} + \beta_7 * SZ_{it} + \beta_8 * LV_{it} + u_{it} \end{aligned}$$

where,

$CSRP_{it}$CSR performance.

BS_{it}Board size.

Table 2 Present operationalization of variables

Variables	Abbreviation	Source	Operationalization
IV			
Board size	BS	CG/CSR/Sustainability Report	Total no. of directors
Board independence	BI	CG/CSR/Sustainability Report	The proportion of total ID
Board meeting frequency	FBM	CG/CSR/Sustainability Report	No. of board meetings held in the year
Gender Diversity	GD	CG/CSR/Sustainability Report	No. of women directors on the board
DV			
CSR Performance	CSRP	CMIE Database	Percentage of profits spend on CSR practices in the FY
CV			
Firms size	SZ	CMIE database	Log. of total sales
Age	AGE	Authors compilation	Years since the incorporation of the firm
Leverage	LEV	CMIE database	Total debt to total assets ratio
Profitability (Return on Assets)	ROA	CMIE database	Net profits after tax to total assets ratio

BI_{it}Board independence.
FBM_{it}board meetings frequency.
GD_{it}Gender diversity.
AGE_{it}Age.
LEV_{it}Leverage.
ROA_{it}Return on assets.
u_{it}Specific error term.
βParameters.
^{it}*i*th observation.

The panel data was short and balanced in the present study. Initially, Breusch-Pagan (LM) test was employed to select among the random effects (RE) model and the ordinary least square (OLS). The results were significant ($p = 0.000$), indicating high heterogeneity in the data. Hence, it was concluded that the OLS model was not suitable for the present data set. Further, F-test was applied to select between the OLS and fixed effects (FE) models. Results were significant ($p = 0.0000$), indicating the OLS model was not an appropriate model for the dataset. The Hausman test was performed to assess the appropriateness of the RE and the FE model, and the outcome was significant ($p = 0.0477$). All the estimated results indicated that the FE model is appropriate for the given dataset. The Bruesch-Godfrey test was conducted to assess the serial correlation and the results showed a significant level of serial correlation

Table 3 Results

Test	Test-statistics	p-value
Test of poolability (Breusch-Pagan) (LM test)	$\chi 2 = 58.71676$	0.0000*
F-test	F = 2.906538	0.0000*
Hausman test (model selection)	$\chi 2 = 15.6486$	0.0477**
Pesaran's CD test (cross-section dependence)	Z = −0.225369	0.8217
Breusch-Godfrey test (serial correlation)	$\chi 2 = 1.86226$	0.0000*
Breusch-Pagan test (heteroskedasticity)	BP = 3932.171	0.0000*

* $p < 0.01$; ** $p < 0.05$; *** $p < 0.1$

(p = 0.0000). The Breusch-Pagan test was performed on the dataset to identify the presence of heteroskedasticity and the results were significant (p = 0.0000). The panel data suffer high spatial correlation due to the non-random selection of cross-section units in this study (Driscoll and Kraay 1998); the Pesaran CD test was used to assess the cross-correlation. The results were insignificant (p = 0.8217). The results are presented in Table 3.

There was an issue of serial correlation and heterogeneity in the data set, which may impact the efficiency of estimators. Data cleaning was performed to solve the issue of heterogeneity. Boxplot was used to remove the extreme values and further, new estimates were calculated. But results indicated no significant improvement and the estimator's direction remained unchanged. The Newey-West estimators are grounded on Newey and West's (1987) std. errors were employed to address the issue of heteroskedasticity and serial correlation (Newey and West 1987; Millo 2017). Table 3 shows result.

4 Discussion

Table 4 provides the descriptive statistics. The dependent variable, CSR performance has a 2.264 mean value, suggesting a marginally better average performance of companies compared to the mandated rule of a minimum of 2% of profits as CSR expenditure. However, CSR performance ranges between 0 to 8.30 percent indicating significant variation in the CSR performance. For independent variables, the range for board size is relatively wider from 4 to 22 members, with an average of 13 members. The FBM average is 4.625, with atleast three meetings and a maximum of ten in a year. This is marginally above the SEBI, India rule of at least 4 meetings in a year. BI has a mean of 6.365, indicating a better representation of independent directors. Their count ranges between 3 to 12 with the SD of 1.567. Firm size mean is 9.63, and ROA ranges between 0.000375 to 1187.566. The average firm age is 45 years where, the oldest company is 114-year-old and they youngest is eight-year-old. Details are provided in Table 4.

Table 4 Descriptive statistics

	CSRP	BS	BIN	FBM	GD	AG	SZ	ROA	LV
Mean	2.264208	12.75335	6.365201	4.625239	1.386233	45.45646	4593.845	73.10492	0.206383
Median	2.000000	13.00000	6.000000	4.000000	1.000000	38.00000	9.533560	0.027975	0.110520
Maximum	8.300000	21.00000	12.00000	10.00000	4.000000	114.0000	1,198,766	1187.566	0.998768
Minimum	0.000000	4.000000	3.000000	3.000000	0.000000	8.000000	4.345666	0.000375	3.238660
SD	0.753181	2.574254	1.567905	0.929041	0.697534	24.25026	74,059.14	172.7230	0.230095
Skewness	2.536679	0.130449	0.287350	1.411056	0.833822	0.749657	16.07805	3.785914	1.218238
Kurtosis	22.58814	4.188539	4.097381	6.272522	4.062257	2.836094	259.5038	19.93602	3.913419
Jarque–Bera	8922.257	32.26673	33.43989	406.9313	85.19293	49.57191	1,456,299	7499.848	147.5459
Probability	0.000000	0.000000	0.000000	0.000000	0.000000	0.000000	0.000000	0.000000	0.000000
Sum	1079.581	6670.000	3329.000	2419.000	725.0000	23,773.73	2,402,581	38,233.87	107.9383
Sum SD	222.7092	3459.182	1283.247	450.5468	253.9809	306,975.1	2.869012	15,572,955	27.63670
Observations	523	523	523	523	523	523	523	523	523

Table 5 provides the results of the correlation. It shows that the CSRP is negatively correlated with the firm size (−0.0417) and leverage (0.103). Whereas, CSR performance was found to be positively correlated with BS, FBM, BI, GD, and ROA.

Table 6 provides the summary of regression results of the DV (CSRP) on IV. The first hypothesis assumed that the BS significantly influences the CSR performance of the companies. It was found that estimated coefficient of BS is positive and statistically significant for the DV. Thus, H1 stands supported. This indicates that more board members in the boardroom are critical to driving the CSR performance of the companies. This finding also corroborates the agency theory perspective, indicating a relatively large-sized board focus on adopting better CSR practices to mitigate the agency problem. Same findings were also reported by Rouf and Hossan (2021) and Cuadrado-Ballesteros et al. (2017). One important justification for this finding could be that large-sized boards take much more rational decisions and address concerns of all the stakeholders in society. Hypothesis 2 predicted that board meetings positively influence the CSR performance of the firm. Contrary to the expectation, the coefficient of FBM suggests an insignificant influence on the FBM on the CSRP. This H2 is not empirically supported. This finding contradicts the results documented by Alshbili et al. (2020) and Kumar et al. (2021). This result is against the expectation that frequent board meetings result in better consideration and monitoring of the social performance of the companies. However, Hazi (2013) Jizi et al (2014) findings also indicated that there is no significant impact of FBM on the CSR practices of the firm.

H3 predicted that BI significantly impacts the CSR practices of the firm. The data indicates that there is a positive and significant correlation between the ID and the success of corporate social responsibility (CSR) initiatives. Thus, H3 stands empirically supported and corroborates the agency theory. Theoretically, this finding suggests that IDs in the board represent the expectations of society and implement better CSR measures. This finding supports the theoretical argument that boards with more independent directors tend to adopt better CSR policies. These results also support the finding of prior studies (Brooks et al. 2009; Post et al. 2015).

For the gender diversity (GD) variable, the result indicates that GD significantly impacts the CSR practices of the companies. Thus, H4 stands supported. This result supports the outcome of prior researchers, especially from the developing economies perspective. It is also aligned with the agency theory viewpoint, indicating women directors in the boardroom improve firms' CSRP. However, it contradicts the outcome by Torchia et al. (2011). This outcome supports the notion that female directors bring a valuable perspective to the board discussion and give impetus to the non-financial performance of the business. This finding also supports the recent trend across the world mandating female directors on the board for better CG mechanisms.

Table 5 Correlation matrix

	CSRP	BS	FBM	BI	GD	AG	SZ	LV	ROA
CSRP	1.00000								
BS	0.17341	1.000000							
FBM	0.10545	0.239229	1.000000						
BI	0.04510	0.733361	0.178307	1.00000					
GD	0.11214	0.119302	0.025724	0.16855	1.00000				
AG	0.03254	−0.041724	0.025759	0.06377	−0.16454	1.000000			
SZ	−0.04407	0.114355	0.025016	0.10421	0.01011	0.120311	1.0000		
LV	−0.0295	−0.057744	0.033194	−0.06064	0.05328	−0.07546	−0.0290	1.000000	
ROA	0.06283	0.088338	−0.03494	0.07456	0.00222	−0.09506	−0.0262	−0.097347	1.000000

Table 6 Hypotheses testing results

Redundant FE tests					
Test cross-section FE					
Effects Test		Statistic	df	p-value	
Cross-section F		2.906538	(74,434)	0.000*	
Cross-section Chi-square		208.10144	74	0.000*	
Fixed Effects Tests					
Variable	*Coefficient*	*Std. Error*	*t-Statistic*	*p-value*	*Hypothesis status*
CSRP	1.173503	0.196030	5.986332	0.0000*	Supported
BS	0.076101	0.016423	4.633916	0.0000*	Supported
BI	−0.087075	0.026808	−3.248146	0.0012*	Supported
FBM	0.046183	0.031006	1.489497	0.1370	Not Supported
GD	0.107970	0.042042	2.568156	0.0105*	Supported
AG	0.002366	0.001195	1.980299	0.0482**	
SZ	−6.07E-07	3.80E-07	−1.595058	0.1113	
LV	−0.062325	0.122061	−0.510604	0.6099	
ROA	0.000217	0.000163	1.333023	0.1831	
R^2	0.369534	Adjusted R^2		0.354881	
F-statistic	4.745407	Durbin-Watson stat		1.257921	
Prob (F-statistic)		0.000012*			

5 Conclusion

There is a emerging body literature based on empirical evidence that effective CG mechanism in the firm improves the nonfinancial performance of the business. To examine this linkage, this research empirically examines the influence of CG attributes on the organizations' CSR performance in the unique setting of India where the legislated CSR approach was adopted in the year 2013. This chapter concentrate on the self-induced CSR performance of companies and its underlying factor for the period from 2014–15 to 2020–21, post major CSR and CG reforms were undertaken by introducing the new Indian Companies Act, of 2013. These findings significantly extend the emerging literature based on CSR and the implications of effective CG mechanisms. The present study offers important insights into the voluntary versus mandatory CSR perspectives. The results revealed that the firm's CSR performance significantly varies despite the enactment of mandatory CSR in India. The implication of these findings propagates the idea that the CSR performance of firms is guided by diverse factors along with mandatory CSR approach enforced by the government.

This study concluded that found that CG attributes like board size, independent directors and women directors significantly impact the firm's CSR performance. However, no significant evidence was found that suggests frequent board meetings significantly improve firms' CSR performance.

This empirical investigation provides several practical implications. This study should be useful for the policymaker, practitioner, regulators, and other stakeholders. The findings enable a better understanding of the various aspects of CG mechanism that lead to better CSR performance. To enhance compliance with the mandatory CSR rule and promote CSR practices, policymakers, and regulators may focus on specific CG attributes to advance the adoption and implementation of CSR practices. This study revealed that organizations should concentrate on bringing more independent directors and female directors to the boardroom. This chapter offer valuable perspective to the debate of the voluntary versus mandatory CSR approach. This empirical investigation provided evidence that a good governance mechanism could improve the firm's CSR performance. The findings provide a crucial assessment of the recent major reforms in the CSR and CG undertaken by the government of India.

This chapter has certain limitations that provide potential avenues for future studies. This study has only considered four CG attributes, future studies in this director may include more CG characteristics such as social responsibility committee and CEO duality. CSR expenditure by the companies is considered as their CSR performance in this study. Further, this study used the sample of only the top seventy-five listed companies listed at NSE, India. The relatively large sample may provide a better insight into the CSR performance of companies in India. Future research in this direction may incorporate these issues to further expand the knowledge of CG and CSR.

References

Abdullah SN, Mohamad NR, Mokhtar MZ (2011) Board independence, ownership and CSR of Malaysian large firms. Corporate Ownership Control 8: 417–431. http://www.virtusinterpress.com/IMG/pdf/10-22495_cocv8i2c4p5.pdf.

Aggarwal VS, Jha A (2019) Pressures of CSR in India: an institutional perspective. J Strateg Manag 12:227–242. https://doi.org/10.1108/JSMA-10-2018-0110

Ahmad NBJ, Rashid A, Gow J (2017) Board independence and corporate social responsibility (CSR) reporting in Malaysia. Australasian Account Bus Financ J 11:61–85. https://doi.org/10.14453/aabfj.v11i2.5

Ahmed K, Hossain M, Adams MB (2006) The effects of board composition and board size on the informativeness of annual accounting earnings. Corporate Govern Int Rev 14:418–431. https://doi.org/10.1111/j.1467-8683.2006.00515.x

Al Fadli A, Sands J, Jones G, Beattie C, Pensiero D (2020) Board independence and CSR reporting: pre and post analysis of JCGC 2009. Int J Law Manag. https://doi.org/10.1108/IJLMA-11-2018-0259

Al Farooque O, Ahulu H (2017) Determinants of social and economic reporting: evidence from Australia, the UK and South African multinational enterprises. Int J Account Inf Manag 25:177–200. https://doi.org/10.1108/IJAIM-01-2016-0003

Alabdullah TTY, Ahmed ER, Muneerali M (2019) Effect of board size and duality on corporate social responsibility: what has improved in corporate governance in Asia?. J Account Sci 3:121–135. https://doi.org/10.21070/jas.v3i2.2810
Almaqtari FA, Al-Hattami HM, Al-Nuzaili KM, Al-Bukhrani MA (2020) Corporate governance in India: a systematic review and synthesis for future research. Cogent Bus Manag 7:1803579. https://doi.org/10.1080/23311975.2020.1803579
Alshbili I, Elamer AA, Beddewela E (2020) Ownership types, corporate governance and corporate social responsibility disclosures: empirical evidence from a developing country. Account Res J. 33:148–166. https://doi.org/10.1108/ARJ-03-2018-0060
Amran A, Lee SP, Devi SS (2014) The influence of governance structure and strategic corporate social responsibility toward sustainability reporting quality. Bus Strateg Environ 23:217–235. https://doi.org/10.1002/bse.1767
Arora A, Sharma C (2016) Corporate governance and firm performance in developing countries: evidence from India. Corp Gov 16:420–436. https://doi.org/10.1108/CG-01-2016-0018
Aswani J, Chidambaran NK, Hasan I (2021) Who benefits from mandatory CSR? Evidence from the Indian Companies Act 2013. Emerg Mark Rev 46:100753
Bacha S, Ajina A, Ben Saad S (2021) CSR performance and the cost of debt: does audit quality matter? Corporate Govern Int J Bus Soc 21:137–158. https://doi.org/10.1108/CG-11-2019-0335
Balon V, Kottala SY, Reddy KS (2022) Mandatory corporate social responsibility and firm performance in emerging economies: an institution-based view. Sustain Technol Entrep 1:100023. https://doi.org/10.1016/j.stae.2022.100023
Barako DG (2007) Determinants of voluntary disclosures in Kenyan companies annual reports. African J Bus Manag 1(5):113–128. https://academicjournals.org/journal/AJBM/article-full-text-pdf/8C361B617434.pdf
Bear S, Rahman N, Post C (2010) The impact of board diversity and gender composition on corporate social responsibility and firm reputation. J Bus Ethics 97:207–221. https://doi.org/10.1007/s10551-010-0505-2
Bhagat S, Bolton B (2008) Corporate governance and firm performance. J Corp Finan 14:257–273. https://doi.org/10.1016/j.jcorpfin.2008.03.006
Brooks A, Oliver J, Veljanovski A (2009) The role of the independent director: evidence from a survey of independent directors in Australia. Aust Account Rev 19:161–177. https://doi.org/10.1111/j.1835-2561.2009.00055.x
Buallay A, Hamdan A, Zureigat Q (2017) Corporate governance and firm performance: evidence from Saudi Arabia. Australasian Account Bus Financ J 11:78–98. https://doi.org/10.14453/aabfj.v11i1.6
Chau G, Gray SJ (2010) Family ownership, board independence and voluntary disclosure: evidence from Hong Kong. J Int Account Audit Tax 19:93–109. https://doi.org/10.1016/j.intaccaudtax.2010.07.002
Cho SJ, Chung CY, Young J (2019) Study on the relationship between CSR and financial performance. Sustainability 11(2):343. https://doi.org/10.3390/su11020343
Crifo P, Forget VD (2015) The economics of corporate social responsibility: a firm-level perspective survey. J Econ Surv 29:112–130. https://doi.org/10.1111/joes.12055
Cuadrado-Ballesteros B, García-Sánchez IM, Martínez-Ferrero J (2017) The impact of board structure on CSR practices on the international scale. Eur J Int Manag 11:633–659. https://doi.org/10.1504/EJIM.2017.087559
Dienes D, Velte P (2016) The impact of supervisory board composition on CSR reporting. Evidence from the German two-tier system. Sustainability 8:63. https://doi.org/10.3390/su8010063
Driscoll JC, Kraay AC (1998) Consistent covariance matrix estimation with spatially dependent panel data. Rev Econ Statist 80:549–560.https://doi.org/10.1162/003465398557825
Dwekat A, Seguí-Mas E, Zaid MA, Tormo-Carbó G (2022) Corporate governance and corporate social responsibility: mapping the most critical drivers in the board academic literature. Meditari Account Res 30:1705–1739. https://doi.org/10.1108/MEDAR-01-2021-1155

Eberhardt-Toth E (2017) Who should be on a board corporate social responsibility committee? J Clean Prod 140:1926–1935. https://doi.org/10.1016/j.jclepro.2016.08.127

Eiselein P, Dentchev NA (2020) Managing conflicting objectives of social enterprises. Soc Enterp J 16(4):431–451

Elmaghrabi ME (2021) CSR committee attributes and CSR performance: UK evidence. Corporate Govern Int J Bus Soc 21:892–919

Elsayih J, Tang Q, Lan Y (2019) Corporate governance and carbon transparency: Australian experience. Account Res J. https://doi.org/10.1108/ARJ-12-2015-0153

Endrikat J, De Villiers C, Guenther TW, Guenther EM (2021) Board characteristics and corporate social responsibility: a meta-analytic investigation. Bus Soc 60:2099–2135. https://doi.org/10.1177/0007650320930638

Fahad P, Rahman PM (2020) Impact of corporate governance on CSR disclosure. Int J Discl Gov 17(2020):155–167. https://doi.org/10.1057/s41310-020-00082-1

García-Sánchez IM, Gómez-Miranda ME, David F, Rodríguez-Ariza L (2019) Board independence and GRI-IFC performance standards: the mediating effect of the CSR committee. J Clean Prod 225:554–562. https://doi.org/10.1016/j.jclepro.2019.03.337

Giannarakis G (2014) The determinants influencing the extent of CSR disclosure. Int J Law Manag. 56:393–416. https://doi.org/10.1108/IJLMA-05-2013-0021

Goering GE (2010) Corporate social responsibility, durable-goods and firm profitability. Manag Decis Econ 31(7):489–496. https://doi.org/10.1002/mde.1508

Guha SK, Samanta N, Majumdar A, Singh M, Bharadwaj A (2019) Evolution of corporate governance in India and its impact on the growth of the financial market: an empirical analysis (1995–2014). Corporate Govern Int J Bus Soc 19(5):945–984. https://doi.org/10.1108/CG-07-2018-0255

Haji A (2013) Corporate social responsibility disclosures over time: evidence from Malaysia. Manag Audit J 28:647–676. https://doi.org/10.1108/MAJ-07-2012-0729

Halliday CS, Paustian-Underdahl SC, Fainshmidt S (2021) Women on boards of directors: a meta-analytic examination of the roles of organizational leadership and national context for gender equality. J Bus Psychol 36:173–191. https://doi.org/10.1007/s10869-019-09679-y

Hickman LE, Iyer SR, Jadiyappa N (2021) The effect of voluntary and mandatory corporate social responsibility on earnings management: evidence from India and the 2% rule. Emerg Mark Rev 46:100750. https://doi.org/10.1016/j.ememar.2020.100750

Jensen MC (1993) The modern industrial revolution, exit, and the failure of internal control systems. J Financ 48:831–880. https://doi.org/10.1111/j.1540-6261.1993.tb04022.x

Jizi MI, Salama A, Dixon R, Stratling R (2014) Corporate governance and corporate social responsibility disclosure: evidence from the US banking sector. J Bus Ethics 125:601–615. https://doi.org/10.1007/s10551-013-1929-2

Jo H, Harjoto MA (2012) The causal effect of corporate governance on corporate social responsibility. J Bus Ethics 106:53–72. https://doi.org/10.1007/s10551-011-1052-1

Kah S, Akenroye T (2020) Evaluation of social impact measurement tools and techniques: a systematic review of the literature. Soc Enterp J 16(4):381–402. https://doi.org/10.1108/SEJ-05-2020-0027

Kakabadse NK, Figueira C, Nicolopoulou K, Hong Yang J, Kakabadse AP, Özbilgin MF (2015) Gender diversity and board performance: Women's experiences and perspectives. Hum Resour Manage 54:265–281. https://doi.org/10.1002/hrm.21694

Kapoor GK, Dhamija S (2017) Mandatory CSR spending-Indian experience. Emerg Econ Studies 3:98–112. https://doi.org/10.1177/2394901517696645

Kathy Rao K, Tilt CA, Lester LH (2012) Corporate governance and environmental reporting: an Australian study. Corporate Govern Int J Bus Soc 12:143–163. https://doi.org/10.1108/14720701211214052

Kent P, Monem R (2008) What drives TBL reporting: good governance or threat to legitimacy? Aust Account Rev 18:297–309. https://doi.org/10.1111/j.1835-2561.2008.0036.x

Kirsch A (2018) The gender composition of corporate boards: a review and research agenda. Leadersh Q 29:346–364. https://doi.org/10.1016/j.leaqua.2017.06.001
Kolk A, Pinkse J (2010) The integration of corporate governance in corporate social responsibility disclosures. Corp Soc Responsib Environ Manag 17:15–26. https://doi.org/10.1002/csr.196
Kumar K (2022a) Emerging phenomenon of corporate sustainability reporting: evidence from top 100 NSE listed companies in India. J Public Aff 22(1):2368. https://doi.org/10.1002/pa.2368
Kumar K (2022b) The uptake of sustainability disclosure practices by environmentally polluting industries in India. Crowther D, Seifi S (Ed) The equal pillars of sustainability. Develop Corporate Govern Respons 17:199–224. https://doi.org/10.1108/S2043-052320220000017010.
Kumar K, Kumari R, Kumar R (2021) The state of corporate sustainability reporting in India: evidence from environmentally sensitive industries. Bus Soc Rev 126(4):513–538. https://doi.org/10.1111/basr.12247
Kumar K, Prakash A (2021) An assessment of mandatory corporate social responsibility rule. Econ Politically Weekly 56:21–24
Kumar K, Prakash A (2020) Managing sustainability in banking: extent of sustainable banking adaptations of banking sector in India. Environ Dev Sustain 22:5199–5217
Li F, Li T, Minor D (2016) CEO power, corporate social responsibility, and firm value: a test of agency theory. Int J Manag Financ 12(5):611–628. https://doi.org/10.1108/IJMF-05-2015-0116
Liu X, Zhang C (2017) Corporate governance, social responsibility information disclosure, and enterprise value in China. J Clean Prod 142:1075–1084. https://doi.org/10.1016/j.jclepro.2016.09.102
M Shamil, M, Shaikh MJ, Ho PL, Krishnan A (2014). The influence of board characteristics on sustainability reporting: empirical evidence from Sri Lankan firms. Asian Rev Account 22:78–97https://doi.org/10.1108/ARA-09-2013-0060
Millo G (2017) Robust standard error estimators for panel models: a unifying approach. J Stat Software 82:1–27. https://doi.org/10.18637/jss.v082.i03
Muller A, Kolk A (2009) CSR performance in emerging markets evidence from Mexico. J Bus Ethics 85:325–337. https://doi.org/10.1007/s10551-008-9735-y
Newey WK, West KD (1987) A simple, positive semi-definite, heteroskedasticity and autocorrelation consistent covariance matrix. Econometrica 55:703–708. https://www.nber.org/system/files/working_papers/t0055/t0055.pdf.
Ng M (2022) The impact of corporate social responsibility expectations on purchase intention of social enterprise products. Soc Enterp J 18(4):585–604. https://doi.org/10.1108/SEJ-01-2022-0001
Nielsen S, Huse M (2010) The contribution of women on boards of directors: going beyond the surface. Corporate Govern Int Rev 18:136–148. https://doi.org/10.1111/j.1467-8683.2010.00784.x
NSE. NIFTY100 index (2019) Retrieved from: www1.nseindia.com/products/content/equities/indices/nifty_100.htm
Olateju DJ, Olateju OA, Adeoye SV, Ilyas IS (2021) A critical review of the application of the legitimacy theory to corporate social responsibility. Int J Manag Studies Res 9(3):1–6. https://doi.org/10.20431/2349-0349.0903001
Orazalin N, Mahmood M (2019) Economic, environmental, and social performance indicators of sustainability reporting: evidence from the Russian oil and gas industry. Energy Policy 121:70–79. https://doi.org/10.1016/j.enpol.2018.06.015
Orazalin N, Mahmood M (2020) Determinants of GRI-based sustainability reporting: evidence from an emerging economy. J Account Emer Econ 10(1):140–164. https://doi.org/10.1108/JAEE-12-2018-0137
Partalidou X, Zafeiriou E, Giannarakis G, Sariannidis N 2020. The effect of corporate social responsibility performance on financial performance: the case of food industry. Benchmarking Int J 27:2701–2720. https://doi.org/10.1108/BIJ-11-2019-0501

Patelli L, Prencipe A (2007) The relationship between voluntary disclosure and independent directors in the presence of a dominant shareholder. European Account Rev 16:5–33. https://doi.org/10.1080/09638180701265820

Patten DM (2020) Seeking legitimacy. Sustain Account Manag Policy J 11(6):1009–1021. https://doi.org/10.1108/SAMPJ-12-2018-0332

Pérez A (2015) Corporate reputation and CSR reporting to stakeholders: gaps in the literature and future lines of research. Corporate Commun Int J 20:11–29. https://doi.org/10.1108/CCIJ-01-2014-0003

Post C, Rahman N, McQuillen C (2015) From board composition to corporate environmental performance through sustainability-themed alliances. J Bus Ethics 130:423–435. https://doi.org/10.1007/s10551-014-2231-7

Prakash A, Kumar K, Khan W, Siddiquei MI (2023) Self-induced versus structured corporate social responsibility: the Indian Context. Measuring sustainability and CSR: From reporting to decision-making. Springer International Publishing, Cham, pp 167–178

Reddy S, Jadhav AM (2019) Gender diversity in boardrooms-a literature review. Cogent Econ Financ 7:1644703. https://doi.org/10.1080/23322039.2019.1644703

Rouf MA, Hossan MA (2021) The effects of board size and board composition on CSR disclosure: a study of banking sectors in Bangladesh. Int J Ethics Syst 37:105–121. https://doi.org/10.1108/IJOES-06-2020-0079

Seifi S, Crowther D (2018) The need to reconsider CSR. In: Redefining corporate social responsibility, pp 1–11. Emerald Publishing Limited

Silva S (2021) Corporate contributions to the sustainable development goals: an empirical analysis informed by legitimacy theory. J Clean Prod 292:125962. https://doi.org/10.1016/j.jclepro.2021.125962

Sufian MA, Zahan M (2013) Ownership structure and corporate social responsibility disclosure in Bangladesh. Int J Econ Financ Issues 3:901–909. https://dergipark.org.tr/en/pub/ijefi/issue/31960/351965

Suman S, Singh S (2021) Corporate governance mechanisms and corporate investments: evidence from India. Int J Product Perform Manag 70(3):635–656. https://doi.org/10.1108/IJPPM-09-2019-0453

Torchia M, Calabrò A, Huse M (2011) Women directors on corporate boards: from tokenism to critical mass. J Bus Ethics 102:299–317. https://doi.org/10.1007/s10551-011-0815-z

Usman B (2020) CSR performance, firm's attributes, and sustainability reporting. Int J Bus Soc 21:521–539. https://doi.org/10.33736/ijbs.3269.2020

Uyar A, Kilic M, Koseoglu MA, Kuzey C, Karaman AS (2020) The link among board characteristics, corporate social responsibility performance, and financial performance: evidence from the hospitality and tourism industry. Tourism Manag Perspect 35:100714. https://doi.org/10.1016/j.tmp.2020.100714

Velte P (2022) Meta-analyses on corporate social responsibility (CSR): a literature review. Manag Rev Quart 72:627–675. https://doi.org/10.1007/s11301-021-00211-2

Wang Z, Hsieh TS, Sarkis J (2018) CSR performance and the readability of CSR reports: too good to be true? Corp Soc Responsib Environ Manag 25:66–79. https://doi.org/10.1002/csr.1440

Webb E (2004) An examination of socially responsible firms' board structure. J Manage Govern 8:255–277. https://doi.org/10.1007/s10997-004-1107-0

Welford R (2007) Corporate governance and corporate social responsibility: issues for Asia. Corp Soc Responsib Environ Manag 14:42–51. https://doi.org/10.1002/csr.139

Withisuphakorn P, Jiraporn P (2016) The effect of firm maturity on corporate social responsibility (CSR): do older firms invest more in CSR? Appl Econ Lett 23(4):298–301. https://doi.org/10.1080/13504851.2015.1071464

Unmasking the Masks of Profits: Analyzing Corporate Social Responsibility Through an Environmental Justice Framework

Amal Narayanan, Radhey Shyam Sharma, and Vandana Mishra

Abstract Current study examines whether Corporate Social Responsibility (CSR) initiatives can genuinely address the environmental and social injustices within the global fashion supply chain, using an environmental justice framework. We analyzed CSR policies of leading fashion brands to identify areas of environmental injustice and evaluate the overall justice these policies embody. Our research highlights that the voluntary compliance mechanisms often fail to protect workers and the environment in low- and middle-income countries. Despite CSR efforts, the fast fashion model perpetuates exploitation and environmental degradation, evidenced by events such as the Rana Plaza collapse and extensive pollution in production regions. The vibrant marketing strategies of fashion brands mask the external costs of production, challenging the effectiveness of CSR. Therefore, the fast fashion industry's current approach to CSR is insufficient in addressing the deep-seated issues of environmental justice. Only by recognizing and rectifying these externalities and exploitations can meaningful progress be made towards a more sustainable and equitable fashion industry. This research calls for a re-evaluation of CSR practices and urges stakeholders to consider the broader implications of their actions on both the environment and vulnerable communities worldwide. The study underscores the need for more robust, legally binding regulations and international oversight to ensure credible and effective CSR practices. Our findings provide critical insights for policymakers,

A. Narayanan · R. S. Sharma · V. Mishra (✉)
Department of Environmental Studies, Bioresources and Environmental Biotechnology Laboratory, University of Delhi, Delhi 110007, India
e-mail: vmishra@es.du.ac

R. S. Sharma
Delhi School of Climate Change and Sustainability, Institute of Eminence, University of Delhi, Delhi, India

V. Mishra
Centre of Interdisciplinary Studies of Mountain and Hill Environment (CISMHE), University of Delhi, Delhi 110007, India

DDA Biodiversity Parks Program, CEMDE, University of Delhi, Delhi 110007, India

P. Singh et al. (eds.), *Integrating Environmental Sustainability and Corporate Social Responsibility: A Move Beyond Profit*, World Sustainability Series,
https://doi.org/10.1007/978-3-031-81469-3_3

industry stakeholders, and researchers, advocating for a shift towards more equitable and sustainable practices in the fashion industry.

Keywords Corporate social responsibility · Environmental sustainability · Fast fashion · Supply chain · Sustainable development · Worker exploitation · Environmental degradation

1 Introduction

The past three decades have witnessed a significant transformation in the fashion industry, often referred to as the "democratization of fashion." This phenomenon has seen the proliferation of new fashion brands that have made fashion accessible to the general public (Papasolomou et al. 2023). This shift has fuelled the rise of the fast fashion business model, which has simultaneously driven the industry's massive growth and its substantial environmental and social costs. As illustrated in Fig. 1, the fashion industry is now the third-largest manufacturing sector globally, renowned for its problematic supply chain. Contributing approximately $2.5 trillion USD to the global economy and providing over 300 million jobs worldwide, the industry also accounts for up to 10% of global CO_2 emissions and is the second-largest consumer of water, significantly contributing to water pollution (Garg et al. 2020; Rawat et al. 2016; UN Climate Change 2018).

Globalization has dispersed textile production to low- and middle-income countries (LMICs), creating a complex and fragmented global fashion supply chain (Niinimäki et al. 2020; Bick et al. 2018). This division reinforces the global north–south divide: labor-intensive production is concentrated in the global south, while design and administration remain in the global north. This disparity perpetuates economic inequalities, visible even in the disposal phase of apparel, where items are incinerated, landfilled, or exported to developing countries, particularly in Africa (Sandin and Peters 2018; Brooks and Simon 2012). Such conditions highlight the fashion industry's environmental injustices, necessitating an analysis through the lens of environmental justice—a movement advocating for equitable environmental treatment regardless of race, color, class, nationality, or gender.

CSR initiatives, including codes of conduct, sustainability reporting, and global partnerships, aim to create a sustainable supply chain (Perry and Wood 2019; Cheung et al. 2009). Promoted as an antidote to business injustices, CSR is intended to enhance the capacity and living standards of stakeholders and society (Lindgreen and Swaen 2010; Watts and Holme 1999). However, CSR often fails to alleviate the moral irresponsibility within the industry, particularly in developing nations where regulatory frameworks are weaker compared to developed countries (Niinimäki et al. 2020; Kuznetsov et al. 2009). In context to the fashion industry, CSR often fails to address the deeper environmental and social injustices embedded in the global south's production landscapes, which further results in growing resentment among the least heard stakeholders—workers and local communities.

Fig. 1 The impacts of fashion industry across the globe

This paper investigates whether CSR can genuinely address these injustices within the global fashion supply chain. By applying an environmental justice framework, we scrutinize the CSR policies of fashion brands, identifying pockets of environmental injustice and evaluating the overall justice embodied in these policies. The current study: (i) analyze CSR and environmental justice concepts; (ii) examine the institutional design of the global fashion supply chain and associated injustices; (iii) juxtapose CSR policies against environmental justice principles. Reflecting on the findings, this research advocates for the development of empirical models to evaluate CSR policies. These models would provide valuable insights for policymakers, industry stakeholders, and researchers, promoting more equitable and sustainable practices in the fashion supply chain.

2 Co-evolution of Principles of Corporate Social Responsibility and Environmental Justice

2.1 Historical Roots and Evolution of Corporate Social Responsibility (CSR)

In 1851, Daniel Salt and Sons, a major textile business in Bradford, England, relocated their industrial operations to the outskirts, establishing a planned industrial community named Saltaire. This initiative aimed to address the severe pollution and dire living conditions in Bradford, then the world's wool textile capital (Smith 2003). Within the capitalist socio-economic order, businesses produce goods and services for profit, a reward granted by society for their capabilities. This profit incentive has been scrutinized since the inception of capitalist production. Despite debates about the motives of Daniel Salt and Sons, their actions demonstrated a clear sense of social responsibility, setting a precedent in their industry.

The scholarly concept of Corporate Social Responsibility (CSR) emerged after Howard R. Bowen's 1953 publication, "Social Responsibilities of the Businessman," earning him the title "Father of CSR" (Carroll 1999). Initially a marginal concept in business schools (Shamir 2011), CSR evolved through contributions from numerous scholars (Fig. 2). Early CSR theories emphasized a normative approach, advocating that corporations should exceed economic gains to address ethical considerations for societal well-being. Corporations, as holders of production means, were expected to embody societal moral values (Smith 2003; McGuire 1963).

Since the 1980s, CSR has become foundational in various business management themes, such as corporate social performance, stakeholder theory, and corporate citizenship. CSR is now seen as a broad concept encompassing corporate social performance, responsiveness, citizenship, governance, accountability, sustainability, and social entrepreneurship (Freeman et al. 2010). The European Commission redefined CSR as "the responsibility of enterprises for their impacts on society," directing corporations to integrate social, environmental, ethical, human rights, and consumer concerns into their core strategies (European Commission 2011). CSR thus entails acting ethically and contributing to societal development, improving the quality of life for stakeholders, including workers, their families, and local communities (Watts and Holme 1999). CSR initiatives encourage investment in human development, environmental enrichment, and resolving social issues (Tai and Chuang 2014). CSR also includes a business case approach, where corporations prioritize self-interest while maintaining moral responsibility to society, thereby achieving a unique status (Smith 2003). Corporations are encouraged to foster stakeholder relationships to pursue common goals (Andriof and Waddock 2017; Maak 2007). The conceptual phase of CSR concluded as corporations faced increasing pressure for transparency. Following the fall of the Berlin Wall, the global economy's expansion and the proliferation of free-market principles reinforced CSR, driven by global NGOs, weak governance, and technological advancements. In this context, CSR is viewed through two

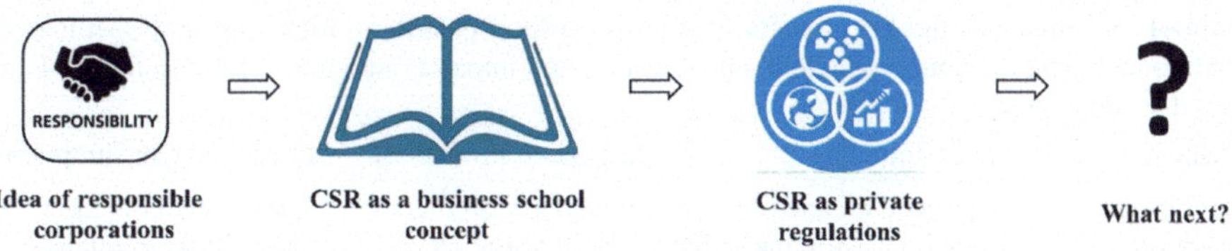

Fig. 2 Evolution of corporate social responsibility

paradigms: the world capitalist paradigm and the world culture paradigm. The former sees CSR as a tool to portray corporations as ethically responsible, avoiding legally binding regulations through voluntary compliance. The latter suggests that CSR is a cultural model disseminated by authoritative global actors, forcing corporations to adopt ethical practices (Shamir 2011) (Fig. 2).

Debates on CSR's nature in the European Union highlighted lobbying by corporations and advocacy groups. Corporates argued that obligatory CSR would undermine business spirit, advocating for voluntary CSR instead. Advocacy groups called for mandatory social and environmental performance reporting. Ultimately, voluntary CSR prevailed (Shamir 2011). During 1970's different authors emphasized the voluntary nature of CSR, arguing that moral responsibilities should not be imposed by regulations (Manne and Wallich 1972; Davis 1973). However, scholars question the effectiveness of voluntary CSR in a competitive free-market environment (Tai and Chuang 2014). The lack of a uniform CSR implementation standard raises concerns about the authenticity of CSR data and stakeholder engagement (Smith 2003; Wood 2010). The global capitalist regime's expansion exacerbates economic disparities, particularly between developed and underdeveloped nations, hegemonically necessitating CSR as a tool for addressing weak governance and exploitation (Baughn et al. 2007). This shift towards private governance compromises state power and alters societal power dynamics. Ultimately, the evolution of CSR reflects an ongoing dialogue about the role and responsibilities of businesses in society (Table 1). As CSR continues to adapt to global economic and cultural changes, its effectiveness in promoting ethical practices and reducing inequalities remains a critical area for further research and development.

2.2 The Evolution and Impact of Environmental Justice

Bullard (1993), a pioneer of the environmental justice movement, stated, "*Whether by conscious design or institutional neglect, communities of color in urban ghettos, rural 'poverty pockets,' or economically impoverished Native American reservations face some of the worst environmental devastations in the nation.*" The environmental justice movement in the United States emerged from the Civil Rights Movement of the late 1950s and early 1960s, which included leaders like Martin Luther King

Table 1 Summary of the key aspects of the fast fashion paradigm, highlighting the shift from traditional to fast fashion, production, and consumption impacts, and resulting environmental and social consequences

Aspect	Description	Impact on Production	Impact on Consumption
Traditional Fashion	Operated on fixed schedules with marketing strategies like trade fairs and fashion shows	Predictable and less frequent production cycles	Consumers bought apparel less frequently and at higher prices
Democratic Fashion	Fast fashion made apparel affordable and accessible by reducing retail prices and increasing external costs	Continuous introduction of new designs, driving frequent production	Encouraged frequent and often reckless consumption
Planned Obsolescence	Reduced product lifespan (absolute obsolescence) and promoted new trends (relative obsolescence)	Focused on short-lived apparel and frequent new styles	Consumers frequently updated wardrobes due to short product lifespan
Supply Chain Changes	Responsive to emerging trends, increasing fashion seasons and demand growth	Faster, flexible production, often relocated to low-cost countries, leading to sweatshops	Constantly new fashion, increased textile waste, worker exploitation
Consumer Pressure	Intense demand for low-cost fashion impacts entire supply chain	Intense competition, labor exploitation in sweatshops	Consumers contribute to labor exploitation and environmental damage

Jr. This movement highlighted "ecological distribution conflicts" such as the deliberate dumping of chemically polluted waste in predominantly minority counties in North Carolina (Martinez-Alier et al. 2016; Mohai et al. 2009; Cutter 1995). Studies confirmed that race and class significantly influenced disproportionate exposure to environmental hazards (Brown 1995; Faber and Kreig 2002). Environmental justice analyses extended beyond health to institutional standards of living, such as property values, education quality, and equity (Fig. 3). Reports from the Environmental Protection Agency in the 1990s also supported these findings.

While the environmental justice movement began as a fight against racism in North America, it expanded to address various forms of discrimination, incorporating factors like class and gender. It evolved into a broader analysis of political

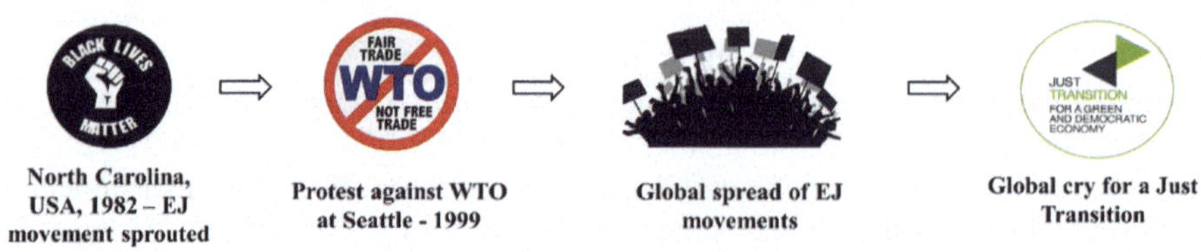

Fig. 3 Evolution of the concept of environmental justice

economy, emphasizing the social costs of environmental degradation (Barca 2012; Cutter 1995). As global political economy expanded through trade agreements, production activities and their environmental impacts were outsourced to developing economies, leading to "ecological distribution conflicts" (Martinez-Alier et al. 2016). This outsourcing sparked the "jobs vs. environment" debate and the globalization of environmental justice movements, exemplified by the protests against WTO globalization policies in Seattle, known as the Battle of Seattle (Mohai et al. 2009).

The environmental justice movement also birthed concepts like "environmentalism of the poor," addressing issues faced by impoverished rural communities deprived of ecosystem rights (Guha and Alier 2013). This aligned with "environmentalism of the dispossessed," which critiques the global capitalist production paradigm for dispossessing people of their resource rights (Martinez-Alier et al. 2016). Other forms of environmental justice include climate justice, biopiracy, and ecological debt. "Working-class environmentalism" emerged as a lens to analyze business value chains (Table 1), addressing workers' struggles and ecological rights (Barca 2012). Health concerns, such as the Black Lung movement in the USA, spurred labor reforms and integrated labor movements with environmental activism.

These seemingly contradictory movements have evolved into the concept of "Just Transition," addressing the "jobs vs. environment" dilemma (Wilgosh et al. 2022). Environmental justice, when challenging societal power relations, divides into affirmative and transformative streams. Transformative ecological justice, advocating radical changes in state, capital, and labor relations, emphasizes a socio-ecological approach while centering human interests (Stevis and Felli 2015). As Just Transition expands to include global energy mixes, it incorporates various socio-political theories, from reactionary to revolutionary. Expanding these discussions across all economic sectors is crucial to addressing power structures and holding those responsible for injustices accountable. This continuous evolution emphasises the importance of environmental justice as a dynamic and integral part of global socio-economic development.

2.3 The Framework of Two Contradictory Phenomena: CSR and Environmental Justice

Despite their divergent trajectories, CSR and Environmental Justice intersect and evolve synonymously with changing socio-political contexts. CSR is often presented as a comprehensive solution to issues arising from corporate activities. However, it lacks a uniform global blueprint. Conversely, environmental justice addresses the external costs and ecological distribution conflicts stemming from corporate activities, demanding accountability, and resolution.

The disparate application of CSR between the Global North and Global South highlights the economic and socio-political divide, constituting the first point of

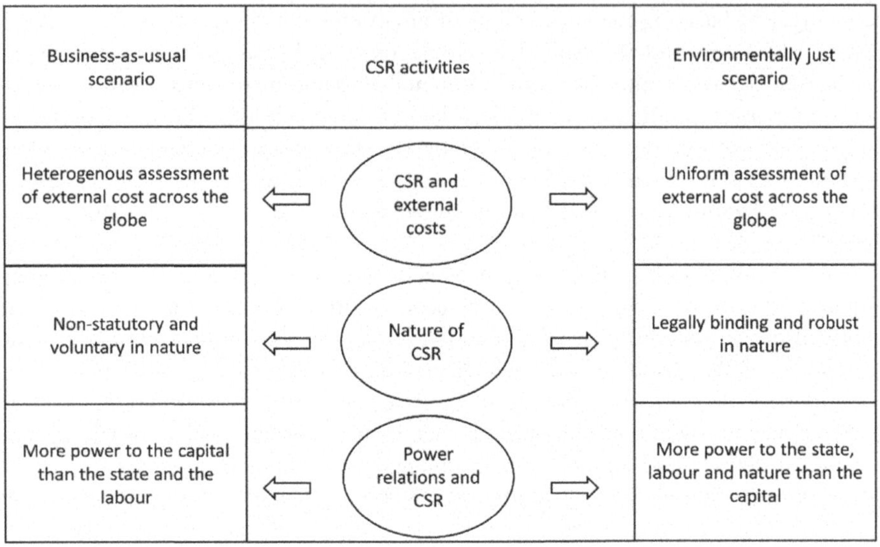

Fig. 4 Schematic representation of framework to analyse CSR policies using the lens of environmental justice

contradiction (Fig. 4). CSR's heterogeneous nature underscores the differing scales used to assess external costs in these regions. When CSR and environmental justice movements globalized, corporations faced immense pressure to be transparent, leading to non-statutory and voluntary compliance mechanisms. The authenticity of these mechanisms was subsequently questioned by environmental justice movements, marking the second point of contradiction.

Both CSR and environmental justice challenge underlying societal power relations. CSR, whether as a 'world culture paradigm' or a 'world capitalist paradigm,' promotes private governance under the guise of social responsibility, often at the expense of state authority. In contrast, environmental justice critiques neoliberal social relations and advocates for increased power for the state, labor, and nature. This forms the third major point of contradiction. As shown in Fig. 4, these three contradictory points can be reflected in a single CSR policy, necessitating a detailed analysis to untwine them. Analyzing CSR policies of fashion retailers through the lens of environmental justice helps assess the degree of justice inherent in these policies.

3 'Bread and Butter' or 'Bread and Grease': The Employment Dilemma

3.1 The Transformation and Impact of Fast Fashion

Fashion is a fundamental aspect of human civilization, deeply embedded in daily life (Mishra et al. 2020). Under the current socio-economic order, the fashion apparel market satisfies this demand through various means of production and distribution. Historically, fashion markets operated with strategies based on sales data, trade fairs, fashion shows, and fabric events, typically limited to two seasons per year (Birtwistle et al. 2003). However, the rise of fast fashion has significantly altered the industry. Fast fashion democratizes fashion by making apparel accessible to a broader audience, a result of the expansionist strategies of certain fashion corporations (Bruce and Daly 2006; Buckley and Clark 2012). As N. Anguelov notes in "The Dirty Side of the Garment Industry," this was achieved by reducing retail prices and increasing external costs, leading to reckless and frequent consumption (Anguelov 2015).

The primary goal of the fashion industry is to continuously launch new designs to attract consumers (Barnes and Lea-Greenwood 2006). This frequent consumption is driven by planned obsolescence, both absolute and relative. Absolute obsolescence dates back to 1939 when DuPont modified their durable nylon thread to be less long-lasting to increase sales (Baša and Farkas 2022). In fast fashion, this results in apparel with a significantly reduced shelf life, often only wearable ten times or even less (Ghemawat et al. 2003; Bhardwaj and Fairhurst 2010; Niinimäki 2015). Relative obsolescence, a key aspect of consumer society, requires the industry to produce and distribute innovative designs to create a constant desire for "newness" (Fels et al. 2016; Barnes and Lea-Greenwood 2006).

Due to this paradigm shift, the fashion supply chain has changed to a system which is highly responsive to emerging trends rather than a system of predetermined projections and calculations. This framework of fast fashion rooted in planned obsolescence which resulted in an increase in the number of fashion seasons per year—Zara's brand reported to have 20 seasons per year, and which brought about a 2% growth in fashion demand per year (Fletcher 2016), proliferated with a detrimental cost of piling up of the textile waste due to the early disposal.

This framework of demand and supply for fashion, consciously made by the corporations to enhance their profits, exerted huge pressure on the global fashion supply chain throughout its length that a highly competitive environment of business emerged out of it which forced the industries to translocate the production units to low and middle income countries in order to bring up the apparels for lowest feasible price and this competition, sometimes known as "the race to the bottom", holds to be liable for the emergence of sweatshops—a term used to denote the spaces of labour exploitation (Portway 2019; Bhardwaj and Fairhurst 2010; Adams 2002). The pressure exerted by consumers ripples through the entire supply chain, impacting workers and communities at every level (Fig. 5). As the industry continues to evolve,

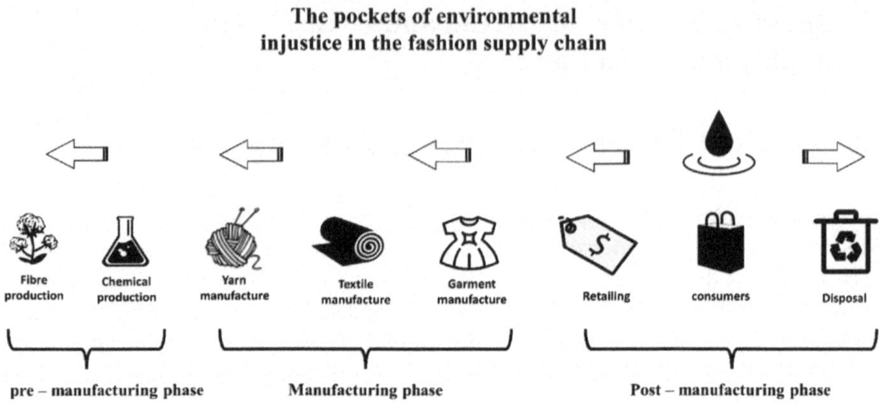

Fig. 5 The ripple effect across the global fashion supply chain

addressing these ethical and environmental challenges remains critical for creating a more sustainable and equitable fashion future.

3.2 *The Hidden Costs of Globalized Fashion Supply Chains*

Globalization has significantly impacted the fashion supply chain, transplanting its roots to the third world, particularly in fiber production and manufacturing (Bick et al. 2018). This industrialization, fuelled by foreign direct investment, has driven economic growth in these countries but has also caused severe environmental degradation (Eweje 2006). To reduce production costs, companies often relocate to nations with less stringent environmental regulations and without mandatory pollution control technologies. This outsourcing is key to the low prices of apparel. The supply chain is structured to minimize lead times (– the time required for a process from its beginning to its end) and maximize responsiveness to consumer demand (Christopher et al. 2004; Barnes and Lea-Greenwood 2006).

Fashion industry utilises various strategies to cope up with the consumer demand and build up a strong and strategic supply chain which is fit for competition, since it is the supply chain that is competing in the modern world of business (Hines 2007). Key strategies include just-in-time sourcing, quick response systems, and rapid supply chain management, all designed to reduce lead times and inventory delays (Bruce et al. 2004; MacCarthy and Jayarathne 2013). In case of Just-in-time sourcing, the merchandise is supplied to the retailers according to the immediate demands encountered by the retailers where the inventory will not be carried forward in time. In quick response system, there will be strong and rapid ripple effect of the consumer demand throughout the supply chain to steer clear of the delays rising out of the fabric and inventory delays and it takes an integrative approach to pass down the arising demands and to appropriate small and endless orders. The rapid supply chain management involves effective structuring of supply chain to harmonize the demand driven process utilizing various information technology in order to minimize the lead times.

These strategies have increased the number of suppliers, who must be flexible to meet retailers' demands, often at the expense of compliance with standards (Barnes and Lea-Greenwood 2006). For instance, Nike sources from around 800 suppliers in over 50 countries, employing more than 600,000 people (Smith 2003). This globally dispersed and complex supply chain contributes to the unsustainability of fast fashion, reflected in environmental and human health deterioration and human rights violations. H&M's sustainability report, "Conscious Action," acknowledges that garment production often occurs in countries with high human rights risks and low environmental awareness (H&M 2011).

The Rana Plaza tragedy in Dhaka, Bangladesh, on April 24, 2013, highlighted these issues when over 1,100 people died, and many more were injured in a factory collapse. Despite global outcry, significant improvements in worker conditions in low- and middle-income countries remain elusive (Taplin 2014). These hidden external costs threaten the reputations of leading brands, compelling them to adopt responsible and sustainable practices to maintain their competitive positions (Turker and Altuntas 2014). Strategies include networking with international or national institutions, projecting a socially responsible image, and other CSR activities. These conditions necessitate a focused environmental justice lens on the hidden costs within the global fashion supply chain, to evaluate the true impact and justice of these CSR policies. Addressing these challenges is essential for creating a more ethical and sustainable fashion industry.

3.3 Diving Deep: The Pockets of Environmental Injustice

The globally fragmented and scattered supply chain of the fashion industry involves various stages and processes, with different industries disproportionately sharing benefits and burdens (Fig. 6). These stages include the petrochemicals and agriculture industries, the design and manufacturing sectors, and the logistics sector. For the current analysis, the global supply chain is divided into three phases: pre-manufacturing (fiber and chemical production), manufacturing (yarn, textile, and garment production), and post-manufacturing (retailing, consumption, and disposal). Each phase involves multiple goods and service providers such as retailers, distributors, logistics companies, designers, merchandisers, yarn and fabric producers, trim producers, garment manufacturers, and embellishment service providers. Analyzing the impacts throughout this supply chain is a complex task due to its large-scale and disintegrated nature. Here, we present a travelogue of apparel from cradle to grave.

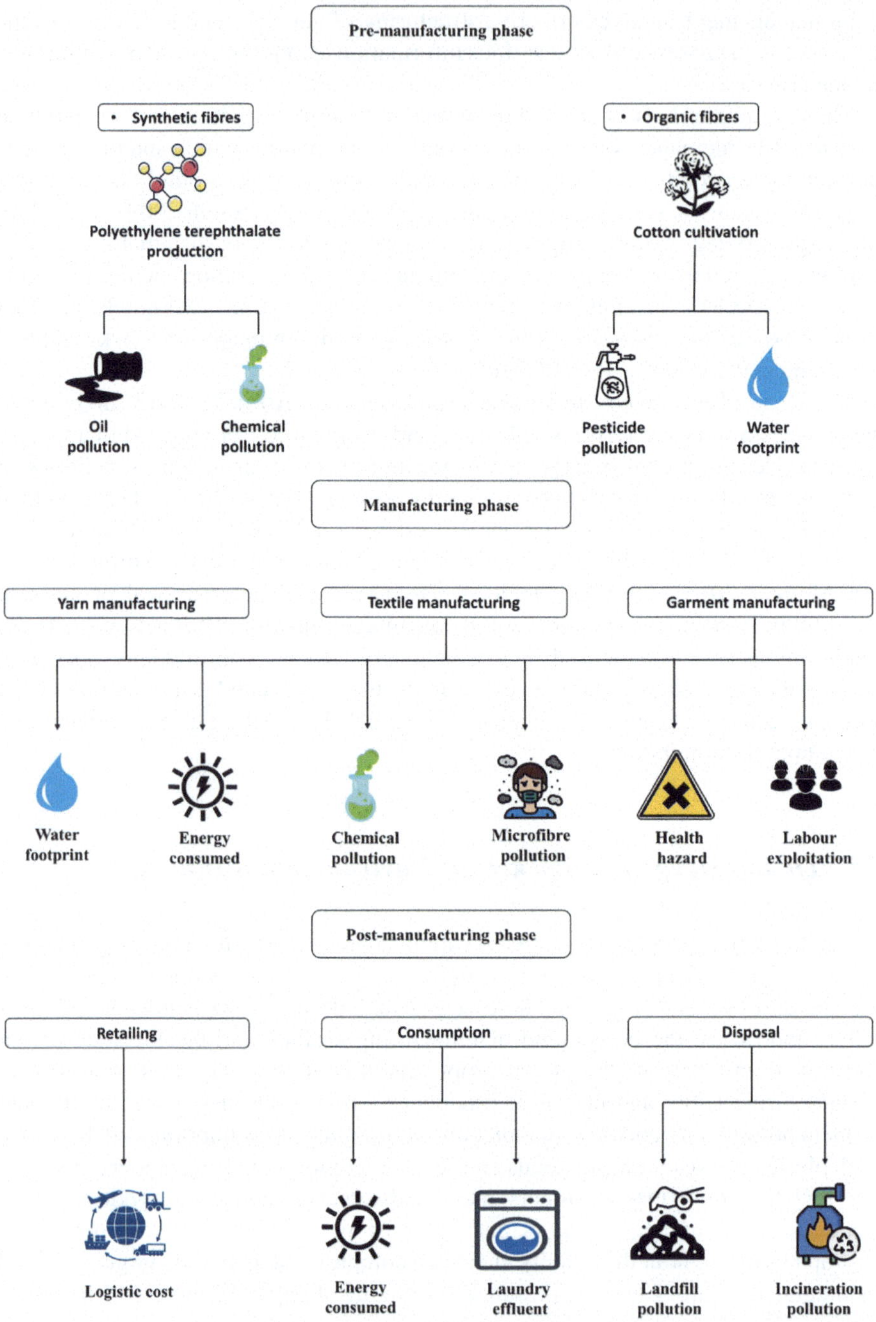

Fig. 6 The global fashion supply chain spans multiple industries and stages, from petrochemicals and agriculture to design, manufacturing, and logistics, involving diverse stakeholders across pre-manufacturing, manufacturing, and post-manufacturing phases

3.3.1 Environmental and Social Impacts of Pre-Manufacturing Phase

The pre-manufacturing phase primarily consists of fiber production, which is spatially scattered and diverse (Table 2). The volume of fiber produced for the textile industry skyrocketed from around 40 million metric tonnes in 1990 to 110 million metric tonnes in 2018, with the majority being oil-based synthetic fibers, mainly polyester (The Fibre Year 2019). The next major share is plant-based cotton fiber, both of which entail substantial environmental and social costs.

Synthetic Fiber Production: Understanding the Real Cost

Polyester Production and Environmental Impact: Polyester fiber dominates the synthetic fibers market, with polyethylene terephthalate (PET) as its primary component. PET, a polymer resin from the polyester family, is produced through a series of processes beginning in the oil industry, where ethylene glycol and terephthalic acid (TPA) are derived from crude oil (Palacios-Mateo et al. 2021). The extraction, refining, and processing phases release harmful toxins, including BTEX compounds (benzene, toluene, ethylbenzene, xylene), particulate matter, NOx, SO_2, and CO, many of which are potential carcinogens (Demirel et al. 2014). This chemical-intensive process significantly contributes to environmental pollution and human health risks (Table 2).

Oil Spills and Environmental Degradation: The environmental impact of polyester production is further compounded by oil spills, a common occurrence in major oil drilling regions. In 2022, the Center for Western Priorities reported 2,449 oil spills in the USA, posing significant threats to local environments and communities (Center for Western Priorities 2023). In the Niger Delta, extensive oil extraction has severely impacted agriculture, fisheries, ecosystems, and human health, with high levels of petroleum hydrocarbons and other trace elements detected in the environment and food (Linden and Palsson 2013). This degradation has led to social unrest, highlighted by the 1995 arrest and execution of Ken Saro-Wiwa and his Ogoni comrades.

PET Production and Environmental Persistence: Once produced, ethylene glycol and terephthalic acid are processed to create PET, which is then used to manufacture yarn (Palacios-Mateo et al. 2021). PET's large aromatic rings make it durable but also persistent in the environment, contributing to long-term pollution (Mandal and Dey 2019). Despite its potential for depolymerization, PET remains a significant environmental concern due to its resistance to degradation.

Rayon Production and Health Risks: Rayon, a semi-synthetic fiber, presents additional challenges. Though derived from natural sources like bamboo and wood pulp, its production involves extensive chemical processing, particularly with carbon disulfide, a solvent known to pose reproductive risks to humans (Mukherjee 2015). The production of 1 kg of rayon releases approximately 251 g of carbon disulfide, making the rayon manufacturing sector the largest anthropogenic source of this chemical (Lee and Brimblecombe 2016). This poses significant health risks to workers and local communities exposed to the manufacturing process.

Table 2 Environmental and social impacts at each stage of the global fashion supply chain, highlighting the pockets of environmental injustice from pre-manufacturing to post-manufacturing phases

Phase	Stage	Process	Environmental Impact	Social Impact
Pre-manufacturing	Synthetic Fibers	Production of PET from crude oil	Releases toxins: BTEX compounds, potential carcinogens	Oil spills impacting agriculture, fisheries, and health (Niger Delta)
		Production of Rayon	Releases carbon disulfide, a reproductive risk chemical	Health risks to workers and local communities
	Organic Fibers	Cotton production	High water usage, pesticide pollution	Water scarcity, salinization, forced labor (Uzbekistan)
Manufacturing	Yarn Production	Conversion of fibers to yarns (spinning, wet processing)	High CO_2 emissions, water contamination	Health hazards: respiratory issues, toxic exposure
	Textile Manufacturing	Weaving/ knitting yarns into fabrics	Chemical pollution (dyes, solvents)	Worker exploitation, poor working conditions
	Garment Manufacturing	Assembling fabrics into garments	Energy-intensive processes, effluent discharge	Sweatshop conditions, low wages, gender discrimination
Post-manufacturing	Distribution	Global transportation and logistics	CO_2 emissions from transport	Global warming impacting agriculture, fisheries etc
	Consumption	Use-phase care: washing, drying, ironing	CO_2 emissions, microfiber pollution	Health risks from detergents, microplastics in food
	Disposal	Textile waste management (landfilling, incineration)	Long degradation time, toxic leachates, air pollution	Waste impacts poorer communities, incineration releases harmful gases

Environmental Foot Print of Cotton Production

Environmental and Social Costs of Cotton Production: The environmental and social impacts of fiber production are not confined to synthetic fibers. Plant-based cotton fiber, a major component of the textile industry, also carries substantial costs. In Central Asia, water diversion from rivers to Uzbekistan's cotton fields has caused significant social and environmental costs, notably shrinking the Aral Sea to 30% of its original size by the 2000s (Loh and Wackernagel 2004). The Environmental Justice Foundation (EJF) highlights the severe impacts on the livelihoods of the indigenous Karakalpakstan people, who have relied on fisheries and subsistence agriculture for 2000 years. Intensive pesticide use, severe water scarcity, and poverty have caused devastating health impacts, turning local inhabitants into environmental refugees, with estimates of around 100,000 individuals, though official data is lacking (EJF 2012). Additionally, Uzbekistan's cotton industry has been plagued by forced and child labor under Islam Karimov's dictatorship. Despite pledges from leading brands to avoid Uzbek cotton, it is still exported to China and Bangladesh, the world's leading textile producers.

Pesticide Use in Cotton Cultivation: Cotton is the largest consumer of pesticides globally, using 11% of all pesticides while occupying only 2.3% of the world's arable land (FAO 2015). Excessive agrochemical use poses significant threats to local ecosystems through eutrophication, nitrate contamination, and soil salinity (Bevilacqua et al. 2014). India is particularly notable for its high water usage and pesticide application, with many pesticides classified as moderately to highly hazardous by the WHO (Khan and Damalas 2015). The Food and Agriculture Organization (FAO) reports extensive use of highly hazardous pesticides like aldicarb, chlorpyrifos, methamidophos, and methomyl in cotton cultivation, which should be phased out but remain in use in regions with weak regulatory frameworks.

Health Impacts on Farmers: Occupational exposure to these pesticides affects many farmers in developing countries, leading to severe health impacts such as damage to the central and peripheral nervous systems, cardiac arrest, and even death. Some pesticide formulations contain carcinogenic, genotoxic, reproductive, or endocrine-disruptive properties, causing severe genetic defects in offspring (Mancini et al. 2005). In this context, India is particularly affected, with extensive pesticide application leading to severe health issues among farm workers, described as a genocide due to the lack of caution in pesticide use (EJF 2007).

The pre-manufacturing phase of fiber production reveals significant environmental and social challenges for both synthetic and organic fibers. The dominance of synthetic fibers like polyester and rayon, combined with the complex and chemically intensive processes involved in their production, highlights the urgent need for more sustainable practices in the textile industry. Similarly, the environmental and social costs of cotton production underscore the necessity for improved regulatory frameworks and responsible sourcing practices. Addressing these issues is crucial for mitigating adverse impacts on ecosystems and human health, ensuring a more responsible and ethical approach to fashion production.

3.3.2 Manufacturing Phase

The manufacturing phase represents the most complex and labor-intensive segment of the fashion supply chain (Table 2). This phase involves converting natural and synthetic fibers into yarns of various lengths and colors through processes such as spinning and, if necessary, wet processing. For polyester fibers, this requires melt spinning, where PET is melted and extruded through small holes to produce yarns with specific qualities. These yarns are then woven or knitted into fabrics, which may undergo additional processes such as dyeing, printing, and finishing to achieve desired properties like color, strength, and design. The fabrics are subsequently assembled into garments in labor-intensive manufacturing units and then sent to retailers, primarily in Western markets.

Environmental Impact: Despite the vibrant colors of the fabrics, the industrial environment where garment production occurs is often dull and grey. This energy-intensive process results in significant CO_2 emissions, largely dependent on the energy sources used for electricity generation (Clarke-Sather and Cobb 2019). Additionally, the textile manufacturing process consumes vast amounts of water, both as a resource and as a medium to absorb industrial effluents, leading to water contamination, especially in countries with lax environmental regulations (Mukherjee 2015). Wet processing, encompassing dyeing, printing, and finishing, is particularly water-intensive and involves extensive use of chemicals. The industry uses around 15,000 different chemicals, many of which are employed in wet processing, including bleaches, surfactants, softeners, dyestuffs, and durable water repellents (Mukherjee et al. 2024a, b; Yadav et al. 2022). Approximately 20% of dyes used are washed away, discharging about 50,000 tonnes of dyes into freshwater sources annually, predominantly in developing countries. Additional chemicals used in spinning and weaving, such as lubricants, accelerators, and solvents, further pollute water bodies (Mishra et al. 2023). The effluents from textile factories often contain toxic heavy metals, bicarbonates, nitrates, dyestuffs, detergents, fats, oils, and various other harmful substances (Gebre-Mariam and Desta 2002; Kumar et al. 2022; Kumar et al. 2023; Madhav et al. 2018). Some dyes are carcinogenic and hormone-disrupting, while others, like azo dyes, can be transformed into dangerous compounds under certain environmental conditions (Rawat et al. 2018, 2022, 2023). The significant environmental impact of textile manufacturing is compounded by the health hazards faced by workers, who are exposed to a variety of harmful substances and conditions during the manufacturing process.

Health Hazards and Worker Safety: Another critical issue is the release of microfibers, a subset of microplastics, into the air, along with toxic gases such as nitrogen and sulfur oxides, hydrocarbons, formaldehyde, acids, and volatile organic compounds (Kaur et al. 2022; Palacios-Mateo et al. 2021; Mukherjee 2015). Polyester microfibers, combined with industrial pollutants, can penetrate human organs, crossing barriers such as the blood–brain and blood-placental barriers, and cause cytotoxicity, tissue damage, fibrosis, carcinogenesis, and other serious health disorders (Palacios-Mateo et al. 2021; Wright and Kelly 2017).

Beyond environmental pollution, the manufacturing phase has significant social costs. Factors contributing to the rise of sweatshops include corporate greed, lax environmental policies, the pursuit of the lowest prices, the proliferation of intermediaries, and pressure to cut costs throughout the supply chain (Adams 2002). These characteristics are intrinsic to the fast fashion industry, making sweatshops prevalent (Bhardwaj and Fairhurst 2010; Berkey 2021). The term "sweatshop" gained prominence in the 1990s when the Kathie Lee Gifford line of clothing in the USA was exposed for inhumane working conditions, including denial of civil rights, unhealthy environments, forced labor, and child labor (Adams 2002). Connecting these health hazards, the exploitation of labor in sweatshops further exacerbates the detrimental effects on workers' wellbeing.

Labor Exploitation and Working Conditions: Sweatshop conditions persist, particularly in developing countries with weak environmental regulations. Bangladesh, a leading manufacturer of ready-made garments, employs about 4.2 million workers in the fashion manufacturing sector (OVI 2017). Despite the significant workforce, it took the Rana Plaza tragedy, which claimed 1,138 lives, to prompt the implementation of the Bangladesh Labor Act of 2006 in 2015, allowing workers to form trade unions (Uddin et al. 2021). However, the partial presence of trade unions means workers in Bangladesh and other developing countries still face severe exploitation. Workers endure physical assault, unfair discrimination, monetary penalties, psychological and sexual harassment, long working hours, and inadequate wages (Zafar and Bhuiyan 2012; Mukherjee 2015). Gender discrimination is also prevalent, with many female workers denied paid maternity leave, contributing to malnutrition and anemia (Mahmud and Afrin 2017; Akhter et al. 2010). The lack of trade unions and worker education exacerbates these issues. The prevalence of these harsh working conditions necessitates a closer examination of corporate social responsibility initiatives aimed at improving worker wellbeing.

CSR and Worker Wellbeing: CSR initiatives have made some efforts to address these problems, but the neglect of workers' social reproduction costs remains a significant issue. Many workers live in substandard conditions, lacking proper sanitation and facing numerous health hazards. The textile manufacturing sector is associated with physical, fire, explosion, chemical, ergonomic, mechanical, and psychological hazards, leading to a wide range of diseases and disorders, including respiratory issues, musculoskeletal disorders, cancer, endocrine dysfunction, accidental injuries, and death (Sant'Ana and Kovalechen 2012; Akther et al. 2017; Wright and Kelly 2017). Workers also suffer from renal failure, cytotoxicity, genotoxicity, mutagenicity, and other health problems due to exposure to harmful chemicals (Bhattacharya et al. 2016; Lellis et al. 2019). Chemical dyes can cause skin and eye irritation, respiratory problems, and other health issues (Tang et al. 2018; Hassan and El-Nemr 2017; El-Hadidy 2022; Zele et al. 2021; Liaqat et al. 2009). This dire situation, coupled with poor living conditions, leads to widespread malnutrition among workers. To address these extensive issues, both consumers and corporations must emphasize accountability and transparency in apparel manufacturing.

Accountability and Transparency in Apparel Manufacturing: All these externalities will continue to proliferate unless significant improvements are made in apparel manufacturing, especially given the projected increase in clothing production in the near future (Remy et al. 2016). This context should prompt consumers to ask critical questions about the origins and production methods of their clothing, as Clarke-Sather and Cobb (2019) suggest: "Where, how, and by whom are the apparels manufactured?". Apparel corporations and state authorities must be held accountable for environmental and social conditions in their industrial operations. They have a responsibility to transparently disclose these aspects to their customers and the broader public. The manufacturing phase of the fashion supply chain is fraught with environmental and social challenges impacting the sustainability of textile industry.

3.3.3 Post-Manufacturing Phase

The post-manufacturing phase of the fashion supply chain extends beyond the creation of garments to include distribution, consumer use, and the end-of-life management of textiles. Each of these stages has significant environmental and social implications (Table 2).

Distribution and Consumer Impact: The distribution centres of fashion apparels are spread across the globe, primarily concentrated in developed nations such as the European Union, the USA, and Japan. These products are packed and shipped from countries in Southeast Asia, including Cambodia and Vietnam, incurring substantial, distribution costs (Quantis 2018). Apparel products are distributed to consumers through major retail outlets in developed nations. At this stage, consumers play a crucial role in either exacerbating or alleviating the environmental impact through their choices in apparel care, such as washing frequency, temperature settings, and ironing habits, all of which affect energy consumption and CO2 emissions (Moazzem et al. 2021a). The distribution and consumer impact phase ties directly to the environmental consequences of how consumers care for their garments, influencing the broader ecological footprint of fashion.

Environmental Impact of Laundry Effluents: Laundry effluent poses a significant threat to freshwater ecosystems and oceans, with microfibers accounting for around 35% of microplastic pollution in the oceans (Boucher and Friot 2017). Microfibers are now ubiquitous, even detected in the food chain through fruits and vegetables (Oliveri Conti et al. 2020). These microplastics threaten food security and biodiversity by severely affecting soil properties on farmlands (Wang et al. 2019; Corradini et al. 2019). The use of sewage sludge as manure is a key source of land pollution by microfibers (Nizzetto et al. 2016; Corradini et al. 2019). Besides the environmental threat, microfibers pose direct health risks to humans based on their physical and chemical properties. Detergents, another major component of laundry effluent, can harm the heart, lungs, kidneys, and skin due to their ingredients, including linear alkyl benzene sulphonates, phenols, and artificial fragrances. While some countries with strong environmental regulations have banned phosphate-containing detergents (Lee and Jones 1986), developing nations often lack proper treatment facilities, exacerbating the pollution from detergent use (Jardak et al. 2016). These environmental impacts from laundry effluents highlight the critical need for better wastewater management and consumer awareness, connecting directly to the broader challenges of textile waste.

Textile Waste Management and Environmental Injustice: In the fast fashion paradigm, the end phase of apparel products is crucial as the amount of textile waste exceeds the capacity of ecosystems to assimilate it. Globally, an average person produces about 17.5 kg of fashion waste per year (Moazzem et al. 2021b). Much of this waste is discarded due to poor fit, changing preferences, and minimal wear. The culture of planned obsolescence drives this massive accumulation of textile waste (Niinimäki et al. 2020). A significant portion of textile waste ends up in landfills or is incinerated, with only a small fraction being recycled. This disposal method wastes valuable resources and labor embedded in discarded products while necessitating the appropriation of more natural resources and labor for new production (Palacios-Mateo et al. 2021). This significant textile waste burden illustrates the urgent need for improved waste management strategies and highlights the issue of environmental injustice.

A significant issue is deadstock, the leftover garments in retailer inventories, which are often incinerated (Niinimäki et al. 2020). Effective textile waste management has the potential to recover much of the value within discarded textiles, as many are suitable for direct reuse (Nørup et al. 2019). However, textile waste management is inadequate even in developed countries, often resulting in waste being exported to third-world countries as second-hand clothing and eventually ending up in their landfills (Moazzem et al. 2021b; Mukherjee 2015). This situation represents environmental injustice, as developing nations bear the brunt of waste generated by developed countries. Efforts by some East African nations to restrict the import of used clothing faced resistance from the U.S. Trade Representative's office, highlighting the complexities of international trade and environmental justice (Brooks and Simons 2012). The problem of deadstock connects to the broader issues of textile waste and the inequitable distribution of its impacts, which also encompasses health risks.

Health and Environmental Risks of Textile Waste: Landfills pose significant risks to human health and nearby ecosystems. Polyester textiles, in particular, are non-biodegradable, producing toxic leachates and air pollutants during their long end-of-life phase (Palacios-Mateo et al. 2021; Bick et al. 2018; Aiama et al. 2016). Regions with less developed waste management systems are particularly affected. Studies in Florida, USA, have shown that those generating the most waste are often not the ones facing its environmental and social impacts, further illustrating environmental injustice (Devoy et al. 2021). Incineration of textile waste also poses pollution risks, emitting harmful gases and toxic ash. Incinerating polyester garments releases various harmful gases such as CO, benzene, benzoic acid, and acetaldehyde (Moltó et al. 2006). The health and environmental risks associated with textile waste emphasize the necessity for recycling efforts, which are currently underutilized.

Recycling Challenges and Potential: Recycling is the most preferable, yet least utilized, option for textile waste management. Recycling can recover value from discarded textiles by transforming them into secondary raw materials for production. Despite the existence of various recycling technologies, such as mechanical, thermal, chemical, and enzymatic bio-recycling, the heterogeneous nature of discarded textiles hinders successful value recovery. Additionally, the lack of markets for products made from recycled raw materials limits recycling efforts (Remy et al. 2016). Consequently, less than 15% of globally discarded textile waste is recycled. These challenges in recycling are intertwined with the broader need for addressing the hidden external costs of the fashion industry.

The fast fashion supply chain is fraught with external costs that are often hidden behind the vibrant marketing strategies of fashion brands. Exposing these costs to the public is challenging, as it risks tarnishing the image of fashion brands. Although corporate social responsibility (CSR) initiatives aim to address these issues, their effectiveness in tackling the external costs of fashion remains questionable.

4 Juxtaposing Environmental Justice and CSR in the Fast Fashion Industry

Analyzing the evolution of environmental justice and corporate social responsibility (CSR) concepts on a spatial and temporal scale provides a framework to evaluate CSR in the fashion sector, particularly in fast fashion (Fig. 4). This framework places CSR activities between the typical business-as-usual scenario and an environmentally just scenario, enabling us to assess the degree of justice achieved by corporations through their CSR efforts. This comprehensive evaluation underscores the need to bridge the gap between current corporate practices and genuine environmental justice.

4.1 CSR and the Spatial Disparity in the External Cost

Retailers adopt sustainable supply chain management strategies to mitigate risks across their supply chains. Turker and Altuntas (2014) propose a holistic framework addressing product lifecycle impacts and supplier performance within dimensions of sustainable development. However, these strategies vary significantly due to disparities in regulatory standards and codes of conduct between corporations in developing and developed nations (Perry and Wood 2019), hindering efforts to mitigate environmental degradation and improve human welfare in the global South (Table 3).

In the fast fashion paradigm, Bick (2018) underscores environmental injustices, particularly the transfer of environmental externalities and social costs to the global

Table 3 Evolution of environmental justice and CSR concepts over time and space and their contrasting and contradictory features

Concept	Key Points	Examples / Observations
CSR and the Spatial Disparity in the External Cost	Retailers use sustainable supply chain management to mitigate risks, but disparities in regulatory standards persist between developed and developing nations	Varying environmental standards, labor laws, and incidents like the Spectrum Building Collapse in Bangladesh
The Voluntary and Non-Statutory Nature of CSR	Partnerships for sustainability in competitive markets face challenges due to monitoring complexities	Monitoring by multinational corporations, voluntary compliance mechanisms like the Global Reporting Initiative, and concerns over worker wages and safety
CSR and the Power Relations of Society	CSR initiatives influence societal power dynamics, impacting governance structures and worker rights	Conflicts between private and national regulations, impact of CSR on labor standards, and critiques of multi-stakeholder initiatives like Accord and Alliance

South. Niinimäki et al. (2020) observe significant disparities in statutory environmental standards between low- and middle-income countries and developed economies.

Zara, a leading fast fashion brand, outlines in its sustainability policy the requirement for supplier compliance with labor laws, International Labour Organization conventions, and human rights principles (Inditex 2022). However, implementation varies widely, with home nations often maintaining stricter labor laws compared to outsourced nations. Weak enforcement of other production-related laws in apparel-producing nations is evident, exemplified by the Spectrum building collapse in Bangladesh in 2005, which resulted in 64 deaths due to illegal construction on unstable land (Barraud de Lagerie 2016).

The spatial disparity among suppliers is apparent in H&M's operations, where major production occurs in the global South with varying union presence—from 100% in Vietnam to 0% in Pakistan, with some nations lacking ratification of ILO conventions on organizing and collective bargaining (H&M 2024). Brands like Zara, operating through 1,733 suppliers and 8,123 factories globally, predominantly focus their policy reports on tier 1 production, often neglecting lower tiers and the associated external costs (Inditex 2022, 2023). This complexity results in varied production costs across nations, prompting brands to shift production to regions with lower operational costs, primarily in the global South.

4.2 The Voluntary and Non—statutory Nature of CSR

Cooperative partnerships between suppliers and retailers are pivotal for enhancing environmental sustainability along the supply chain. However, integrating sustainability into a competitive business environment necessitates continuous monitoring and auditing of supply chain entities, which runs counter to the fast fashion model focused on rapid production (Cheung et al. 2009). This irony is compounded by the fact that these monitoring and auditing processes are often overseen by the same multinational corporations that establish the standards and guidelines for compliance. For example, the Global Reporting Initiative, a major sustainability framework funded by corporations like Ford, General Motors, Nike, and Shell (Shamir 2011), illustrates this dual role. The involvement of Royal Dutch Shell, a significant player in corporate sustainability yet embroiled in numerous global lawsuits, underscores the complexities of voluntary compliance mechanisms (Martine-Alier et al. 2016).

The inherent uncertainty of these mechanisms stems from their lack of standardized scope and global coverage. Concerns also arise regarding the ambiguous treatment of worker wages. H&M, for instance, outsources production and claims no responsibility for garment workers' salaries while setting vague standards for supplier wage distribution (H&M 2024). The collapse of the Spectrum building in Bangladesh highlighted the ineffectiveness of corporate CSR policies and triggered global calls for legally binding CSR standards (Barraud de Lagerie 2016).

Efforts to introduce transparency in fashion supply chains include initiatives like sustainability tagging, where garments are labeled with universal green tags to inform consumer choices (Remy et al. 2016). However, the proliferation of green tags has raised concerns about their credibility and potential for marketing exploitation (Portway 2019). Furthermore, multi-stakeholder initiatives such as Accord and Alliance, aimed at improving safety and working conditions in Bangladesh's Ready-Made Garment (RMG) sector, have faced criticism for lacking adequate worker representation, allowing dominant stakeholders like multinational corporations and allied NGOs to wield disproportionate influence (Mausumi and Rahman 2018).

As non-statutory and voluntary CSR practices continue to proliferate, they weave a complex web of corporate social responsibility, where entities at each node offer varied goods and services, creating an exclusive realm within corporate business structures.

4.3 CSR and the Power Relations of Society

The proliferation of CSR initiatives, comprising multiple nodes of entities, into the socio-political domains of nations alters societal relations and weakens underlying governance structures. Persistent industrial accidents in apparel production units in the global south, compounded by pressure from Western consumer organizations on retailer brands, have placed worker communities in a dilemma between state-led labor regulations and private regulations imposed by overseas retailers (Barraud de Lagerie 2016). Analyzing the aftermath of the 2005 Spectrum collapse and subsequent improvements, in 2016, tensions escalated due to conflicts between private and national regulations, where workers feared losing their autonomy under regulations imposed by foreign retail brands. Union representatives' express apprehensions over the concentration of responsibilities in a few global actors, thereby limiting the role of the democratic state.

Bartley and Egels-Zanden (2015) examine the impact of codes of conduct in industrial units in Indonesia, producing apparel, electronics, and footwear, on labor standards. They differentiate between impacts on globally recognized standards such as health, safety, and employment, and less emphasized standards like the right to organize and collective bargaining. Their analysis supports earlier literature, indicating that retailer-imposed codes of conduct can influence worker health and safety and formalize the workforce, though outcomes are contingent on their alignment with retail brands' cost structures and administrative controls. However, these codes often fail to foster equitable labor relations across the value chain, particularly when they prioritize their exclusive authority upon the value chain over workers' rights (Bartley and Egels-Zanden 2015).

Issues of labor representation also arise in multi-stakeholder initiatives like Accord and Alliance, where worker involvement remains unclear, and accountability to legal authorities is minimal, undermining the initiative's effectiveness (Mausumi and Rahman 2018).

The profit-driven institutional design of CSR under a business-case approach seeks to reshape societal and governance power dynamics by marginalizing worker and community agency. This approach contradicts environmental justice movements advocating for greater empowerment of states, workers, and communities in decision-making processes concerning nature.

Alessandro Nicita's empirical analysis of Madagascar's clothing industry for the World Bank reveals potential for significant increases in the purchasing power of the working population (Nicita 2007). However, the analysis overlooks the widespread casualization of the workforce in fashion manufacturing units across the Global South, a trend embraced by state authorities despite its long-term social costs and implications for job creation.

The complexities and disparities highlighted in this discussion underscore the formidable challenges facing the fast fashion industry in achieving genuine environmental justice and effective CSR. Addressing these challenges demands a comprehensive approach involving stricter regulations, enhanced transparency, and a shift towards sustainable business practices. By acknowledging and confronting these issues, the fashion industry can progress towards a more ethical and sustainable future.

5 Conclusion

This review has critically examined the incursion of the global fashion supply chain into the Global South, highlighting the resulting repercussions for local populations and the effectiveness of CSR policies introduced by fashion brands through the lens of environmental justice. The findings reveal significant shortcomings in current CSR initiatives, which often fail to address the external costs of production effectively. These policies frequently do not align with the principles of environmental justice, as they overlook the varying degrees of external costs across different nations and rely on voluntary compliance mechanisms that lack robustness. The analysis emphasises that the efforts of fashion corporations under the guise of CSR remain inadequate in improving the lives of people in developing nations. Instead, these efforts often perpetuate environmental injustice, as evidenced by the burdens placed on the environment and labourers in countries such as Uzbekistan, Bangladesh and the Niger Delta. The disparity in labour rights and environmental standards between the Global South and developed nations further exacerbates this issue. This research highlights the need for more empirical models to evaluate CSR policies and assess the environmentally just behaviour of fashion brands. Such models are essential for identifying and addressing the pockets of environmental injustice that exist within the global fast fashion supply chain. Additionally, exploring the potential for legally binding regulations and the role of international oversight in ensuring credible and effective CSR initiatives could provide valuable insights.

References

Adams RJ (2002) Retail profitability and sweatshops: a global dilemma. J Retailing Consumer Serv 9(3):147–153. http://nlcnet.org/PRESS/BIOSCharlie.htm

Aiama D, Carbone G, Cator D, Challender D (2016) Biodiversity risks and opportunities in the apparel sector. Int Union Conserv Nat

Akhter S, Salahuddin AFM, Iqbal M, Malek ABMA, Jahan N (2010) Health and occupational safety for female workforce of garment industries in Bangladesh. J Mech Eng 41(1):65–70

Andriof J, Waddock S (2017) Unfolding stakeholder engagement. In: Andriof J, Waddock S, Husted B, Rahman SS (eds) Unfolding stakeholder thinking. Routledge, pp 19–42

Anguelov N (2015) The dirty side of the garment industry: fast fashion and its negative impact on environment and society. CRC Press

Barca S (2012) On working-class environmentalism: a historical and transnational overview. Interface J Soc Movements 4(2):61–80

Barnes L, Lea-Greenwood G, Lane OH (2006) Fast fashioning the supply chain: shaping the research agenda. J Fash Mark Manag 10(3):259–271

Barraud de Lagerie P (2016) Conflicts of responsibility in the globalized textile supply chain: lessons of a tragedy. J Consum Policy 39(4):397–416. https://doi.org/10.1007/s10603-016-9312-1

Bartley T, Egels-Zandén N (2015) Responsibility and neglect in global production networks: the uneven significance of codes of conduct in Indonesian factories. Global Netw 15(S1):S21–S44. https://doi.org/10.1111/glob.12077

Baša K, Farkas S (2022) Buying behaviour and planned obsolescence in the fashion industry. Soc Econ Revue 20(2). https://doi.org/10.52665/ser20220202

Baughn CC, Bodie NL, McIntosh JC (2007) Corporate social and environmental responsibility in Asian countries and other geographical regions. Corp Soc Responsib Environ Manag 14(4):189–205. https://doi.org/10.1002/csr.160

Berkey B (2021) Sweatshops, structural injustice, and the wrong of exploitation: why multinational corporations have positive duties to the global poor. J Bus Ethics 169(1):43–56. https://doi.org/10.1007/s10551-019-04299-1

Bevilacqua M, Ciarapica FE, Mazzuto G, Paciarotti C (2014) Environmental analysis of a cotton yarn supply chain. J Clean Prod 82:154–165. https://doi.org/10.1016/j.jclepro.2014.06.082

Bhardwaj V, Fairhurst A (2010) Fast fashion: response to changes in the fashion industry. Int Rev Retail Distrib Consumer Res 20(1):165–173. https://doi.org/10.1080/09593960903498300

Bick R, Halsey E, Ekenga CC (2018) The global environmental injustice of fast fashion. Environ Health Global Access Sci Source 17(1):1–4. https://doi.org/10.1186/s12940-018-0433-7

Birtwistle G, Siddiqui N, Fiorito SS (2003) Quick response: perceptions of UK fashion retailers. Int J Retail Distrib Manag 31(2):118–128. https://doi.org/10.1108/09590550310462010

Boucher J, Friot D (2017) Primary microplastics in the oceans: a global evaluation of sources. Gland, Switzerland. Int Union Conserv Nat. https://doi.org/10.2305/IUCN.CH.2017.01.en

Brooks A, Simon D (2012) Unravelling the relationships between used-clothing imports and the decline of African clothing industries. Dev Chang 43(6):1265–1290. https://doi.org/10.1111/j.1467-7660.2012.01797.x

Brown P (1995) Race, class, and environmental health: a review and systematization of the literature. Environ Res 69(1):15–30. https://doi.org/10.1006/enrs.1995.1021

Bruce M, Daly L (2006) Buyer behaviour for fast fashion. J Fash Mark Manag 10(3):329–344. https://doi.org/10.1108/13612020610679303

Bruce M, Daly L, Towers N (2004) Lean or agile: a solution for supply chain management in the textiles and clothing industry? Int J Oper Prod Manag 24(2):151–170. https://doi.org/10.1108/01443570410514867

Buckley C, Clark H (2012) Conceptualizing fashion in everyday lives. Design Issues 28(4):79–92. https://about.jstor.org/terms

Bullard RD (1993) Anatomy of environmental racism and the environmental justice movement. Confronting environmental racism: voices from the grassroots, 15–39

Carroll AB (1999) Corporate social responsibility: evolution of a definitional construct. Bus Soc 38(3):268–295. https://doi.org/10.1177/000765039903800303

Centre for Western Priorities (2023) 2022 spills tracker. https://westernpriorities.org/resource/2022-spills-tracker/

Cheung DKK, Welford RJ, Hills PR (2009) CSR and the environment: business supply chain partnerships in Hong Kong and PRDR, China. Corp Soc Responsib Environ Manag 16(5):250–263. https://doi.org/10.1002/csr.208

Christopher M, Lowson R, Peck H (2004) Creating agile supply chains in the fashion industry. Int J Retail Distrib Manag 32(8):367–376. https://doi.org/10.1108/09590550410546188

Clarke-Sather A, Cobb K (2019) Onshoring fashion: worker sustainability impacts of global and local apparel production. J Clean Prod 208:1206–1218. https://doi.org/10.1016/j.jclepro.2018.09.073

Corradini F, Meza P, Eguiluz R, Casado F, Huerta-Lwanga E, Geissen V (2019) Evidence of microplastic accumulation in agricultural soils from sewage sludge disposal. Sci Total Environ 671:411–420. https://doi.org/10.1016/j.scitotenv.2019.03.368

Cutter SL (1995) Race, class and environmental justice. Prog Hum Geogr 19(1):111–122. https://doi.org/10.1177/030913259501900111

Davis K (1973) The case for and against business assumption of social responsibilities. Acad Manag J 16(2):312–322. https://doi.org/10.5465/255331

Demirel G, Özden Ö, Döğeroğlu T, Gaga EO (2014) Personal exposure of primary school children to BTEX, NO2 and ozone in Eskişehir, Turkey: relationship with indoor/outdoor concentrations and risk assessment. Sci Total Environ 473–474:537–548. https://doi.org/10.1016/j.scitotenv.2013.12.034

DeVoy JE, Congiusta E, Lundberg DJ, Findeisen S, Bhattacharya S (2021) Post-consumer textile waste and disposal: Differences by socioeconomic, demographic, and retail factors. Waste Manage 136:303–309. https://doi.org/10.1016/j.wasman.2021.10.009
El-Hadidy NM (2022) Occupational health profile of textile dyeing workers. Egypt J Occup Med 46(2):1–20. https://doi.org/10.21608/ejom.2021.57155.1147
Environmental Justice Foundation (2007) The deadly chemicals in cotton. Environmental Justice Foundation in collaboration with Pesticide Action Network UK
Environmental Justice Foundation (2012) The true costs of cotton: cotton production and water insecurity
European Commission (2011) A renewed EU strategy 2011–14 for corporate social responsibility. https://eur-lex.europa.eu/legal-content/EN/TXT/?uri=CELEX:52011DC0681
Eweje G (2006) The role of MNEs in community development initiatives in developing countries: corporate social responsibility at work in Nigeria and South Africa. Bus Soc 45(2):93–129. https://doi.org/10.1177/0007650305285394
Faber DR, Krieg EJ (2002) Unequal exposure to ecological hazards: environmental injustices in the Commonwealth of Massachusetts. Health Perspect 110(4):339–348. https://doi.org/10.1289/ehp.02110339
Fels A, Falk B, Schmitt R (2016) Social media analysis of perceived product obsolescence. Procedia CIRP 50:571–576. https://doi.org/10.1016/j.procir.2016.04.147
Fletcher K (2016) Craft of use: post-growth fashion. Routledge
Freeman RE, Parmar BL, Harrison JS, Wicks AC, Purnell L, De Colle S (2010) Stakeholder theory: the state of the art. Acad Manag Ann 4(1):403–445. https://doi.org/10.5465/19416520.2010.495581
Garg S, Kumar P, Singh S, Yadav A, Dumée LF, Sharma RS, Mishra V (2020) *Prosopis juliflora* peroxidases for phenol remediation from industrial wastewater—an innovative practice for environmental sustainability. Environ Technol Innov 19:100865. https://doi.org/10.1016/j.eti.2020.100865
Gebre-Mariam Z, Desta Z (2002) The chemical composition of the effluent from Awassa Textile Factory and its effects on aquatic biota. SINET: Ethiopian J Sci 25(2):263–274. https://doi.org/10.4314/sinet.v25i2.18084
Ghemawat P, Nueno JL, Dailey M (2003) ZARA: fast fashion. Harvard Business School, Boston, MA
Guha R, Alier JM (2013) Varieties of environmentalism: Essays North and South. Routledge
H&M Group (2024) Wages. https://hmgroup.com/sustainability/fair-and-equal/wages/
H&M Group (2011) Conscious actions sustainability report 2011. https://hmgroup.com/sustainability/conscious-actions/sustainability-report-2011/
Hines T (2007) The emergence of supply chain management as a critical success factor for retail organisations. In: Fernie J, Sparks L (eds) International retail marketing: A case study approach. Routledge, pp 108–122
Inditex Group (2022) Sustainability policy report. https://www.inditex.com/itxcomweb/api/media/db9dc1cf-b01a-47c0-afa0-f66aa33361d8/inditex_sustainability_policy.pdf?t=1667565990404
Inditex Group (2023) Community investment policy. https://www.inditex.com/itxcomweb/api/media/5534ba9a-e8aa-43d1-b7a3-300cebf6cd09/Inditex+Community+Investment+Policy.pdf?t=1695715303861
Jardak K, Drogui P, Daghrir R (2016) Surfactants in aquatic and terrestrial environment: occurrence, behaviour, and treatment processes. Environ Sci Pollut Res 23(4):3195–3216. https://doi.org/10.1007/s11356-015-5803-x
Kaur H, Rawat D, Poria P, Sharma U, Gibert Y, Ethayathulla AS, Dumée LF, Sharma RS, Mishra V (2022) Ecotoxic effects of microplastics and contaminated microplastics–Emerging evidence and perspective. Sci Total Environ 841:156593
Khan M, Damalas CA (2015) Occupational exposure to pesticides and resultant health problems among cotton farmers of Punjab, Pakistan. Int J Environ Health Res 25(5):508–521. https://doi.org/10.1080/09603123.2014.980781

Kumar P, Mishra V, Bhardwaj S, Garg S, Dumée LF, Sharma RS (2023) Five-wheel framework for system-based monitoring of heavy metal pollution in rivers. Environ Qual Manag 32(4):9–17. https://doi.org/10.1002/tqem.21740

Kumar P, Mishra V, Yadav S, Yadav A, Garg S, Poria P, Farooqi F, Dumée LF, Sharma RS (2022) Heavy metal pollution and risks in a highly polluted and populated Indian river–city pair using the systems approach. Environ Sci Pollut Res 29(40):60212–60231. https://doi.org/10.1007/s11356-022-20092-w

Kuznetsov A, Kuznetsova O, Warren R (2009) CSR and the legitimacy of business in transition economies: the case of Russia. Scand J Manag 25(1):37–45. https://doi.org/10.1016/j.scaman.2008.11.008

Lee C-L, Brimblecombe P (2016) Anthropogenic contributions to global carbonyl sulfide, carbon disulfide and organosulfides fluxes. Earth Sci Rev 160:1–18. https://doi.org/10.1016/j.earscirev.2016.06.005

Lee GF, Jones RA (1986) Detergent phosphate bans and eutrophication. Environ Sci Technol 20(4):330–331. https://doi.org/10.1021/es00146a603

Lellis B, Fávaro-Polonio CZ, Pamphile JA, Polonio JC (2019) Effects of textile dyes on health and the environment and bioremediation potential of living organisms. Biotechnol Res Innov 3(2):275–290. https://doi.org/10.1016/j.biori.2019.09.001

Liaqat I, Arshad M, Arshad R, Arshad N (2009) Exposure to textile chemicals leads to microcytic anemia and hypersensitivity in textile workers. Pak J Zool 41(5):385–393

Lindén O, Pålsson J (2013) Oil contamination in Ogoniland. Niger delta. AMBIO 42(6):685–701. https://doi.org/10.1007/s13280-013-0412-8

Lindgreen A, Swaen V (2010) Corporate social responsibility. Int J Manag Rev 12(1):1–7. https://doi.org/10.1111/j.1468-2370.2009.00277.x

Loh J, Wackernagel M (2004) World Wide Fund for Nature, WWF, UNEP World Conservation Monitoring Centre, & Global Footprint Network. (2004). Living planet report: 2004. WWF-World Wide Fund for Nature

Taplin IM (2014) Who is to blame? Crit Perspect Int Bus 10(1/2):72–83. https://doi.org/10.1108/cpoib-09-2013-0035

Maak T (2007) Responsible leadership, stakeholder engagement, and the emergence of social capital. J Bus Ethics 74(4):329–343. https://doi.org/10.1007/s10551-007-9510-5

MacCarthy BL, Jayarathne PGSA (2010) Fast fashion: achieving global quick response (GQR) in the internationally dispersed clothing industry. In: Chitkara R (ed) Fast fashion. Springer, pp 37–60

MacCarthy BL, Jayarathne PGSA (2013) Supply network structures in the international clothing industry: differences across retailer types. Int J Oper Prod Manag 33(7):858–886. https://doi.org/10.1108/IJOPM-12-2011-0478

Madhav S, Ahamad A, Singh P, Mishra PK (2018) A review of textile industry: wet processing, environmental impacts, and effluent treatment methods. Environ Qual Manage 27(3):31–41. https://doi.org/10.1002/tqem.21538

Mahmud MI, Afrin S (2017) Sustainability of female workforce in the ready-made garment industries of Bangladesh. Int J Sci Qual Anal 3(5):49–54

Mancini F, Van Bruggen AH, Jiggins JL, Ambatipudi AC, Murphy H (2005) Acute pesticide poisoning among female and male cotton growers in India. Int J Occup Environ Health 11(3):221–232. https://doi.org/10.1179/oeh.2005.11.3.221

Mandal S, Dey A (2019) PET chemistry. In: Recycling of polyethylene terephthalate bottles, pp 1–22. Elsevier. https://doi.org/10.1016/B978-0-12-811361-5.00001-8

Manne HG, Wallich HC (1972) The modern corporation and social responsibility. Am J Sociol 77(6):1479–1480

Martinez-Alier J, Temper L, del Bene D, Scheidel A (2016) Is there a global environmental justice movement? J Peasant Stud 43(3):731–755. https://doi.org/10.1080/03066150.2016.1141198

Mausumi N, Rahman M (2018) Impact of stakeholder initiatives: Bangladesh garment industry. SCMS J Indian Manag 15(2):31–41

Mishra V, Mudgal N, Rawat D, Poria P, Mukherjee P, Sharma U, Kumria P, Pani B, Singh M, Yadav A, Farooqi F (2023) Integrating microalgae into textile wastewater treatment processes: Advancements and opportunities. J Water Process Eng 55:104128
Mishra V, Sharma U, Rawat D, Benson D, Singh M, Sharma RS (2020) Fast-changing life-styles and ecotoxicity of hair dyes drive the emergence of hidden toxicants threatening environmental sustainability in Asia. Environ Res 184:109253
Moazzem S, Crossin E, Daver F, Wang L (2021a) Assessing environmental impact reduction opportunities through life cycle assessment of apparel products. Sustain Prod Consum 28:663–674. https://doi.org/10.1016/j.spc.2021.06.015
Moazzem S, Wang L, Daver F, Crossin E (2021b) Environmental impact of discarded apparel landfilling and recycling. Resour Conserv Recycl 166:105338. https://doi.org/10.1016/j.resconrec.2020.105338
Mohai P, Pellow D, Roberts JT (2009) Environmental justice. Annu Rev Environ Resour 34:405–430. https://doi.org/10.1146/annurev-environ-082508-094348
Moltó J, Font R, Conesa JA (2006) Study of the organic compounds produced in the pyrolysis and combustion of used polyester fabrics. Energy Fuels 20(5):1951–1958. https://doi.org/10.1021/ef060205e
Mukherjee S (2015) Environmental and social impact of fashion: towards an eco-friendly, ethical fashion. Int J Interdiscip Multidiscip Studies 2(3):22–35
Mukherjee P, Sharma RS, Rawat D, Sharma U, Karmakar S, Yadav A, Mishra V (2024a) Microbial communities drive flux of acid orange 7 and crystal violet dyes in water-sediment system. J Environ Manag 351:119699
Mukherjee P, Sharma RS, Mishra V (2024b) Deciphering the ecological impact of azo dye pollution through microbial community analysis in water–sediment microcosms. Environ Sci Pollut Res 1–19
Nicita A (2007) Who benefits from export-led growth? Evidence from Madagascar's textile and apparel industry. J Afr Econ 17(3):465–489. https://doi.org/10.1093/jae/ejm030
Niinimäki K, Peters G, Dahlbo H, Perry P, Rissanen T, Gwilt A (2020) The environmental price of fast fashion. Nature Rev Earth Environ 1(4):189–200. https://doi.org/10.1038/s43017-020-0039-9
Nizzetto L, Futter M, Langaas S (2016) Are agricultural soils dumps for microplastics of urban origin? Environ Sci Technol 50(20):10777–10779. https://doi.org/10.1021/acs.est.6b04140
Nørup N, Pihl K, Damgaard A, Scheutz C (2019) Quantity and quality of clothing and household textiles in the Danish household waste. Waste Manage 87:454–463. https://doi.org/10.1016/j.wasman.2019.02.020
Oliveri Conti G, Ferrante M, Banni M, Favara C, Nicolosi I, Cristaldi A, Fiore M, Zuccarello P (2020) Micro- and nano-plastics in edible fruit and vegetables: the first diet risks assessment for the general population. Environ Res 187:109677. https://doi.org/10.1016/j.envres.2020.109677
Palacios-Mateo C, van der Meer Y, Seide G (2021) Analysis of the polyester clothing value chain to identify key intervention points for sustainability. Environ Sci Eur 33(1):2. https://doi.org/10.1186/s12302-020-00447-x
Papasolomou I, Melanthiou Y, Tsamouridis A (2023) The fast fashion vs environment debate: Consumers' level of awareness, feelings, and behaviour towards sustainability within the fast-fashion sector. J Mark Commun 29(2):191–209. https://doi.org/10.1080/13527266.2022.2154059
Perry P, Wood S (2019) Exploring the international fashion supply chain and corporate social responsibility: Cost, responsiveness and ethical implications. In: Fernie J, Sparks L (eds) Logistics and retail management: Emerging issues and new challenges in the retail supply chain. Kogan Page, pp 1–26
Portway S (2019) Climate justice isn't sexy: the double failure of sustainable fashion marketing and activism. Fashion, Style and Popular Culture 6(1):49–67. https://doi.org/10.1386/fspc.6.1.49_1

Quantis (2018) Measuring fashion: Environmental impact of the global apparel and footwear industries. https://quantis-intl.com/report/measuring-fashion-2018/
Rawat D, Mishra V, Sharma RS (2016) Detoxification of azo dyes in the context of environmental processes. Chemosphere 155:591–605. https://doi.org/10.1016/j.chemosphere.2016.04.068
Rawat D, Sharma RS, Karmakar S, Arora LS, Mishra V (2018) Ecotoxic potential of a presumably non-toxic azo dye. Ecotoxicol Environ Saf 148:528–537. https://doi.org/10.1016/j.ecoenv.2017.10.049
Rawat D, Sharma U, Poria P, Finlan A, Parker B, Sharma RS, Mishra V (2022) Iron-dependent mutualism between *Chlorella sorokiniana* and *Ralstonia pickettii* forms the basis for a sustainable bioremediation system. ISME Commun 2(1):83
Rawat D, Sharma U, Yadav A, Poria P, Farooqi F, Pani B, Mukherjee P, Singh M, Sharma RS, Mishra V (2023) A bottom-up approach to select microbes from textile wastewater for detoxification of aromatic amine and azo dye in a single stage. J Water Process Eng 53:103668
Remy N, Speelman E, Swartz S (2016) Style that's sustainable: a new fast-fashion formula. McKinsey Global Institute
Sandin G, Peters GM (2018) Environmental impact of textile reuse and recycling: a review. J Clean Prod 184:353–365. https://doi.org/10.1016/j.jclepro.2018.02.266
Sant'Ana MA, Kovalechen F (2012) Evaluation of the health risks to garment workers in the city of Xambrê-PR, Brazil. Work 41:5647–5649.https://doi.org/10.3233/WOR-2012-0906-5647
Shamir R (2011) Socially responsible private regulation: World-culture or world-capitalism? Law Soc Rev 45(2):313–336. https://doi.org/10.1111/j.1540-5893.2011.00441.x
Smith NC (2003) Corporate social responsibility: whether or how? Calif Manage Rev 45(4):52–76. https://doi.org/10.2307/41166188
Stevis D, Felli R (2015) Global labour unions and just transition to a green economy. Int Environ Agreements Politics Law Econ 15(1):29–43. https://doi.org/10.1007/s10784-014-9266-1
Tai F-M, Chuang S-H (2014) Corporate social responsibility. Ibusiness 6(3):117–130. https://doi.org/10.4236/ib.2014.63013
Tang AYL, Lo CKY, Kan C (2018) Textile dyes and human health: a systematic and citation network analysis review. Color Technol 134(4):245–257. https://doi.org/10.1111/cote.12331
Turker D, Altuntas C (2014) Sustainable supply chain management in the fast fashion industry: an analysis of corporate reports. Eur Manag J 32(5):837–849. https://doi.org/10.1016/j.emj.2014.02.001
Uddin M, Ullah MR, Dipto MRR (2021) Assessment of the current working condition of the garment workers and determining the importance of labor union for the improvement of working condition in the RMG industry of Bangladesh. In: Proceedings of the 11th annual international conference on industrial engineering and operations management (Vol 9)
UN Climate Change (2018) UN helps fashion industry shift to low carbon. UN Climate Change News. https://unfccc.int/news/un-helps-fashion-industry-shift-to-low-carbon
Wang J, Liu X, Li Y, Powell T, Wang X, Wang G, Zhang P (2019) Microplastics as contaminants in the soil environment: a mini-review. Sci Total Environ 691:848–857. https://doi.org/10.1016/j.scitotenv.2019.07.209
Watts, & Holme R (1999) Corporate social responsibility: meeting changing expectations. World Business Council for Sustainable Development
Wilgosh B, Sorman AH, Barcena I (2022) When two movements collide: learning from labour and environmental struggles for future just transitions. Futures 137:102903. https://doi.org/10.1016/j.futures.2022.102903
Wood DJ (2010) Measuring corporate social performance: a review. Int J Manag Rev 12(1):50–84. https://doi.org/10.1111/j.1468-2370.2009.00274.x
Wright SL, Kelly FJ (2017) Plastic and human health: a micro issue? Environ Sci Technol 51(12):6634–6647. https://doi.org/10.1021/acs.est.7b00423
Yadav A, Kumar P, Rawat D, Garg S, Mukherjee P, Farooqi F, Roy A, Sundaram S, Sharma RS, Mishra V (2022) Microbial fuel cells for mineralization and decolorization of azo dyes: recent advances in design and materials. Sci Total Environ 826:154038

Zafar M, Bhuiyan A (2012) Present status of garment workers in Bangladesh: An analysis. IOSR J Bus Manag 3(5):43–53. www.iosrjournals.org

Zele YT, Kumie A, Deressa W, Bråtveit M, Moen BE (2021) Registered health problems and demographic profile of integrated textile factory workers in Ethiopia: a cross-sectional study. BMC Public Health 21(1):1526. https://doi.org/10.1186/s12889-021-11556-4

Does CSR Contribute to Sustainable Development? Characterization of CSR in Colombian Companies

Gustavo A. Yepes-López, Julián Mauricio Cruz-Pulido, and José Luis Camarena-Martínez

Abstract It is theorized that corporate social responsibility contributes to sustainable development to the extent that it favors social inclusion, reduces environmental impact, and reinforces governance while at the same time strengthening the economic viability of the company; however, not all corporate social responsibility practices are equal or contribute in the same way to sustainable development. This is because many of these, on the one hand, do not reduce the impacts of the operation of companies. On the other hand, they do not directly assume the associated costs, which is why this article aims to characterize the practice of corporate responsibility in Colombia in terms of its contribution to sustainable development, identifying whether the company itself assumes CSR practices. To achieve this objective, data from 7,233 Colombian companies are examined, performing a two-stage cluster analysis (hierarchical and K-means), resulting in an empirical model of characterization of modalities of social responsibility versus CSR cost externalization, where diverse behaviors were found, which were grouped into five profiles—externalizers, conservatives, politically correct, undecided and pioneers—that account for the different perspectives with which Colombian companies approach their corporate responsibility and their contribution to sustainable development.

Keywords Corporate social responsibility · CSR · Sustainable development · Characterization · Profiles · Colombia · Emerging markets

G. A. Yepes-López (✉) · J. M. Cruz-Pulido · J. L. Camarena-Martínez
School of Management, Universidad Externado de Colombia, Bogotá, Colombia
e-mail: gustavo.yepes@uexternado.edu.co

J. M. Cruz-Pulido
e-mail: julian.cruz@uexternado.edu.co

J. L. Camarena-Martínez
e-mail: Jose.camarena@uexternado.edu.co

P. Singh et al. (eds.), *Integrating Environmental Sustainability and Corporate Social Responsibility: A Move Beyond Profit*, World Sustainability Series,
https://doi.org/10.1007/978-3-031-81469-3_4

1 Introduction

The growing interest in corporate social responsibility evidenced by the increase in practices, standards, reports, and indices (Batista and Francisco 2018; Busch et al. 2024; Halkos and Nomikos 2021; Jan et al. 2023; Lambin and Thorlakson 2018; Velte 2023; Xu et al. 2024) is a clear indication that this significant trend in corporate management is currently strongly influencing managerial thinking and practice around the world, however, it is not clear whether those practices that promise improvements in the social, environmental and governance performance of the organizations that implement them, actually do so (Ali and Kaur 2021; Busch et al. 2024; Halme et al. 2020; Moon 2007; Taran and Mirkin 2020; Wijesundara et al. 2024) or if they are simply communication strategies that seek to improve the image and reputation of companies with light or misleading information (Balluchi et al. 2020; Gatti et al. 2019; Huynh et al. 2024; Oppong-Tawiah and Webster 2023; Šimunović et al. 2024; Skarmeas and Leonidou 2013; Stoeckl and Luedicke 2015).

Various authors have characterized corporate social responsibility practices under different parameters and models (Aguinis et al. 2024; Carroll 1979, 2021; Davis and Blomstrom 1966; Marques and Mintzberg 2015; McAdam 1973; Rangan et al. 2015; Wilson 1975; Zadek 2007) which, despite contributing to knowledge about CSR management within companies and in different sectors of the economy, do not make their contribution to society evident (Christensen and Murphy 2004; Johnston 2011; Newell 2008) or specifically to sustainable development (Attig et al. 2023; Farias et al. 2024; In and Park 2019; Singhal et al. 2024).

In emerging markets, countries with significant economic, social, and environmental demands, particularly CSR, promise to catalyze economic growth, social inclusion, and environmental care (Cezarino et al. 2022). However, most research focuses on developed (Muller and Kolk 2009; Nyabakora and Mohabir 2024), limiting the ability to recognize and clearly establish the similarities, particularities, and effectiveness of CSR performance in this type of market (Boubakri et al. 2021; Jamali and Karam 2018). For these reasons, the objective of this document is to characterize the practice of corporate responsibility in Colombia, a country widely recognized as an emerging market, and to identify the contribution of Colombian companies to the sustainable development of the country since the need to recognize the types of CSR that best contribute to sustainable development has been raised (Freze et al. 2023; Moon 2007; Wang et al. 2024).

To achieve this objective, both the theory associated with CSR is used to determine the modalities and contribution of the practices, and the theory of social costs to identify the appropriation or externalization of the investments associated with these practices, in a sample of 7,233 companies, to which a 2-stage cluster analysis (hierarchical and K-means) is carried out. Next, the reader will find the literature review, the description of the methodology, the results obtained, the discussion and conclusions that we trust will be of practical and theoretical use to the reader.

2 Literature Review

Corporate Social Responsibility and Sustainable Development have been the subject of debate in recent decades, ranging from the lack of unanimity in their definitions, through their nature, the achievements they pursue and the expected scope, the most efficient mechanisms for their implementation and management, to the results obtained.

Apart from these important discussions of concepts at the individual level, a new debate has taken shape associated with the relationship between CSR and Sustainable Development, with proposals that are also dissimilar and with diverse and even contradictory results. While some researchers find that empirically, in particular contexts, there is a great distance between these two concepts (Husser et al. 2012; Simionescu 2015), others warn that despite their different origins, they converge, sharing similar perspectives and even shared measurement mechanisms (Ashrafi et al. 2018; Bansal and Song 2017; Loew et al. 2004). However, longitudinal studies find an increasingly strong relationship between these concepts (Bhargava and Ligade 2023; Sánchez-Teba et al. 2021). Within the framework of this controversial relationship, there is also the debate on the contribution of CSR to Sustainable Development (Moon 2007) or specifically to the United Nations Sustainable Development Goals (de Almeida Barbosa Franco et al. 2024; Mahajan et al. 2024),which are associated, on the one hand, with social problems and the expectations that society has about the behavior of companies and, on the other, the way or mechanisms with which companies respond to these expectations.

In the first place, the various problems and expectations of society towards companies can be catalogued by means of criteria, which can be called alternatives or modalities that from the CSR respond to these problems and social expectations, which can be grouped through different taxonomies or classifications. Some authors look for classifications by the regions or geographical areas where they are applied (Galego-Álvarez et al. 2014; Husted et al. 2016; Masindi et al. 2023; Zamir and Saeed 2020) or by the economic sectors that carry them out (Pillai et al. 2024; Sancho et al. 2017; Song et al. 2020) or for their contribution to the solution of social issues (Kotler and Lee 2005; Lucarelli and Severini 2024; Sheehy 2015). However, in recent years, most agree with including the company's contribution to social expectations, set out in the components of sustainable development—economic, social, and environmental—(May et al. 2007).

The traditional criteria by which CSR was measured were associated with the company's contribution to solving social and community problems (Kassis and Majaj 2012; Kotler and Lee 2005; Sheehy 2015). However, since the 1990s, the elements or criteria that currently take into account such activities are generally found in international mechanisms for the promotion of sustainable development and social responsibility, such as ISO 26000 (International Standards Organization 2010), Global Reporting Initiative (2020), OECD (2011) as well as the UN Global Compact (2021), which are widely used worldwide (Raufflet et al. 2017) and which have become popular in recent years, especially in the financial sector under the name of ESG criteria or risks (Jiang et al. 2024). Secondly, the measures or actions with which companies respond to society's expectations have also been extensively studied and can be categorized according to the attitudes or postures with which companies can approach these social commitments, which occur in terms of stages of evolution or CSR performance profiles.

In this case too, the answers are diverse. For example, Carroll (1979) warns that companies can adopt different types of receptivity to CSR and consolidates the categories with which different authors characterize the social responsiveness of companies between doing nothing and doing much (see Table 1).

Another way of categorizing practices has been using profiles that show the characteristics and traits to measure performance. For instance, Taran and Mirkin (2020) group them into five: improving CSR performance (CS), de-improving CSR performance (SB), limited CSR dimensions (CS), CSR modality approach (SD), expansion of CSR modalities (SE); Farias et al. (2024) propose five profiles called following (1) as usual, (2) hypocritical suitor, (3) hypocritical cooperator, (4) responsible company, and (5) purpose-driven company. In addition, a study in a group of 586 Luxembourg companies proposed two profiles: strategic CSR and responsive CSR (Bocquet et al. 2013).

The company's commitment to social responsibility has also been the subject of analysis. Dowling and Moran warned that social responsibility actions could only be bolted to the company's strategy and not integrated as they should, thus obtaining less coherence. with the main actions of the organization, offering less credibility to its interest groups and the general public (Dowling and Moran 2012; Rahman and Blake 2021).

Table 1 Social responsiveness categories and profiles classified in previous literature

Author(s)	Social responsiveness Categories				
	Do less				Do more
Davis and Blomstrom (1968)	Withdrawal	PR approach	Legal approach	Negotiation	Problem solving
McAdam (1973)	Fight to the end	Do only what is necessary	Be progressive	Lead the industry	
Wilson (1975)	Reaction	Defense	Accommodation	Proaction	
Zadek (2007)	Defensive	Compliance	Management	Strategic	Civilian
NBS (2012)	Operational optimization	Organizational transformation	Systems creation		
Marques and Mintzberg (2015)	Business as usual	Public relations	Stakeholder perspective	CSR business case	Civil
Rangan et al. (2015)	Philanthropic	Operational effectiveness	Business model transformation		

Source own elaboration

In the case of Colombia specifically, the CSR performance was characterized, also under the traits and characteristics of 25 large companies, determining the following four profiles: (1) functionalist, (2) constructivist, (3) culturalist paradigms, and (4) sociopolitical (Villa-Castaño et al. 2023) however, none of them took into account in this evaluation of the contribution of CSR to Sustainable Development, how companies assume or do not assume the costs derived from their CSR actions proposed by Johnston (2011) thus preventing the company from using it as a promotional and market element, without taking care of the adverse effects created in society by itself. Another example includes the sustainability five dimensions model, which characterizes sustainability practices according to five dimensions: (1) global management, (2) competitiveness, (3) stakeholder relationships, (4) business impacts, and (5) accountability and transparency (Yepes-López et al. 2021).

2.1 Proposed Model

Although the contribution to sustainable development, as noted in the previous section, has a relevant convergence between the criteria and the different modalities of the practices used to identify the themes with which companies respond to social expectations, and despite the fact that they have been closely related as in Carroll's pyramid (Carroll 2016; Janowski 2020) their nature and impact are not the same (Margolis and Walsh 2003). Some contributions are associated with cash or kind donations that companies make to contribute to the solution of one or more social problems (Adrian et al. 2013; Montuschi 2009), while others are concerned with the impacts of their operations and decision-making processes (Porter and Kramer 2006; Rangan et al. 2015; Wartick and Cochran 1985; Zollo 2004) and increasingly associated with the concept of sustainable development, as proposed by the most prominent international standards.

In order to group the 14 modalities or alternatives of CSR resulting from the review above, it was considered to divide them into those that contribute to society but are not necessarily associated with the productive activities of the company. These will be cataloged in the first group, Philanthropic CSR (Kotler and Lee 2005). The alternatives or modalities related to the productive activity of the companies that can improve efficiency in terms of resource use, as well as optimize production processes (Porter and Kramer 2006), which could help reduce production costs, will be cataloged in a second group called Operational Efficiency CSR (see Table 2).

On the other hand, the mechanisms or means used to achieve operational efficiency do not have the same affinity, because they are very diverse. According to Carroll (1979), companies can adopt different degrees of receptivity to CSR. After reviewing the literature presented above, an unanalyzed position evaluates the negative or positive effects of companies' CSR actions on third parties or society. Therefore, an alternative means of identifying the firm's contribution level may be the level of assimilation or transfer of costs associated with CSR by firms known as external costs or externalities. The theory of social costs (CST) initially proposed by

Table 2 Alternatives or modalities of CSR by activity group

Group 1 Philanthropic CSR	Group 2 Operational Efficiency CSR
Contributions toward community (financial contribution)	Environment
Social and community projects	Human rights
Donations	Labor Aspects
Social marketing	Transparency and anti-corruption
Commitment to social causes	Freedom of competition
Cause-related marketing	Suppliers and distributors
Volunteering	Customers and consumers

Source Yepes et al. (2023)

Ronald Coase highlights the existence of both positive and negative effects on the economic system, which are not registered in the price mechanism but generate a series of information failures in the system itself, in turn causing inefficiencies in the allocation of resources (Coase 1960). These externalities can arise in every decision that is made in an economic activity and represent everything that society assumes to keep companies functioning, whether positive or negative (Kapp 1950), thus, the proposed way to register this type of corporate stance to address CSR is to associate these actions with accounting records (Yepes-López et al. 2023).

Accounting records in this sense are the alternatives with which a firm can record the disbursement accrued for the SR activity(s) in which it carries out. It should be noted, however, that when a company assigns the type of accounting record to a CSR activity, two situations may occur. First, it is possible that, in the first instance, the activity in question did not involve any financial outlay, so in this case it would not be possible to include it in the accounting system, this would be the case, for example, of employee volunteering practices, i.e. there is no accounting record. The second situation assumes that CSR activity entails a monetary outlay by the organization, obliging companies to record it in their accounts. In Colombia, there is a standardized scheme of accounting records called the Single Plan of Accounts (PUC), through which firms are required to have a uniform record of economic operations under the same structure. This allows for the commensurability and transparency of the accounting information of Colombian companies (Decreto 2650 de 1993 1993).

According to the International Financial Reporting Standards (IFRS) system and the PUC in Colombia, there are the following alternatives to record an economic disbursement: (1) donations, (2) non-operating expenses, (3) operating expenses, (3) cost and (4) reserves. Likewise, for the purposes of this study, the option of not registering the activity is contemplated, and in such a case, it would be interpreted that the cost of said activity would be outsourced in its entirety to a third party (Antheaume 2004; Hoi et al. 2013). Thus, it is proposed that a CSR activity's externalization level should be seen as a continuum, where there is a maximum level of cost externalization at one end, while at the opposite end, there is a minimum level of cost transfer.

2.2 CSR Contribution to Sustainable Development Map

Of the two elements described above, CSR Modalities and the Type of Accounting Record, the CSR Contribution Sustainable Development Map (CCSDM) is proposed. These two components constitute a Cartesian plane where the companies' position in relation to the two variables can be identified. This combination of elements results in a map that can help classify companies, taking as a reference the modality of the CSR carried out and the type of accounting record assigned in the accounting. This will make it possible to determine the collective position of all the companies studied, defined by groupings of individual points in the quadrants of the Cartesian plane. The SDCM (Fig. 1) proposes the possible generic behaviors in which CSR practices can be characterized against the theory of social costs, taking into account the axes based on the variables of Modality and Registration.

The combination of the firms' CSR modality and CSR cost externalization determines the possible characteristic behaviors, which constitute the following four sectors:

First quadrant: the CSR practices of companies in this quadrant are not formally recorded in accounting and are therefore considered external costs (Coase 1988). Likewise, they use the modalities associated with group 1 called Philanthropic CSR, which focuses on the contribution of charitable initiatives and initiatives of social and community interest, which makes this the sector that generates the most negligible social contribution since not only does it outsource its CSR actions, but these do not generate concrete solutions to the causes of social problems associated with the

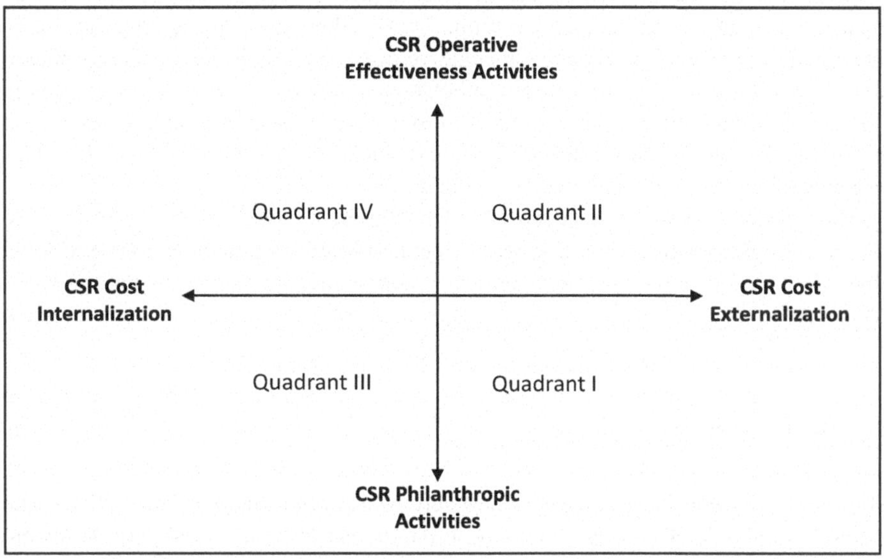

Fig. 1 CSR Contribution to sustainable development Map (CCSDM). *Source* own elaboration

impacts of the company itself (Baudillard 2007). Companies located in this sector primarily focus their efforts on mitigating the symptoms of social problems. Although these types of contributions are perfectly valid, they fail to develop the potential that the company has to contribute from its area of influence to reduce the causes of social problems (Pimonenko et al. 2020).

Second quadrant. This sector is made up of CSR practices that maintain a significant level of cost externalization. However, they use the modalities of group two, which are associated with the company's operation. Therefore, this sector has a relatively better social contribution than the previous one since it seeks to reduce the causes of social problems from its area of influence; though it externalizes the costs associated with its practices, and it is society or a third party that pays for them.

Third quadrant. The third sector groups together CSR practices recorded in its accounting system. However, the fact that they do register them does not mean that this group internalizes all their costs associated with CSR. Additionally, this sector is characterized by a predilection in the use of the practices of the philanthropic group (Kotler and Lee 2005). In other words, this group is more involved in charitable activities, which means that companies miss out on their potential to provide solutions to their own impacts. For this reason, this quadrant generates the second most significant social contribution since, although its CSR practices are inclined towards operational efficiency, it does not ultimately seek to reduce its negative externalities.

Fourth Quadrant. Finally, this sector includes the CSR practices that are effectively recorded in the accounting system and mainly include the modalities associated with reducing the company's own impacts (Porter and Kramer 2006). Thus, this sector is considered to be the one in which the practices and initiatives that contribute the most to companies and society are found. This is because, on the one hand, they make deliberate efforts to reduce the negative impacts of their operation or improve their social contribution, and, on the other hand, they record their CSR actions in accounting, thus reducing the externalization of these costs. Although the mere fact of its registration does not imply the non-existence of cost transfer, formality does presume a better impact and less externalization.

The CSR Contribution to Sustainable Development Map thus serves as a tool to characterize the contribution of CSR to sustainable development, as opposed to an alternative vision in the corporate stance of addressing CSR, such as the theory of social costs.

3 Methodology

This paper is a descriptive and explanatory cross-sectional study (Bryman 2015). The methodological details are set out below. First, the variables involved in the study are presented. Then, the study participants are examined. The study's hypotheses and methods for testing them are then explained. This section ends by narrowing down the theoretical and software tools used.

3.1 Study Variables

The variables analyzed in this study are:

- Registry and modality variables: these are made up of the accounting records corresponding to CSR activities and the modalities in which these activities are carried out. In terms of records, CSR activities can be recorded in reserves (paid by the shareholder), in costs (paid by the consumer), in operating and non-operating expenses (paid by the company), in donations and tax exemptions (paid by the state) or they can be unrecorded. The modalities, on the other hand, are divided into two groups or two types of activities: (1) philanthropic activities related to economic contribution projects, foundations, and donations, social marketing, support for noble causes, cause-related marketing, and volunteering; and (2) operational activities, related to the environment, human rights, labor, transparency, fair competition, suppliers and customers.
- Quadrants of the CSR Contribution to Sustainable Development Map: The proposed theoretical model distributes companies into four quadrants explained in the theoretical framework. The analysis uses this variable to verify the proposed model through actual data.
- CSR business profile: consists of a synthetic variable obtained by applying clustering algorithms to the registration and modality variables.

3.2 Participants

According to the Superintendence of Societies of Colombia, in 2012, were 27,000 companies registered as commercial companies, branches of foreign companies, and sole proprietorships throughout the national territory (Superintendencia de Sociedades 2012). The official instrument for collecting information is Form 31 Business Practices. Based on this instrument, the total number of these companies with CSR practices is estimated. Of the 18,897 companies that responded to the questionnaire adequately, 7,233 companies stated that they carried out CSR activities. According to Colombia's Single Business and Social Registry (RUES), 855,686 companies were formally registered in 2012. It is worth noting that the Superintendence of Societies does not supervise corporations, associative work companies, or financial entities of public services. These entities are, therefore, outside the population universe.

The population universe corresponds to 7,233 companies that appropriately filled out Form 31 and claimed to carry out CSR actions in 2012. The population universe has been selected in this period due to a more robust questionnaire on CSR issues

(Superintendencia de Sociedades 2011). Questionnaire 31 was modified in subsequent years by the Superintendence of Companies, reducing the amount of information collected. Using this source of information provides access to the data of the population universe so that the sample design is the census—incorporation of the entire target population in the sample—(Wu and Thompson 2020). The sample size is 7,233 participating companies; considering that the sample design is census-based, the sample size coincides with the population size; therefore, the sampling error in the analyses is equal to zero. However, it is pertinent to carry out hypothesis tests to reflect on the behavior of hypothetical companies that, meeting the selection criteria, have not been able to enter the population for unknown reasons or companies of the same population under identical conditions in subsequent years.

3.3 Hypotheses

The hypotheses of the study are as follows:

H1: Companies are distributed according to the CSR Contribution to Sustainable Development Map quadrants. To test this hypothesis, the map is implemented and interpreted.

H2: Companies have different behaviors regarding externalization and the relevance of CSR activities. In order to test this hypothesis, the new synthetic variable CSR Business Profile is constructed using algorithms of aggregation to the data of CSR accounting records and CSR modalities; the construction of this particular variable is presented in detail.

H3: The different ways in which companies approach CSR are expressed in the Social Contribution Map. This hypothesis evaluates the relationship between the quadrant variables of the Map of Contribution to Sustainable Development that determines the theoretical approach and the CSR Business Profile, which shows the behaviors of companies according to the data.

3.4 Theoretical and Software Tools

The implementation of the CCSDM is carried out according to a score for the accounting records (reserve = 0, cost = 2, operating expense = 2, non-operating expense = 2, tax discount = 4, non-registration = 4), which is transformed to the scale (0, 1) and a modality score (operational efficiency = 0, Philanthropy = 1). These quadrants can be combined to generate an outsourcing index (Yepes-López et al. 2023).

The construction of the new CSR Business Profile variable is carried out using clustering algorithms. The grouping process is carried out in two stages. The first stage encompasses the agglomerative hierarchical procedure with Ward link, which is used in the second stage by applying the K-Means algorithm with which the final result is obtained. The number of groups is obtained from the dendrogram obtained in the first stage (Everitt et al. 2011; Han et al. 2022).

The hypothesis test starts with descriptive statistics and visualization (H1) and continues by applying the Shapiro–Wilk normality test for the selection of parametric or non-parametric procedures. Then, the silhouette coefficient is calculated on the resulting profiles and Student's or Wilcoxon-Mann–Whitney t-tests for equality to zero in a tail (H2) and chi-square statistics (H3) (Gibbons and Chakraborti 2014; Razali and Wah 2011; Shapiro and Wilk 1965; Thijssen 2016). Access to the data is provided by the Superintendence of Industry and Commerce. Statistical analysis and methodological results are obtained using R software (R Core Team 2023).

4 Results

This section includes the results of the data analysis.

H1: *Companies are distributed according to the quadrants of the Map of Contribution to Sustainable Development.* The quadrant distribution is shown in Fig. 2.

The CCSDM then defines four quadrants that comprise all the possibilities for a company's behaviors. However, companies are mostly concentrated in only one of these quadrants.

H2: *There are different behaviors on the part of companies regarding the outsourcing and relevance of CSR activities.* By applying the aggregation algorithms, a new synthetic variable is obtained that describes the different behaviors of companies concerning CSR in terms of modalities and accounting records. This variable is a central result in this study, called the CSR Business Profile. The data collected provide an overview of the structure of CSR modalities and CSR accounting records of each company. Table 3 shows the percentage values of the use of registries and modalities in CSR.

Consequently, $7 \times 14 = 98$ pieces of data are used for each company to create the profiles. Figure 3 shows five distinct groups by applying the agglomerative hierarchical clustering algorithm with Ward's least variance link.

An optimal group structure is obtained by refining this result using a k-means procedure. The profiles found are shown in Table 4.

The CSR Business Profiles synthetic variable is distributed according to Fig. 4 below.

The calculation of the silhouette coefficient, its mean, and significance support this hypothesis. The value of the average silhouette coefficient is close to zero (0.0987), which indicates a low definition of the groups. According to this value, the groups are diffuse and mixed with each other, generating cases of confusion, where some companies may present characteristics of two or more groups simultaneously. The silhouette coefficient is calculated for each company to test the hypothesis; thus, its

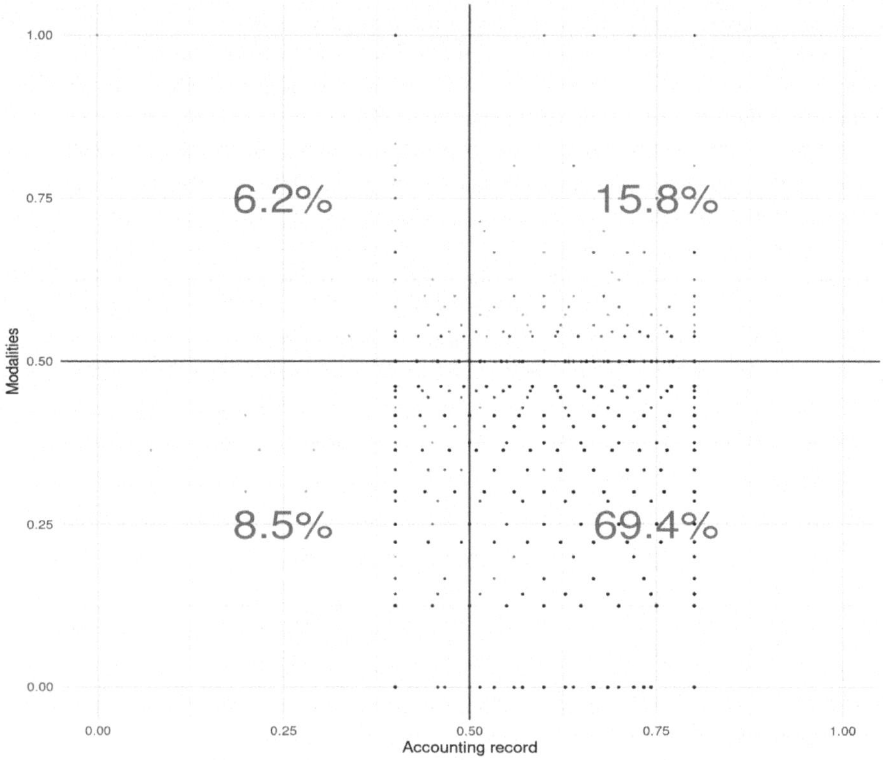

Fig. 2 Percentage of companies located in the quadrants of the CCSDM. *Source* Own elaboration

mean and corresponding significance are calculated, which evaluates as an alternative hypothesis that the global silhouette coefficient is greater than zero. The Shapiro–Wilk test indicates that the individual silhouette coefficient does not follow a normal distribution. The p-value is less than 0.05 (Wilcoxon-Mann–Whitney p-value < 2.2e-16), indicating statistically significant evidence that the value is greater than zero. This means that, although it is a relatively low value, the silhouette coefficient is greater than zero, substantively supporting the existence of the different profiles identified.

Therefore, there are different ways of approaching CSR management in the companies in the sample. This management is affected by several factors, such as the modality in which the activities associated with CSR are carried out. These modalities can be operational or philanthropic. In turn, the accounting record of CSR activities shows the responsibility the company assumes, shared with other stakeholders, or directly transferred to the state. The different forms of CSR management are expressed in the CSR Business Profile variable. These profiles are not well defined, generating cases of confusion, where some companies may present characteristics of two or more groups simultaneously.

Table 3 Percentage of companies that respond favorably for each CSR modality and accounting record

CSR modality	Accounting record type					
	Reserves (%)	Cost (%)	Operating expense (%)	Non-operating expense (%)	Tax deduction (%)	No registry (%)
VOL	0.06	2.36	9.11	5.94	0.08	26.03
CRM	0.10	2.79	5.68	4.15	0.11	21.82
CSC	0.15	2.99	9.46	16.07	0.64	31.95
SMA	0.04	1.98	9.95	5.06	0.11	24.19
DON	0.15	2.52	10.81	30.37	4.22	18.26
SCP	0.14	3.51	9.39	10.41	0.37	35.23
CTC	0.08	3.94	10.06	12.86	0.76	24.62
C&C	0.06	4.11	14.43	4.23	0.06	71.96
S&D	0.06	3.75	11.27	3.75	0.07	73.33
COM	0.08	2.13	9.01	3.87	0.03	72.79
TAC	0.04	2.17	11.97	5.50	0.08	67.34
LAA	0.08	7.92	36.46	7.33	0.11	39.60
HUR	0.07	3.80	14.90	5.97	0.10	70.95
ENV	0.10	14.95	22.31	8.95	0.30	45.53

Notes VOL = volunteering, CRM = cause related marketing, CSC = commitment to social causes, SMA = social marketing, DON = donations, SCP = social and community projects, CTC = contributions toward community, C&C = customers and consumers, S&D = suppliers and distributors, COM = freedom of competition, TAC = transparency and anti-corruption, LAA = labor aspects, HUR = human rights, ENV = environment. Source: Yepes-López et al. (2023)

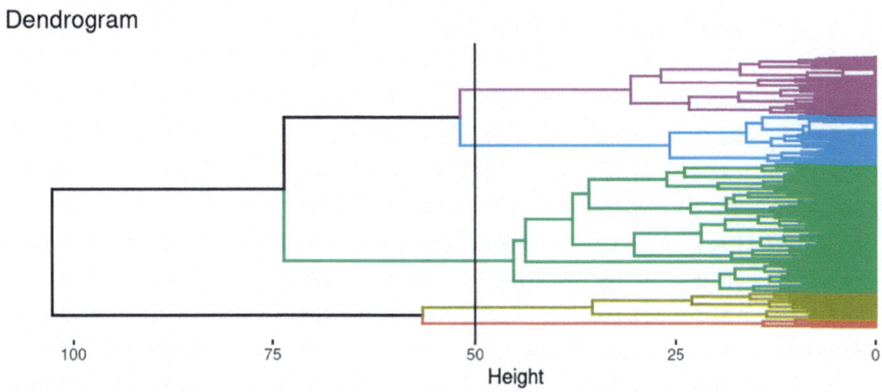

Fig. 3 Dendrogram of the cumulative hierarchical procedure. Note. Five groups are selected. *Source* own elaboration

Table 4 Clusters found in the sample: externalizer, conservative, undecided, politically correct and precursor

Cluster	Account registry type						
	CSR Modality	Reserves (%)	Cost (%)	Operating expense (%)	Non-operating expense (%)	Tax deduction (%)	No registry (%)
Externalizer	VOL	0.00	0.86	2.63	0.92	0.13	69.72
	CRM	0.07	1.05	0.86	0.66	0.00	62.61
	CSC	0.00	0.53	2.17	3.16	0.33	88.74
	SMA	0.07	0.26	2.44	0.99	0.20	74.46
	DON	0.20	1.51	7.83	23.37	4.08	48.58
	SCP	0.07	0.86	1.45	0.92	0.13	88.87
	CTC	0.07	1.71	4.28	5.46	0.46	63.46
	C&C	0.00	0.66	2.17	0.53	0.13	94.73
	S&D	0.00	1.05	1.32	0.33	0.07	94.40
	COM	0.07	0.59	0.46	0.26	0.13	93.94
	TAC	0.00	0.59	2.30	0.46	0.13	90.98
	LAA	0.00	5.27	19.49	3.69	0.26	66.82
	HUR	0.00	0.92	2.30	0.86	0.26	94.01
	ENV	0.20	8.82	10.60	4.34	0.20	71.49
Conservative	VOL	0.05	0.53	1.25	1.68	0.00	13.28
	CRM	0.00	1.58	1.25	1.34	0.05	8.87
	CSC	0.10	1.15	2.68	8.77	0.48	16.54
	SMA	0.00	0.67	3.21	2.01	0.00	6.86
	DON	0.05	1.34	3.98	20.18	3.26	16.73
	SCP	0.05	1.29	1.53	4.03	0.19	22.67
	CTC	0.05	1.05	2.54	5.56	0.43	17.83
	C&C	0.05	1.01	1.34	0.34	0.05	92.33
	S&D	0.05	1.05	0.48	0.14	0.10	89.12
	COM	0.00	0.34	0.14	0.05	0.00	86.05
	TAC	0.00	0.29	0.58	0.34	0.05	83.99
	LAA	0.10	2.01	0.58	2.40	0.05	85.14
	HUR	0.05	0.43	1.05	0.38	0.05	95.78
	ENV	0.00	7.05	8.87	4.12	0.14	72.82
Undecided	VOL	0.04	5.23	9.43	5.15	0.04	16.63
	CRM	0.16	4.98	3.79	2.92	0.08	13.26
	CSC	0.21	6.30	9.43	20.91	0.54	20.79
	SMA	0.08	4.41	9.39	3.75	0.08	14.82
	DON	0.16	4.57	11.49	38.12	4.45	6.30
	SCP	0.21	7.62	8.97	12.39	0.49	24.54

(continued)

Table 4 (continued)

Cluster	Account registry type						
	CSR Modality	Reserves (%)	Cost (%)	Operating expense (%)	Non-operating expense (%)	Tax deduction (%)	No registry (%)
	CTC	0.08	7.74	11.16	16.63	0.91	13.50
	C&C	0.12	8.73	8.81	1.32	0.04	72.00
	S&D	0.12	7.00	2.22	0.66	0.04	78.51
	COM	0.16	4.65	1.28	0.49	0.00	75.46
	TAC	0.12	5.19	7.08	1.77	0.08	66.57
	LAA	0.08	15.97	62.12	7.49	0.08	1.69
	HUR	0.08	8.60	15.15	4.03	0.00	63.73
	ENV	0.04	25.03	29.44	8.32	0.41	24.66
Politically correct	VOL	0.00	1.06	6.61	56.35	0.00	8.20
	CRM	0.00	1.06	5.82	43.12	0.00	6.08
	CSC	0.00	1.06	0.53	78.57	1.32	4.23
	SMA	0.00	1.06	6.35	45.50	0.00	6.88
	DON	0.00	0.53	2.12	71.96	7.41	3.17
	SCP	0.00	1.06	2.91	69.84	0.53	5.56
	CTC	0.00	1.59	3.70	62.43	0.79	4.76
	C&C	0.00	3.97	18.78	62.96	0.00	11.11
	S&D	0.00	4.50	12.70	61.90	0.26	14.81
	COM	0.26	1.85	9.26	63.23	0.00	16.93
	TAC	0.00	1.32	6.61	77.78	0.00	7.67
	LAA	0.00	5.56	30.69	56.88	0.26	2.12
	HUR	0.00	2.38	12.43	72.22	0.00	10.32
	ENV	0.00	8.99	13.76	66.67	0.26	4.76
Precursor	VOL	0.24	1.95	41.29	5.24	0.37	13.64
	CRM	0.24	3.41	31.43	3.41	0.61	11.81
	CSC	0.49	3.29	44.34	15.35	1.58	11.81
	SMA	0.00	1.71	44.34	5.60	0.37	10.96
	DON	0.37	2.19	35.69	27.16	4.75	8.40
	SCP	0.37	3.05	48.23	10.96	0.85	13.15
	CTC	0.24	5.24	39.59	11.08	1.71	12.06
	C&C	0.00	4.75	85.02	2.56	0.00	5.97
	S&D	0.00	5.60	83.19	1.58	0.00	5.85
	COM	0.00	2.19	70.16	2.92	0.00	17.78
	TAC	0.00	1.34	75.76	5.72	0.12	11.08

(continued)

Table 4 (continued)

Cluster	Account registry type						
	CSR Modality	Reserves (%)	Cost (%)	Operating expense (%)	Non-operating expense (%)	Tax deduction (%)	No registry (%)
	LAA	0.24	5.12	85.75	3.29	0.00	2.92
	HUR	0.24	4.14	73.81	4.87	0.24	14.49
	ENV	0.37	19.24	61.02	4.99	0.61	8.65

Note VOL = volunteering, CRM = cause related marketing, CSC = commitment to social causes, SMA = social marketing, DON = donations, SCP = social and community projects, CTC = contributions toward community, C&C = customers and consumers, S&D = suppliers and distributors, COM = freedom of competition, TAC = transparency and anti-corruption, LAA = labor aspects, HUR = human rights, ENV = environment

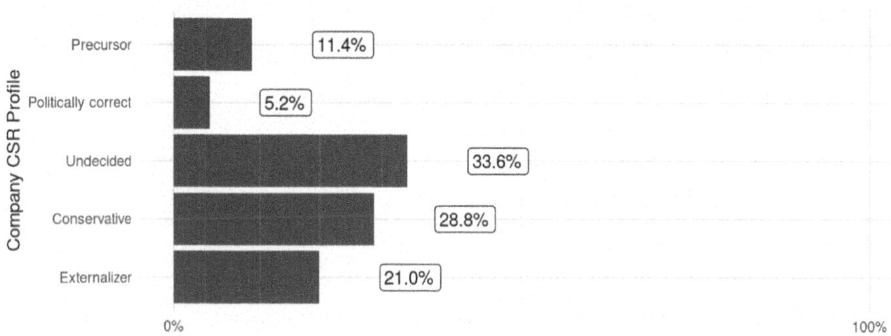

Fig. 4 Distribution of the sample concerning the groups found. *Source* own elaboration

H3: *The different ways in which companies approach CSR are expressed in the CCSDM.* It can be seen that approximately 70% of the companies belong to the lower right quadrant of the Map of Contribution to Sustainable Development. This can be contrasted with the distribution of CSR Business Profiles in the quadrants. Figure 5 shows the distribution of the companies according to the profiles in the defined quadrants.

Although the CCSDM covers all possible behaviors of companies in the four quadrants, there is not enough diversity in companies when implementing CSR. On average, the Externalizers, Conservative and Undecided CSR Business Profiles are located at the lower right quadrant, while the Precursor and Politically Correct CSR Business Profiles are located in the lower left quadrant. The test yields a p-value less than 0.05 (chi-square p-value < 2.2e-16), which implies the rejection of the null hypothesis. This means that there is statistically significant evidence of the relationship between CSR Business Profiles and the Quadrants of the Contribution to Sustainable Development Map. This relationship can observed in Table 5.

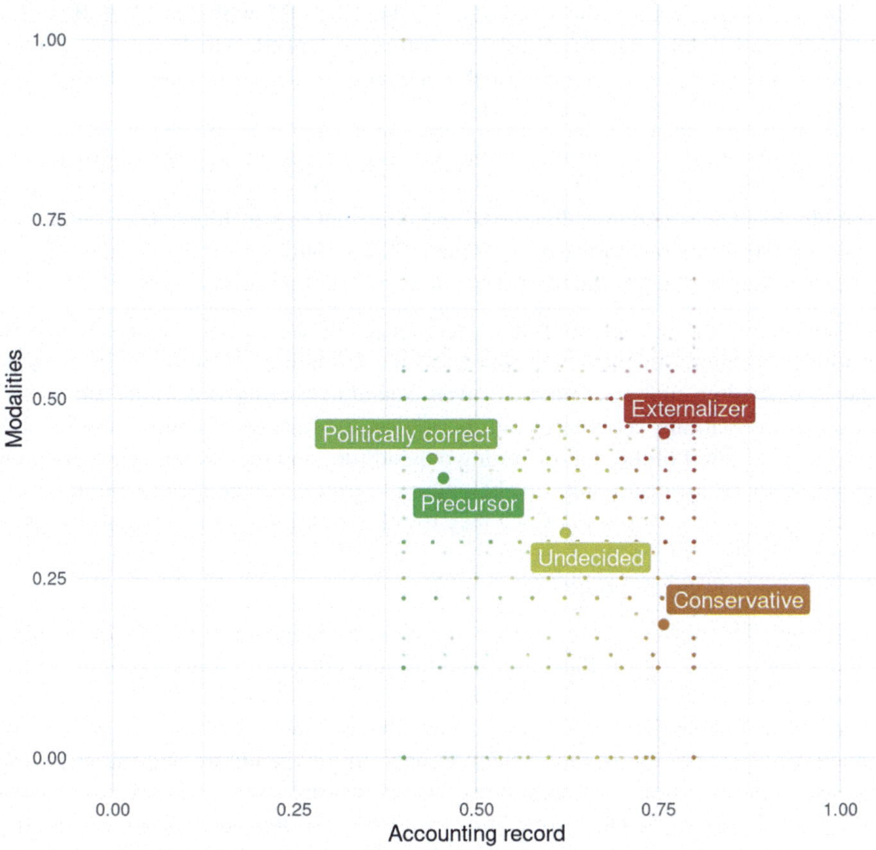

Fig. 5 CSR business profiles located in the quadrants of the CCSDM. *Source* own elaboration

Table 5 Contingency table between the CSR Business Profile and quadrants of the CCSDM

		Company CSR profile				
		Externalizer (%)	Conservative (%)	Undecided (%)	Politically correct (%)	Precursor (%)
Map of contribution to sustainable development quadrant	IV	9.02	0.90	4.62	0.40	0.85
	III	0.00	0.00	0.94	1.89	3.41
	II	12.06	28.04	26.57	0.58	2.19
	I	0.00	0.00	1.22	2.37	4.94

Source own elaboration

Finally, when relating the quadrants of the CCSDM with the CSR Business Profile, a statistically significant relationship is evidenced, which shows a coincidence between the proposed theoretical model and the data collected and allows the five types of profiles to be determined:

Externalizers: In the first profile, companies use almost entirely the alternative of non-registration in all modalities, registering a small percentage of non-operating expenditure in the modality donations and contributions to foundations, and they were called Externalizers since most of their CSR activities are transferred to society or third parties, making it the least convenient profile for society.

Conservatives: The second profile also presents a high concentration in the alternative of non-registration but shows exceptions in the modalities of volunteering, cause-related marketing, support for noble causes, social marketing, donations and contribution to foundations, economic contribution and realization of community interest projects, registered in the non-operating expenditure and the operating expense, all belonging to group 1 and only one belonging to group 2, care for the environment and the name Conservatives is since they use the traditional group modalities, they register timidly more and use several alternatives to register, which makes it the second least convenient profile for society.

Undecided: The third profile maintains the majority behavior of the use of the non-registry alternative, but in this case, also the modalities of group 1 (volunteering, cause-related marketing, support for noble causes, social marketing, donations and contribution to foundations, economic contribution and realization of projects of community interest) registered in the non-operating expenditure and the operating expense, labor, environment, human rights and transparency modalities associated with group 2 are recorded in the operating expenditure, followed by the non-operating expenditure, and they were named undecided, although their behavior is not ideal, it is the one that comes closest to this since they are the ones that outsource the least, making it the most convenient profile for society.

Politically correct: The companies in the fourth profile, on the other hand, concentrate the records of all modalities in the non-operating expense alternative, followed by a small percentage that uses the operating expense alternative and the no registry alternative. It is worth noting that in the modality of donations and contributions to foundations, a small proportion of the registration alternative called donation is presented, and they were called Politically Correct since it combines several registration alternatives, focusing on the modalities of community support and internal control, which are the modalities that offer the most visibility, making it the third least suitable profile for society.

Precursors: Finally, the fifth profile has a higher concentration of the operating expense registration alternative in all modalities, followed by the cost, non-registration alternatives, and a lower percentage in non-operating expenditure and the name of precursors were given to them since, although they are beginning to register, they have not yet clearly defined a position or strategy than the second most convenient profile for society.

5 Discussion

Considering the hypotheses raised in this study, it can be seen how the contribution to sustainable development by companies has different perspectives with which their performance can be characterized, and although they agree on aspects such as the mechanisms with which the contribution to sustainable development is evaluated according to international standards (e.g., ISO26000, GRI, UN Global Compact, OECD, etc.). which are associated with the company's value chain, seeking to reduce or improve the negative impact of companies, as well as increase efficiency in their operation for society or the environment (Raufflet et al. 2017). Also present are practices that include economic contributions to the community, participation in social and community projects, donations, social marketing campaigns, commitment to social causes, marketing campaigns with a cause, and volunteer programs (Kotler and Lee 2005) that, as mentioned above, these types of activities seek to contribute to a good cause, they do not necessarily reduce the impact of the operation of companies on society and the environment (Yepes-López et al. 2023).

On the other hand, according to Carroll (1979) companies can adopt different degrees of receptivity to social responsibility, such as the level of externalization of costs, taking as a reference the type of accounting record, with some similarities such as the donation record is similar in the category of the public relations approach of Davis and Blomstrom (1966); since a company that makes donations, a philanthropic activity, usually seeks to position its brand or improve the public's perception of it.

Likewise, the non-operating expense type of accounting record can relate to the reactive category of social responsiveness suggested by Wilson (1975). For example, penalties accrued for an excessive pollution fine would be considered a non-operating expense. So would be the actions and disbursements a company makes after the damage caused to mitigate negative externalities, i.e., a reactive CSR (Chang 2015; Rim and Ferguson 2020).

On the other hand, the operating expense option could be tied to Wilson's *defensive* McAdam's *do only what is required,* and Davis and Blomstrom's *legal approach* categories. This is because the activities in this category are those that are part of the day-to-day operation of companies. Thus, the costs produced, for example, in purchasing safety equipment (e.g., fire extinguishers, first aid kit) for installations that could be mandatory for the company would fall into this category, as well as the purchase of machinery and equipment (e.g., energy-saving light bulbs, sheets of recycled paper) for the company's daily activities. Also, providing minimum employment benefits by law, such as enrolling employees in the health care system, could be classified here.

Likewise, the accounting option, cost, can be compared to the *be progressive* category mentioned by McAdam (1973) and that of *accommodation* by Wilson (1975).

This is concerning the fact that the registration of a social responsibility activity as a cost would imply considering it as part of an investment, which would be progressing towards a strategic vision of CSR by the company (Afrin 2013) where the costs of its social actions are assumed with a minimum level of externalization. Thus, each proposed accounting record alternative could be linked to a level of outsourcing in a similar way to the categories of "social responsiveness" analyzed by Carroll (1979). The results obtained with the five profiles—externalizers, conservatives, politically correct, undecided, and precursors—can also resemble a type of evolution of CSR management as proposed by Zadek (2007), Marques and Mintzberg (2015) or Rangan et al. (2015) with the differences in terms of evaluation criteria.

The CSR Contribution to Sustainable Development Map is an alternative way of categorizing CSR actions, which not only allows us to see the attitude and commitment that companies assume towards CSR, but also the level of evolution of the practices used, through profiles that, although different results identified in the literature (Bocquet et al. 2013; Farias et al. 2024; Taran and Mirkin 2020) respond to particular concerns about companies' performance, which can occur in different contexts. The CCSDM may be used in emerging markets, which can be very useful for public institutions, civil society organizations, research institutions, and even individual private companies since it allows them to know the contribution to the sustainable development of the sample or unit evaluated beyond their effort in carrying out well-intentioned actions, identifying their commitment, reduction or compensation of the impact generated by their operation and identifying a stage of evolution or level of maturity. The CCSDM could also be helpful in the development of public policies or business guidelines, especially these days when the obligation to comply with the objectives and goals of the United Nations 2030 Agenda and the worrying situation that the world is experiencing with the systematic increase of greenhouse gases and global warming (Calvin et al. 2023).

6 Conclusions

Although CSR and Sustainable Development are concepts that have different origins, are getting closer every day. At its beginnings, CSR had been widely associated with social contribution and support activities, known as charitable or philanthropic, since the 90s; however, this association has slowly been linked to companies' impacts on society and the environment. In other words, we have gone from associating CSR with seeking ways to reverse the causes of many of society's most significant problems and not the manifestations of these. Consequently, measurement of CSR performance has been standardized and organized under parameters, guides, and standards established at the international level.

Different methodologies have been proposed to carry out these measurements, ranging from the simple reporting of performance against international criteria through sectoral and regional studies to the characterization of companies through more sophisticated mechanisms, which seek to identify the type of receptivity or attitude that companies assume towards CSR and the companies' levels of progress or maturity towards CSR, or the characteristics and traits with which companies approach CSR. An original approach with which CSR performance can be measured is the concept of social costs, initially promoted by Ronald Coase, which helped determine the company's willingness and commitment to assume CSR-related costs.

In order to conceptualize this alternative approach, a theoretical model called the CSR Contribution to Sustainable Development Map is proposed. The CCSDM is made up of four quadrants in which all the possibilities for a company's CSR behaviors are considered. The model was tested in a sample of 7,233 companies operating in Colombia, using scores on the accounting records and modalities proposed by Yepes-López et al. (2023). The results showed the sampled companies heterogeneously distributed: 69.4% in the first quadrant, 15.8% in the second quadrant, 8.5% in the third quadrant, and 6.2% in the fourth quadrant, showing that almost 70% of Colombian firms' practices are linked to social contribution or philanthropic activities and externalize the associated costs; while only 6.2% of the sample carry out CSR actions linked their operational activities and assume or internalize their costs.

The theoretical model's empirical validation was carried out employing clustering algorithms, intending that the different behaviors of companies in practice should be reflected on the CCSDM. When the algorithms were applied, five types of behavior of companies concerning CSR manifested: Externalizers, conservatives, indecisive, politically correct, and precursors. The identification of these profiles not only allows us to identify the behaviors through the modalities used and the types of accounting records used to formalize CSR practices but also to identify the level of convenience of these behaviors for society, which can be ordered as follows: (1) Precursors, (2) Undecided, (3) Politically Correct, (4) Conservative, and (5) Externalizers. However, even though these profiles determine the behaviors of the sample in Colombia, the results show that the situation is far from ideal. The most desirable profile, precursors, is located in the lower right of the model, showing the distance from the ideal CSR practices that the model proposes—located at the top left of the model—inviting us to redouble efforts to promote a much more responsible type of CSR and consistent with the challenges that Sustainable Development imposes on contemporary society.

Within the study's limitations, it is worth noting that the companies' behaviors are very similar. Very few companies assume CSR costs and carry out operational efficiency-related CSR practices, limiting their analysis due to information scarcity. On the other hand, the modifications made in recent years by the Colombian Superintendence of Societies, form 31 survey—the source of the information in this study—do not allow for more up-to-date data. Finally, although the sample design is the census, the target population is short compared to the total number of companies at the national level; due to the limited coverage of the Superintendence of Companies, other types of companies were left out. The findings of this study may have implications both in the promotion activities of institutions that promote CSR in Colombia and in the development of public policies. By facilitating the design of more focused strategies so that CSR definitions and practices are not so open to interpretation and to benefit society and the environment more significantly, contributing in a more decisive way to sustainable development since they provide a tool for the study and analysis of CSR.

Within the lines of future research, the continuation and sophistication of the proposed model and its use in other contexts (countries or regions) come to mind. This would allow interesting comparative analyses between developed countries and emerging markets such as Colombia and between different emerging markets to make conceptual adjustments and collective efforts of promotion and research on the subject in different contexts. On the other hand, future research might inquire into mechanisms that make it possible to equate the effects and possible positive and negative effects of business operations on the environment, society, or the economy in order to know the true impact of a company or a general sector, which in turn would allow for more constructive and purposeful productive sectors. Another future field of research may involve research projects that seek to identify the causality between CSR and the economic performance of companies (e.g., whether or not CSR generates greater profitability for a company or a group of companies) or the relationship between the profiles and characterization variables of companies such as size, age, economic sector. Other studies could explain the relationship between the profiles and the efforts to formalize CSR in companies, such as investment, policies, and CSR reports. Finally, it is proposed that research on corporate social responsibilities should focus on expanding the scope of these responsibilities to society and the environment, which can facilitate overcoming the still persistent dilemma between the maximization of profits and the responsibilities of companies to society.

References

Adrian M, Phelps L, Gatte A (2013) Philanthropy and corporate social responsibility: is giving, enough to truly be ethical? Int J Acad Bus World 7(1):83–90

Afrin S (2013) Traditional vs strategic corporate social responsibility: in pursuit of supporting sustainable development. J Econ Sustain Dev 4(20):153–157

Aguinis H, Rupp DE, Glavas A (2024) Corporate social responsibility and individual behaviour. Nat Hum Behav 8(2):219–227. https://doi.org/10.1038/s41562-023-01802-7

Ali SS, Kaur R (2021) Effectiveness of corporate social responsibility (CSR) in implementation of social sustainability in warehousing of developing countries: a hybrid approach. J Clean Prod 324:129154. https://doi.org/10.1016/j.jclepro.2021.129154

Antheaume N (2004) Valuing external costs–from theory to practice: implications for full cost environmental accounting. Europ Account Rev 13(3):443–464. https://doi.org/10.1080/0963818042000216802

Ashrafi M, Adams M, Walker TR, Magnan G (2018) How corporate social responsibility can be integrated into corporate sustainability: a theoretical review of their relationships. Int J Sust Dev World 25(8):672–682. https://doi.org/10.1080/13504509.2018.1471628

Attig N, Hu W, Rahaman MM, Zaman AA (2023) Overselling corporate social responsibility. Financ Manage 52(3):573–610. https://doi.org/10.1111/fima.12434

Balluchi F, Lazzini A, Torelli R (2020). CSR and greenwashing: a matter of perception in the search of legitimacy. In: Del Baldo M, Dillard J, Baldarelli M-G, Ciambotti M (Eds), Accounting, accountability and society: trends and perspectives in reporting, management and governance

for sustainability (pp 151–166). Springer International Publishing. https://doi.org/10.1007/978-3-030-41142-8_8
Bansal P, Song H-C (2017) Similar but not the same: differentiating corporate sustainability from corporate responsibility. Acad Manag Ann 11(1):105–149
Batista AA da S, Francisco AC de (2018) Organizational sustainability practices: a study of the firms listed by the corporate sustainability index. Sustainability 10(1):226.https://doi.org/10.3390/su10010226
Baudillard J (2007) La Sociedad del Consumo, sus mitos y estructuras. Siglo XXI editores
Bhargava A, Ligade P (2023) Corporate social responsibility (CSR) and sustainability—a scientometric analysis of the interrelationship. J Indian Bus Res 15(1):110–124. https://doi.org/10.1108/JIBR-04-2022-0096
Bocquet R, Le Bas C, Mothe C, Poussing N (2013) Are firms with different CSR profiles equally innovative?: empirical analysis with survey data. Eur Manag J 31(6):642–654. https://doi.org/10.1016/j.emj.2012.07.001
Boubakri N, El Ghoul S, Guedhami O, Wang H (Helen) (2021) Corporate social responsibility in emerging market economies: determinants, consequences, and future research directions. Emerg Markets Rev 46:100758. https://doi.org/10.1016/j.ememar.2020.100758
Bryman A (2015) Social research methods (4th ed.). Oxford University Press
Busch T, Barnett ML, Burritt RL, Cashore BW, Freeman RE, Henriques I, Husted BW, Panwar R, Pinkse J, Schaltegger S, York J (2024) Moving beyond "the" business case: How to make corporate sustainability work. Bu Strategy Environ 33(2):776–787. https://doi.org/10.1002/bse.3514
Calvin K, Dasgupta D, Krinner G, Mukherji A, Thorne PW, Trisos C, Romero J, Aldunce P, Barrett K, Blanco G, Cheung WWL, Connors S, Denton F, Diongue-Niang A, Dodman D, Garschagen M, Geden O, Hayward B, Jones C, … HM (2023) IPCC, 2023: climate change 2023: synthesis report. Contribution of Working Groups I, II and III to the Sixth assessment report of the intergovernmental panel on climate change [Core Writing Team, H. Lee and J. Romero (eds.)]. IPCC, Geneva, Switzerland. (Arias P, Bustamante M, Elgizouli I, Flato G, Howden M, Méndez-Vallejo C, Pereira JJ, Pichs-Madruga R, Rose SK, Saheb Y, Sánchez Rodríguez R, Ürge-Vorsatz D, Xiao C, Yassaa N, Romero J, Kim J, Haites EF, Jung Y, Stavins R, … Péan C, Eds.). https://doi.org/10.59327/IPCC/AR6-9789291691647
Carroll AB (1979) A three-dimensional conceptual model of corporate performance. Acad Manag Rev 4(4):497–505. https://doi.org/10.5465/AMR.1979.4498296
Carroll AB (2016) Carroll's pyramid of CSR: taking another look. Int J Corp Soc Responsib 1(1):3. https://doi.org/10.1186/s40991-016-0004-6
Carroll AB (2021) Corporate social responsibility: perspectives on the CSR construct's development and future. Bus Soc 60(6):1258–1278. https://doi.org/10.1177/00076503211001765
Cezarino LO, Liboni LB, Hunter T, Pacheco LM, Martins FP (2022) Corporate social responsibility in emerging markets: opportunities and challenges for sustainability integration. J Clean Prod 362:132224. https://doi.org/10.1016/j.jclepro.2022.132224
Chang C-H (2015) Proactive and reactive corporate social responsibility: antecedent and consequence. Manag Decis 53(2):451–468. https://doi.org/10.1108/MD-02-2014-0060
Christensen J, Murphy R (2004) The Social Irresponsibility of corporate tax avoidance: taking CSR to the bottom line. Development 47(3):37–44. https://doi.org/10.1057/palgrave.development.1100066
Coase RH (1960) The problem of social cost. In: GC (Ed.), Classic papers in natural resource economics, pp 87–137. Palgrave Macmillan. https://doi.org/10.1057/9780230523210_6
Coase RH (1988) The firm, the market, and the law. University of Chicago Press
Davis K, Blomstrom RL (1966) Business and its environment. McGraw-Hill
de Almeida Barbosa Franco J, Franco Junior A, Battistelle RAG, Bezerra BS (2024). Dynamic capabilities: unveiling key resources for environmental sustainability and economic sustainability, and corporate social responsibility towards sustainable development goals. Resources 13(2):22.https://doi.org/10.3390/resources13020022

Decreto 2650 de 1993 (1993). https://incp.org.co/Site/productosyservicios/legislativa/2650.htm
Dowling G, Moran P (2012) Corporate reputations: built in or bolted on? Calif Manage Rev 54(2):25–42. https://doi.org/10.1525/cmr.2012.54.2.25
Everitt BS, Landau S, Leese M, Stahl D (2011) Cluster analysis (5th ed.). John Wiley & Sons, Ltd. https://doi.org/10.1002/9780470977811
Farias G, Landrum NE, Farias C, Krysa I (2024) Explorations in organized hypocrisy and a proposed direction for a sustainable future. Sustain Account Manag Policy J 15(1):1–22. https://doi.org/10.1108/SAMPJ-05-2022-0250
Freze T, Korneev A, Krayneva R, Oruch T, Kandalov W, Strielkowski W (2023) Business leadership and corporate social responsibility in the post-COVID era. Economies 11(3):98. https://doi.org/10.3390/economies11030098
Galego-Álvarez I, Formigoni H, Antunes MTP (2014) Corporate social responsibility practices at Brazilian firms. Revista De Administração De Empresas 54:12–27. https://doi.org/10.1590/S0034-759020140103
Gatti L, Seele P, Rademacher L (2019) Grey zone in—greenwash out. A review of greenwashing research and implications for the voluntary-mandatory transition of CSR. Int J Corporate Soc Respon 4(1):6. https://doi.org/10.1186/s40991-019-0044-9
Gibbons JD, Chakraborti S (2014) Nonparametric statistical inference: Revised and Expanded (4th ed.). CRC Press. https://books.google.com/books?id=_J6RDwAAQBAJ
Global Reporting Initiative (2020) GRI Standards. https://www.globalreporting.org/how-to-use-the-gri-standards/resource-center/
Halkos G, Nomikos S (2021) Corporate social responsibility: trends in global reporting initiative standards. Econ Anal Policy 69:106–117. https://doi.org/10.1016/j.eap.2020.11.008
Halme M, Rintamäki J, Knudsen JS, Lankoski L, Kuisma M (2020) When is there a sustainability case for CSR? pathways to environmental and social performance improvements. Bus Soc 59(6):1181–1227. https://doi.org/10.1177/0007650318755648
Han J, Pei J, Tong H (2022) Data mining: concepts and techniques (4th ed.). Morgan kaufmann
Hoi CK, Wu Q, Zhang H (2013) Is corporate social responsibility (CSR) associated with tax avoidance? Evidence from Irresponsible CSR activities. Account Rev 88(6):2025–2059. http://www.jstor.org/stable/23525962
Husser J, André J, Barbat G, Lespinet-Najib V (2012) CSR and sustainable development: are the concepts compatible? Manag Environ Qual Int J 23(6):658–672. https://doi.org/10.1108/14777831211262936
Husted BW, Jamali D, Saffar W (2016) Near and dear? The role of location in CSR engagement. Strateg Manag J 37(10):2050–2070. https://doi.org/10.1002/smj.2437
Huynh A-T, Knápková A, Bui T-D, Nguyen T-T-H (2024) The impact of institutional pressures on corporate social responsibility and green marketing adoption: an empirical approach in Vietnam banking industry. Int J Bank Market 42(3):620–641. https://doi.org/10.1108/IJBM-04-2023-0228
In SY, Park KY (2019) When do firms oversell or undersell environmental sustainability?: an empirical analysis of sustainability communications. Stanford Global Projects Center Working Paper Series. https://doi.org/10.2139/ssrn.3264923
International Standards Organization (2010) ISO 26000—social responsibility. Standards. https://www.iso.org/iso-26000-social-responsibility.html
Jamali D, Karam C (2018) Corporate social responsibility in developing countries as an emerging field of study. Int J Manag Rev 20(1):32–61. https://doi.org/10.1111/ijmr.12112
Jan A, Rahman HU, Zahid M, Salameh AA, Khan PA, Al-Faryan MAS, Che Aziz RB, Ali HE (2023) Islamic corporate sustainability practices index aligned with SDGs towards better financial performance: Evidence from the Malaysian and Indonesian Islamic banking industry. J Cleaner Prod 405:136860. https://doi.org/10.1016/j.jclepro.2023.136860
Janowski A (2020) Philanthropy and the contribution of Andrew Carnegie to corporate social responsibility. Sustainability 13(1):155. https://doi.org/10.3390/su13010155

Jiang P-C, Feng G-F, Wang H-J, Chang C-P (2024) CSR from different perspectives: the global ESG indexes updated. Corporate Soc Respons Environ Manag n/a(n/a). https://doi.org/10.1002/csr.2814

Johnston A (2011) Facing up to social cost. Griffith Law Review 20(1):221–244. https://doi.org/10.1080/10383441.2011.10854696

Kapp KW (1950) The social costs of private enterprise. Harvard University Press

Kassis S, Majaj Y (2012) CSR and philanthropy: different forms of effective social investment. In: Jamali D, Sidani Y (Eds), CSR in the middle east: fresh perspectives, pp 11–24. Palgrave Macmillan UK. https://doi.org/10.1057/9781137266200_2

Kotler P, Lee N (2005) Corporate social responsibility: doing the most good for your company and your cause. John Wiley & Sons

Lambin EF, Thorlakson T (2018) Sustainability standards: interactions between private actors, civil society, and governments. Annu Rev Environ Resour 43(1):369–393. https://doi.org/10.1146/annurev-environ-102017-025931

Loew T, Ankele K, Braun S, Clausen J (2004) Significance of the CSR debate for sustainability and the requirements for companies. https://www.ioew.de/fileadmin/_migrated/tx_ukioewdb/future-IOEW_CSR-Study_Summary.pdf

Lucarelli C, Severini S (2024) Anatomy of the chimera: environmental, Social, and Governance ratings beyond the myth. Bus Strat Environ n/a(n/a). https://doi.org/10.1002/bse.3688

Mahajan R, Kumar S, Lim WM, Sareen M (2024) The role of business and management in driving the sustainable development goals (SDGs): current insights and future directions from a systematic review. Bus Strat Environ n/a(n/a), 1–37. https://doi.org/10.1002/bse.3712

Margolis JD, Walsh JP (2003) Misery loves companies: rethinking social initiatives by business. Adm Sci Q 48(2):268–305. https://doi.org/10.2307/3556659

Marques J, Mintzberg H (2015) Why corporate social responsibility isn't a piece of cake. MIT Sloan Manag Rev 56(4):1–6. http://sloanreview.mit.edu/article/why-corporate-social-responsibility-isnt-a-piece-of-cake/

Masindi D, Singh P, Msweli P (2023) ESG in the developing economies–does it matter. Int J Innov Res Multidiscip Educ 2:74–83. https://doi.org/10.58806/ijirme.2023.v2i3n01

May SK, Cheney G, Roper J (2007) The debate over corporate social responsibility. Oxford University Press

McAdam TW (1973) How to put corporate responsibility into practice. Bus Soc Rev/Innov 6:8–16

Montuschi L (2009) Consideraciones respecto de la ética en los negocios, la responsabilidad social empresaria y la filantropía estratégica. Serie Documentos de Trabajo, Universidad del CEMA: Área: negocios. https://ideas.repec.org/p/cem/doctra/414.html

Moon J (2007) The contribution of corporate social responsibility to sustainable development. Sustain Dev 15(5):296–306. https://doi.org/10.1002/sd.346

Muller A, Kolk A (2009) CSR performance in emerging markets evidence from Mexico. J Bus Ethics 85(2):325–337. https://doi.org/10.1007/s10551-008-9735-y

Newell P (2008) CSR and the limits of capital. Dev Chang 39(6):1063–1078. https://doi.org/10.1111/j.1467-7660.2008.00530.x

Nyabakora WI, Mohabir SE (2024) Corporate social responsibility knowledge base: a bibliometric analysis. Modern Finance 2(1). https://doi.org/10.61351/mf.v2i1.123

OECD (2011) OECD guidelines for multinational enterprises. http://www.oecd.org/daf/inv/mne/48004323.pdf

Oppong-Tawiah D, Webster J (2023) Corporate sustainability communication as 'Fake News': Firms' greenwashing on Twitter. Sustainability 15(8):6683. https://doi.org/10.3390/su15086683

Pillai R, Islam MdA, Sreejith S, Al-Malkawi HA (2024) Comparative analysis of environmental, social and governance (ESG) ratings: do sectors and regions differ? J Manage Governance 1–41. https://doi.org/10.1007/s10997-023-09692-7

Pimonenko T, Bilan Y, Horák J, Starchenko L, Gajda W (2020) Green brand of companies and greenwashing under sustainable development goals. Sustainability 12(4):1679. https://doi.org/10.3390/su12041679

Porter ME, Kramer MR (2006) Strategy and society: the link between competitive advantage and corporate social responsibility. Harv Bus Rev 84(12):78–92. https://hbr.org/2006/12/strategy-and-society-the-link-between-competitive-advantage-and-corporate-social-responsibility
R Core Team (2023) R: a language and environment for statistical computing. https://www.R-project.org/
Rahman N, Blake L (2021) A review of CSR classification schemes and the operationalization of bolted-on vs. built-in CSR. Bus Ethics Environ Respons 30(3):248–261. https://doi.org/10.1111/beer.12345
Rangan K, Chase L, Karim S (2015) The truth about CSR. Harv Bus Rev 93(1–2):41–49. https://hbr.org/2015/01/the-truth-about-csr
Raufflet E, Portales Derbez L, García de la Torre C, Lozano Aguilar JF, Barrera Duque E (2017) Responsabilidad, ética y sostenibilidad empresarial (1st ed.). Pearson Educación
Razali NM, Wah YB (2011) Power comparisons of shapiro-wilk, kolmogorov-smirnov, lilliefors and anderson-darling tests. J Stat Modeling Analyt 2(1):21–33
Rim H, Ferguson MAT (2020) Proactive versus reactive CSR in a crisis: an impression management perspective. Int J Bus Commun 57(4):545–568. https://doi.org/10.1177/2329488417719835
Sánchez-Teba EM, Benítez-Márquez MD, Bermúdez-González G, Luna-Pereira M del M (2021) Mapping the knowledge of CSR and sustainability. Sustainability 13(18):10106.https://doi.org/10.3390/su131810106
Sancho MPL, Jorge ML, Madueño JH (2017) Characterisation of CSR practices in Spanish SMEs through a cluster analysis. Int J Manag Enterprise Dev 16(4):308–336. https://doi.org/10.1504/IJMED.2017.086912
Shapiro SS, Wilk MB (1965) An analysis of variance test for normality (complete samples). Biometrika 52(3/4):591–611. https://doi.org/10.2307/2333709
Sheehy B (2015) Defining CSR: problems and solutions. J Bus Ethics 131(3):625–648. https://doi.org/10.1007/s10551-014-2281-x
Simionescu LN (2015) The relationship between corporate social responsibility (CSR) and sustainable development (SD). Int Audit Risk Manag 10(2):179–190. https://search.ebscohost.com/login.aspx?direct=true&db=bsu&AN=108555377&lang=es&site=ehost-live
Šimunović D, Murtarelli G, Romenti S (2024) Sustainability data distortions: the use of visual impression management techniques in corporate sustainability reporting. J Commun Manag 28(1):93–109. https://doi.org/10.1108/JCOM-12-2022-0134
Singhal N, Paul P, Giri S, Taneja S (2024) Corporate social responsibility: impact on firm performance for an emerging economy. J Risk Financ Manag 17(4):171. https://doi.org/10.3390/jrfm17040171
Skarmeas D, Leonidou CN (2013) When consumers doubt, Watch out! The role of CSR skepticism. J Bus Res 66(10):1831–1838. https://doi.org/10.1016/j.jbusres.2013.02.004
Song B, Wen J, Ferguson MA (2020) Toward effective CSR communication in controversial industry sectors. J Mark Commun 26(3):243–267. https://doi.org/10.1080/13527266.2018.1536887
Stoeckl VE, Luedicke MK (2015) Doing well while doing good? An integrative review of marketing criticism and response. J Bus Res 68(12):2452–2463. https://doi.org/10.1016/j.jbusres.2015.06.032
Superintendencia de Sociedades (2011) Guia para diligenciar el formulario informe de prácticas empresariales. Informe 31. https://docplayer.es/16199213-Guia-para-diligenciar-el-formulario-informe-de-practicas-empresariales-informe-31.html
Superintendencia de Sociedades (2012) Sistema Integrado de Información Societaria. https://siis.ia.supersociedades.gov.co/#
Taran Z, Mirkin B (2020) Exploring patterns of corporate social responsibility using a complementary K-means clustering criterion. Bus Res 13(2):513–540. https://doi.org/10.1007/s40685-019-00106-9
Thijssen J (2016) A Concise introduction to statistical inference. CRC Press. https://books.google.com/books?id=cuiVDQAAQBAJ

UN Global Compact (2021) Act globally engage locally. Colombia. https://www.unglobalcompact.org/engage-locally/latin-america/colombia
Velte P (2023) Which institutional investors drive corporate sustainability? A systematic literature review. Bus Strat Environ 32(1):42–71. https://doi.org/10.1002/bse.3117
Villa-Castaño LE, Perdomo-Ortiz J, Dueñas-Ocampo S (2023) Business–society interface: an exploration of a paradigmatic heuristic model of corporate social responsibility in Colombia. Corp Soc Responsib Environ Manag 30(6):3203–3221. https://doi.org/10.1002/csr.2547
Wang X, Chaiyasat C, Vorapuksirikool V (2024) Ethics and CSR in business: a review and future research. In: Russ-Eft DF, Alizadeh A (Eds), Ethics and human resource development: societal and organizational contexts, pp 279–297. Springer International Publishing. https://doi.org/10.1007/978-3-031-38727-2_14
Wartick SL, Cochran PL (1985) The evolution of the corporate social performance model. Acad Manag Rev 10(4):758–769. https://doi.org/10.2307/258044
Wijesundara CNR, Khatibi A, Tham J (2024) A critique on theoritical framework for adoption of corporate sustainability practices: iintegrating institutional motives, stakeholder motives, and managerial motives. Int J Professional Bus Rev 9(1):1–15. https://doi.org/10.26668/businessreview/2024.v9i1.4202
Wilson I (1975) What one company is doing about today's demands on business. In: Steiner GA (Ed), Changing business-society interrelationships. Los Angeles Graduate School of Management, UCLA
Wu C, Thompson ME (2020) Sampling theory and practice (J. Chen & (Din) Ding-Geng Chen, Eds.). Springer. https://doi.org/10.1007/978-3-030-44246-0
Xu L, Xie L, Mei S, Hao J, Zhang Y, Song Y (2024) Corporate sustainability reporting and stakeholders' interests: evidence from China. Sustainability 16(8):3443. https://doi.org/10.3390/su16083443
Yepes-López GA, Bello-Pérez CJ, Cruz JM, Camarena JL (2021) El modelo pentadimensional aplicado a las empresas de Bogotá, Colombia. In: García de la Torre C, Portales L, Yepes-López GA (Eds), Sostenibilidad en las pymes de América Latina y el Caribe (1st ed., pp 101–138). Universidad Externado de Colombia
Yepes-López GA, Camarena JL, Cruz J (2023) Who pays for corporate social responsibility?: Proposal for an externalization index of CSR costs. In: Singh P, Milshina Y, Batalhão A, Sharma S, Hanafiah MM (Eds), The route towards global sustainability: challenges and management practices (pp 195–228). Springer International Publishing. https://doi.org/10.1007/978-3-031-10437-4_10
Zadek S (2007) The path to corporate responsibility. In: Zimmerli WC, Holzinger M, Richter K (Eds), Corporate ethics and corporate governance (pp 159–172). Springer Berlin Heidelberg. https://doi.org/10.1007/978-3-540-70818-6_13
Zamir F, Saeed A (2020) Location matters: Impact of geographical proximity to financial centers on corporate social responsibility (CSR) disclosure in emerging economies. Asia Pacific J Manag 37(1):263–295. https://doi.org/10.1007/s10490-018-9619-3
Zollo M (2004) Philanthropy or CSR: a strategic choice. A special report by European Business Forum. https://www.johnelkington.com/archive/ebf_CSR_report.pdf

Role of Corporate Social Responsibility in Promoting Green Growth Amidst a Competitive Business Environment

Kshitiz Jha, Akanksha Bhardwaj, Kapinder, and Archna Kumar

Abstract In a competitive business environment, companies might tend to invest more on new R&D initiatives and technologies or employ the latest available technologies, in order to achieve cost savings, quality improvements, and production efficiency. Businesses can gain a competitive edge in the market and improve their financial health by projecting an image of being a green company that cares about the environment, which will make employees and customers prefer the company's products over others. Industries are adopting sustainable practices at an increasing rate due to the challenges posed by global environmental issues and the growing awareness of corporate responsibility. In the present scenario, managing industrial waste is one way that corporate social responsibility contributes significantly to green growth. For instance, improving the state of the soil, giving it all the nutrients it needs for a healthy crop yield, and reducing loss are all important for raising crop productivity. The impact of bio magnification of these chemicals released by industries causing disruption in the food chain However, there are several researches and studies done to prove alternatives that can be used for environmental improvement and cause negligible harm. An upsurge is the adoption of waste recovery technologies, which minimize, recycle, and anticipate responsibility for turning dispose of into byproducts. Steel industry byproducts, such as steel slag, are an outstanding illustration of the "wastes" generated by the steel industry that are turned into goods with an extensive range of potential uses. Steel slag offers an opportunity to be an economical nutrient substitute in agriculture. Through employing these recycled materials, natural resources are conserved, landfill waste is avoided, and CO2 emissions are decreased. This chapter serves as a demonstration of the significance of corporate social responsibility in strengthening green growth within the commercial segment.

Keywords Corporate social responsibility · Byproduct management · Green growth

K. Jha · A. Bhardwaj · A. Kumar (✉)
Amity Institute of Biotechnology, Amity University, Noida 201313, Uttar Pradesh, India
e-mail: akumar21@amity.edu

Kapinder
Department of Zoology, University of Allahabad, Prayagraj 211002, Uttar Pradesh, India

P. Singh et al. (eds.), *Integrating Environmental Sustainability and Corporate Social Responsibility: A Move Beyond Profit*, World Sustainability Series,
https://doi.org/10.1007/978-3-031-81469-3_5

1 Introduction

Industries are adopting sustainable practices at an increasing rate due to the challenges posed by global environmental issues and the growing awareness of corporate responsibility.

It represents an important intersection of economic and ecological interests as businesses come under increasing pressure to strike a balance between environmental stewardship and profitability.

Corporate Social Responsibility (CSR) is a business approach in which a company prioritizes' positive impact on society beyond financial goals. It entails consideration of Environmental, Social, and Governance factors in business operations to contribute to the improvement of society and the environment.

CSR encompasses environmental impact, ethical responsibility, philanthropy, and financial responsibilities. This involves open disclosure, policy development, and training to enhance CSR performance. Companies that implement CSR practices can benefit from lower costs, better risk management, improved employee retention, increased customer loyalty, and access to a larger addressable market (Hamidu et al. 2015; Lim and Greenwood 2017).

Implementing a CSR strategy can benefit businesses in a number of ways, including:

- Enhancing employee performance and effective risk management.
- Promoting the growth of local communities.
- Improving Financial Performance
- Reducing operating costs
- Enhancing customer loyalty.
- Enhancing workplace diversity and inclusion
- Improving product durability and functionality.
- Having improved product safety
- Increasing employee retention.
- Improving brand image (Lim and Greenwood 2017).

2 Green Growth

Green growth means promoting economic growth and development while ensuring that natural assets continue to provide the resources and environmental services required for our well-being (Fig. 1). It entails a shift to a more sustainable and environmentally conscious approach to economic development, focusing on the long-term health of the planet and its resources (Soundarrajan and Vivek 2016).

Green growth strategies prioritize maximizing natural assets economic potential without compromising their ability to provide essential services like clean air and water, as well as resilient biodiversity for food production and human health. Natural

Fig. 1 Green growth strategies

assets are not infinitely substitutable, and green growth policies consider that (Ali et al. 2021; Cheng et al. 2020).

- It entails reconsidering growth strategies to promote environmentally sustainable and socially inclusive development.
- Green growth focuses on reducing carbon footprints and greenhouse gas emissions.
- Increasing energy efficiency and switching to renewable sources
- Developing green technologies and industries (Hao et al. 2021; Soundarrajan and Vivek 2016).

3 Crucial Role of Corporate Social Responsibility (CSR) in Promoting Green Growth

Initiatives help companies adopt green technologies, improve energy efficiency, and reduce carbon footprints, minimizing their environmental impact (Fernando and Hor 2017).

- Promotes the growth of green industries, including renewable energy, manufacturing, and agriculture, which are crucial for a sustainable economy.
- Encourages companies to create eco-friendly products, services, and processes, supporting the transition to a green economy.
- Aims to promote sustainable practices and raise environmental awareness by engaging stakeholders such as employees, customers, suppliers, and communities.
- Initiatives often align with the Sustainable Development Goals (SDGs), which aim to achieve sustainable development and promote green growth.
- Adopting CSR practices can help companies improve their reputation, attract eco-conscious consumers, and reduce operational costs through efficient resource use, giving them a competitive advantage (Rathore 2017).
- Promotes sustainable consumption and production, including responsible sourcing, sustainable packaging, and recycling/reuse (Prabhu and Aithal 2023).
- Investing in Natural Capital: CSR initiatives often prioritize protecting and restoring ecosystems to maintain ecological balance and promote green growth (AyuPurnamawati et al. 2023).

4 Application of CSR in Industries

Corporate Social Responsibility (CSR) can be instilled in various industries based on their unique characteristics, stakeholder expectations, and potential risks and opportunities. Here are some examples of how different industries can promote CSR:

- Retailers and consumer goods companies can encourage CSR by promoting sustainable practices such as reducing packaging waste, using environmentally friendly materials, and supporting local suppliers. By implementing these CSR initiatives, retailers can help to ensure environmental sustainability, support local communities, and meet the growing consumer demand for environmentally friendly products and practices.

For example, a retailer may invest in sustainable packaging, source products from local suppliers, or promote recycling programs (Naidoo and Gasparatos 2018; Beitzen-Heineke et al. 2017).

- Automotive companies can encourage CSR by promoting sustainable practices such as lowering emissions, using environmentally friendly materials, and supporting environmental initiatives. (Lee 2019; Jaderná and Přikrylová 2018)

For example, a car manufacturer may invest in electric vehicles and reduce waste during production.

5 Contributions of Steel Industries in Promoting Green Growth

The steel industry has an essential role in the growth of the world economy as it supplies raw materials for several industries, however, the traditional methods of handling, storage, and disposal of hazardous steel industry byproducts has been a matter of concern.. Improper handling, storage, and disposal of hazardous steel industry byproducts can contaminate the soil, making the area unusable for farming or habitation and can also disrupt the food chain causing various health hazards. High concentrations of pollutants in effluent discharges by steel plants have the potential to contaminate water sources and landmasses and also harming aquatic life and the nearby communities that depend on these resources. Therefore it is necessary to ensure appropriate management of steel slag with minimal risk. Adopting waste recovery technologies that reduce and reuse "waste" internally in the plant and take responsibility for the conversion of waste to by-products is becoming a growing trend. By using these recycled materials, landfill waste is avoided, CO_2 emissions are decreased, and natural resources are preserved.

5.1 Case Study

China being a major producer of steel slag (100 million tons of slag per annum) has only been able to utilize 29.5% which result in causing harm to environment and economy of the country (Guo et al. 2018). However in past few years the slag utilization Rate has increased in various parts of the world.

Slag utilization rate

- Japan—98.4%
- Europe—87%
- USA—84%

These countries adapted various ways of Industrial waste management due to which the damage caused to the environment is minimized and also used as an asset in promoting green growth. The use of these byproducts in agriculture can extensively prevent the accumulation of harmful substances in the environment and also helps in increasing plant growth and improve soil properties by neutralizing acidic soil, enhance the nutrient content of soil and improve the water holding capacity of soil.

6 Steel Slag

The main residual material formed in basic oxygen furnace by converting iron into steel or steel manufacturing in electric arc furnaces by the process of scrap melting by melting scrap in electric arc furnace is called steel slag. It is a major byproduct of steel producing industries has shown significant application in a various work fields, one of them is the agricultural field (O'Connor et al. 2021).

6.1 Chemical Composition of Steel Slag

The major chemical components of steel slag (Fig. 2) are:

- Calcium oxide (CaO)
- Phosphorus pent oxide (P2O5)
- Silicon dioxide (SiO)
- Magnesium oxide (MgO)
- Manganese (II) oxide (MnO)
- Ferric oxide and Ferrous oxide (FeO and Fe_2O_3) (Navarro et al. 2010; Yi et al. 2012) (Liu et al. 2019; Sinha et al. 2013)

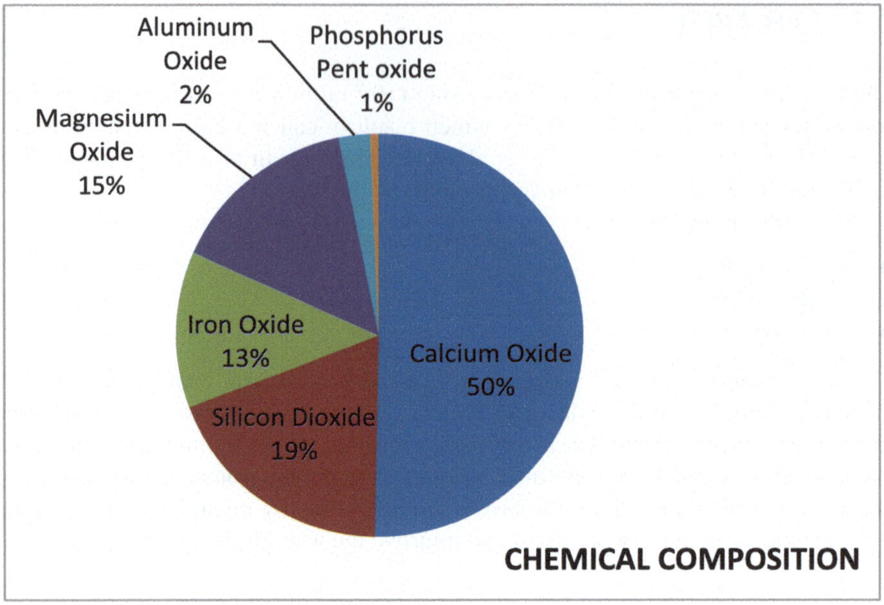

Fig. 2 The major chemical components of steel slag

6.2 Harmful Affects of Steel Slag

Various harmful effect of steel slag is summarized in Table 1 and Fig. 3.

6.3 Beneficial Aspects of Using Steel Slag

Steel slag has significant amounts of priceless minerals and metals. By using general, physical or chemical methods of processing minerals, such as crushing, grinding, classification, hydro cyclone, magnetic separation, flotation, leaching, or roasting, can recover certain values. These solid wastes must be converted from one form to another in order to be reused, either by the same production unit or by a new industrial installation. This is crucial for both environmental protection and resource conservation of metals and minerals.

The following are recent developments in the treatment of steel slag:

- The environmental effect of the steel industry has significantly decreased as a result of increased reuse of iron and steelmaking slag (Kumar 2019).
- Blast furnace slag, ladle slag, and electric arc furnace slag are some possibilities for managing steelmaking slags (Aziz et al. 2022).

Fig. 3 Various conducts of steel slag disposal from a steel producing plant

Table 1 Heavy metals present in steel slag with their harmful effects on living beings

Heavy metal	Probable harmful effects	References
Lead	• Neurotoxicity • Developmental disorders • Kidney damage • Impaired cognitive functions	Karcioglu and Arslan (2019) Mahurpawar (2015) Fu and Xi (2020) Mitra et al. (2022)
Chromium	• Carcinogenicity • DNA damage • Respiratory problems • Dermatitis	
Cadmium	• Kidney damage, • Bone demineralization • Cancer	
Arsenic	• Carcinogenicity • Skin lesions • Respiratory problems	
Nickel	• Carcinogenicity • Respiratory problems • Skin allergies	
Copper	• Liver damage • Gastrointestinal issues • Developmental disorders	
Zinc	• Reduced plant growth • Phytotoxicity • Soil contamination	

6.4 Use of Steel Slag for Improving Soil Quality

Chemical fertilizers have gained massive demand since the last century in the field of agriculture due it's readily serving mineral and nutrient capability. But the use of chemical based fertilizers has reduced significantly due to its low micronutrient content as well as due to the human health concerns. Therefore, it is very crucial to find new substitute materials to enhance agricultural yields (Foley et al. 2011; Li et al. 2007).

Steel slag, which is created in large quantities by industries that make steel, can be used to improve soil quality and crop productivity. Steel slag use in agriculture can increase crop yields while reducing the amount of waste that can build up in the environment. Depending on the slag's composition and the soil type, steel slag may have different effects on soil fertility and plant growth. Therefore, before applying the slag to the soil, it is crucial to understand its chemical and physical characteristics (Deus et al. 2018; O'Connor et al. 2021).

6.5 Use of Slag for Nutritent Fullfillment of Soil

There are high mineral content and acceptable low levels of heavy metals present in steel slag, which, when used properly, can improve the soil's chemical properties by increasing the content of silicon, phosphorus, calcium, and magnesium as well as other advantageous elements lead and cadmium (Gao et al. 2020).

Rice is a typical Si hyper-accumulating plant species, with shoots containing up to 10% Si on a dry weight basis. Steel slags as a low-cost source of silicon can be used as a silicate fertilizer (Alcarde and Rodella 2003; Das et al. 2007).

6.6 Use of Slag as Soil Acidity Corrector

Acidity of the soil restricts crop growth and yield in many parts of the world. The most typical use of lime is to neutralize acidic soil. Since it is only weakly soluble in water, it is advised that it be incorporated into the soil to ensure a better correction. Alternatives to lime for reducing soil acidity include steel slags. Calcium and magnesium silicates, and the SiO32 base gives a neutralizing effect slag Therefore, it can also be used in agriculture as fertilizers and soil acidity correctors. The release of OH- ions when CaO and MgO dissolve in water causes steel slag to be added to soil to raise the pH of acidic soils (Islam et al. 2022).

Based on the amount of steel slag added to the soil, it can either have a positive or negative impact on plant growth. A study found that soil with low levels of steel slag (0.2% to 1%) had progressive effects on the growth of plant, mineral uptake and bacterial biomass, whereas large amount of steel slag addition (Above 1%) can decrease the mineral uptake capacity as well as the microbial activity and plant growth in the soil (Islam et al. 2022; O'Connor et al. 2021). Lead (Pb) and cadmium (Cd) are two heavy metals present in steel slag in low concentrations that are suitable for agricultural use. However, the presence of other heavy metals in steel slag, like chromium (Cr) and nickel (Ni), may pose a risk to the health of the soil (Garcia-Guinea et al. 2010; Ning et al. 2016).

To prevent potential harm to soil health and human health, it is crucial to monitor the amount of heavy metals in steel slag and limit its application to soil (O'Connor et al. 2021).

6.7 Steel Slag Use for Heavy Metal Stabilization

Heavy metal like Zn, Cu, Pb and Cd that are present in soils can be solidified or stabilized using steel slag. As a substance for stabilizing heavy metals in contaminated soil, steel slag has shown promising results. Heavy metals in soil can be stabilized with the use of steel slag, by lowering their bioavailability and potential for harming the soil (Jiang et al. 2022; Ning et al. 2016).

7 Conclusion

In the changing business environment, Corporate Social Responsibility (CSR) emerges as a critical strategy for fostering long-term growth while increasing competitive advantage. As global environmental challenges become more pressing and societal expectations for responsible business practices rise, industries are forced to innovate and implement green technologies and sustainable practices. CSR initiatives not only assist businesses in addressing these challenges, but also allow them to meet the growing demands of environmentally conscious consumers and stakeholders. A prime example of CSR's transformative potential is the steel industry's innovative use of a byproduct—steel slag—as a bio-fertilizer. Steel slag, previously considered waste, is now being repurposed to improve soil health and agricultural productivity. Companies that actively engage in sustainable practices can strengthen their market position, increase brand loyalty, and gain the trust of environmentally conscious consumers. This dedication to sustainability may also attract investors and partners who value long-term environmental stewardship. Companies that integrate sustainability into their core operations can achieve regulatory compliance, reduce operational risks, foster innovation, and drive growth. Furthermore, CSR initiatives allow businesses to fulfill their responsibility to a greener future, ensuring that economic growth does not come at the expense of the planet's health. This balanced approach to growth and sustainability is critical to the long-term success of both industries and the global community.

References

A Aziz I, Abdullah MMAB, Mohd Salleh MAA, Ming L, Li L, Vizureanu P, Nemes O, Mahdi S (2022) Recent developments in steelmaking industry and potential alkali activated based steel waste: a comprehensive review. Materials 15:1948. https://doi.org/10.3390/ma15051948

Alcarde JC, Rodella A (2003) Qualidade e legislação de fertilizantes e corretivos. Tópicos Em Ciência Do Solo 3:291–334

Ali EB, Anufriev VP, Amfo B (2021) Green economy implementation in Ghana as a road map for a sustainable development drive: a review. Scientific African 12:e00756. https://doi.org/10.1016/j.sciaf.2021.e00756

Ayu Purnamawati IG, Yuniarta GA, Jie F (2023) Strengthening the role of corporate social responsibility in the dimensions of sustainable village economic development. Heliyon 9(4):e15115. https://doi.org/10.1016/j.heliyon.2023.e15115

Beitzen-Heineke EF, Balta-Ozkan N, Reefke H (2017) The prospects of zero-packaging grocery stores to improve the social and environmental impacts of the food supply chain. J Clean Prod 140:1528–1541. https://doi.org/10.1016/j.jclepro.2016.09.227

Cheng Z, Li L, Liu J (2020) Natural resource abundance, resource industry dependence and economic green growth in China. Resour Policy 68:101734. https://doi.org/10.1016/j.resourpol.2020.101734

Das B, Prakash S, Reddy P, Misra VN (2007) An overview of utilization of slag and sludge from steel industries. Resour Conserv Recycling 50:40–57.https://doi.org/10.1016/j.resconrec.2006.05.008

Deus ACF, Bertani RM de A, Meirelles GC, Soares A de AVL, Moreira LLQ, Fernandes LTB, DM, Deus ACF, Bertani RM de A, Meirelles GC, Soares A de AVL, Moreira LLQ, Fernandes LT, DM (2018) The comprehensive utilization of steel slag in agricultural soils. In: Recovery and utilization of metallurgical solid waste. IntechOpen. https://doi.org/10.5772/intechopen.81440

Fernando Y, Hor WL (2017) Impacts of energy management practices on energy efficiency and carbon emissions reduction: a survey of malaysian manufacturing firms. Resour Conserv Recycl 126:62–73. https://doi.org/10.1016/j.resconrec.2017.07.023

Foley J, Ramankutty N, Brauman K, Cassidy E, Gerber J, Johnston M, Mueller N, O'Connell C, Ray D, West P, Balzer C, Bennett E, Carpenter S, Hill J, Monfreda C, Polasky S, Rockström J, Sheehan J, Siebert S, Zaks D (2011) Solutions for a cultivated planet. Nature 478:337–342. https://doi.org/10.1038/nature10452

Fu Z, Xi S (2020) The effects of heavy metals on human metabolism. Toxicol Mech Methods 30(3):167–176. https://doi.org/10.1080/15376516.2019.1701594

Gao D, Wang F-P, Wang Y-T, Zeng Y-N (2020) Sustainable utilization of steel slag from traditional industry and agriculture to catalysis. Sustainability 12:9295. https://doi.org/10.3390/su12219295

Garcia-Guinea J, Correcher V, Recio-Vazquez L, Crespo-Feo E, Gonzalez-Martin R, Tormo L (2010) Influence of accumulation of heaps of steel slag on the environment: determination of heavy metals content in the soils. An Acad Bras Ciênc 82:267–277. https://doi.org/10.1590/S0001-37652010000200003

Guo J, Bao Y, Wang M (2018) Steel slag in China: treatment, recycling, and management. Waste Manage 78:318–330. https://doi.org/10.1016/j.wasman.2018.04.045

Hamidu A, Haron M, Amran A (2015) Corporate social responsibility: a review on definitions, core characteristics and theoretical perspectives (SSRN Scholarly Paper 2861506). https://papers.ssrn.com/abstract=2861506

Hao L-N, Umar M, Khan Z, Ali W (2021) Green growth and low carbon emission in G7 countries: how critical the network of environmental taxes, renewable energy and human capital is? Sci Total Environ 752:141853. https://doi.org/10.1016/j.scitotenv.2020.141853

Islam Z, Tran QT, Koizumi S, Kato F, Ito K, Araki KS, Kubo M (2022) Effect of steel slag on soil fertility and plant growth. J Agric Chem Environ 11(3):Article 3. https://doi.org/10.4236/jacen.2022.113014

Jaderná E, Přikrylová J (2018) Green solutions in automotive industry. | Marketing Science & Inspirations | EBSCOhost. https://openurl.ebsco.com/contentitem/gcd:129596807?sid=ebsco:plink:crawler&id=ebsco:gcd:129596807

Jiang Q, He Y, Wu Y, Dian B, Zhang J, Li T, Jiang M (2022) Solidification/stabilization of soil heavy metals by alkaline industrial wastes: a critical review. Environ Pollut (Barking, Essex: 1987) 312:120094. https://doi.org/10.1016/j.envpol.2022.120094

Karcioglu O, Arslan B (2019) Poisoning in the modern world: new tricks for an old dog? BoD—Books on Demand

Kumar S (2019) Recent trends in slag management & utilization in the steel industry 1:94–102

Lee JW (2019) Global corporate citizenship: cross-cultural comparison of best practices in the global automotive industry (SSRN Scholarly Paper 4035167). https://papers.ssrn.com/abstract=4035167

Li BY, Zhou DM, Cang L, Zhang HL, Fan XH, Qin SW (2007) Soil micronutrient availability to crops as affected by long-term inorganic and organic fertilizer applications. Soil Tillage Res 96(1):166–173. https://doi.org/10.1016/j.still.2007.05.005

Lim JS, Greenwood CA (2017) Communicating corporate social responsibility (CSR): Stakeholder responsiveness and engagement strategy to achieve CSR goals. Public Relations Rev 43(4):768–776. https://doi.org/10.1016/j.pubrev.2017.06.007

Liu W, Li H, Zhu H, Xu P (2019) Properties of a steel slag-permeable asphalt mixture and the reaction of the steel slag-asphalt interface. Materials 12(21):3603. https://doi.org/10.3390/ma12213603

Mahurpawar M (n.d.) Effects of heavy metals on human health

Mitra S, Chakraborty AJ, Tareq AM, Emran TB, Nainu F, Khusro A, Idris AM, Khandaker MU, Osman H, Alhumaydhi FA, Simal-Gandara J (2022) Impact of heavy metals on the environment and human health: Novel therapeutic insights to counter the toxicity. J King Saud Univ Sci 34(3):101865. https://doi.org/10.1016/j.jksus.2022.101865

Naidoo M, Gasparatos A (2018) Corporate environmental sustainability in the retail sector: drivers, strategies and performance measurement. J Clean Prod 203:125–142. https://doi.org/10.1016/j.jclepro.2018.08.253

Navarro C, Díaz M, Villa-García MA (2010) Physico-chemical characterization of steel slag. Study of its behavior under simulated environmental conditions. Environ Sci Technol 44(14):5383–5388. https://doi.org/10.1021/es100690b

Ning D, Liang Y, Liu Z, Xiao J, Duan A (2016) Impacts of steel-slag-based silicate fertilizer on soil acidity and silicon availability and metals-immobilization in a paddy soil. PLoS ONE 11(12):e0168163. https://doi.org/10.1371/journal.pone.0168163

O'Connor J, Nguyen TBT, Honeyands T, Monaghan B, O'Dea D, Rinklebe J, Vinu A, Hoang SA, Singh G, Kirkham MB, Bolan N (2021) Production, characterisation, utilisation, and beneficial soil application of steel slag: a review. J Hazard Mater 419:126478. https://doi.org/10.1016/j.jhazmat.2021.126478

Prabhu N, Aithal PS (2023) Inbound corporate social responsibility model for selected Indian banks and their proposed impact on attracting and retaining customers—a case study (SSRN Scholarly Paper 4715266). https://papers.ssrn.com/abstract=4715266

Rathore B (2017) Aligning profitability and environmental responsibility: a study on sustainable marketing strategies. Eduzone Int Peer Rev/Refereed Multidiscip J 6(2), Article 2

Sinha A, Havanagi V, Mathur S (2013) Steel slag waste material for the construction of road. Indian Highways 41:15–22

Soundarrajan P, Vivek N (2016) Green finance for sustainable green economic growth in India. Agric Econ (Zemědělská Ekonomika) 62(1):35–44. https://doi.org/10.17221/174/2014-AGRICECON

Yi H, Xu G, Cheng H, Wang J, Wan Y, Chen H (2012) An overview of utilization of steel slag. Procedia Environ Sci 16:791–801. https://doi.org/10.1016/j.proenv.2012.10.108

Are Firms Strategically Engaging in CSR? An Exploratory Study of SCSR Dimensions in an Emerging Market

José Luis Camarena, Gustavo Adolfo Yepes López, Julián Mauricio Cruz Pulido, and Maria Teresa Camacho Ríos

Abstract This article aims to determine whether companies that carry out CSR actions do so strategically. Following (Burke and Logsdon, Long Range Plan 29:495–502, 1996) model of strategic CSR [centrality, specificity, proactivity, visibility, and voluntarism] and based on data from 5,136 companies surveyed by the Superintendence of Companies of Colombia in 2017, a principal components analysis was carried out to explore which items of the Superintendence survey aligned with the theoretical model, additionally, a k means cluster analysis was conducted to group companies according to their SCSR levels. The results of the PCA analysis made it possible to expand the original five dimensions by adding two subdimensions. These findings showed that the specificity dimension not only benefits the company but also benefits its stakeholders. Additionally, the visibility subdimension focuses on appreciating a company's CSR efforts by its supply chain members, namely suppliers and distributors. The cluster analysis yielded two profiles: (1) one with a high degree of alignment of its strategy with CSR practices called *strong SCSR* and (2) another with a lack of interest in these actions, denominated *weak SCSR*.

Keywords Strategic CSR · Strategic CSR dimensions · Colombia · Cluster analysis

---

J. L. Camarena (✉) · G. A. Yepes López · J. M. Cruz Pulido · M. T. Camacho Ríos
Universidad Externado de Colombia School of Management, Bogotá, Colombia
e-mail: jose.camarena@uexternado.edu.co

G. A. Yepes López
e-mail: gustavo.yepes@uexternado.edu.co

J. M. Cruz Pulido
e-mail: julian.cruz@uexternado.edu.co

P. Singh et al. (eds.), *Integrating Environmental Sustainability and Corporate Social Responsibility: A Move Beyond Profit*, World Sustainability Series,
https://doi.org/10.1007/978-3-031-81469-3_6

1 Introduction

As of writing, the world is facing overwhelming negative news, including the rise in natural disasters since the turn of the century (Alimonti and Mariani 2023) and the records of high temperatures being broken year after year, for example (Cheng et al. 2021; The Guardian 2023); Significant inequality in the distribution of wealth (Oxfam International 2024), or the growing warnings of the effects of climate change (Calvin et al. 2023), among other aspects strongly related to the economic, social, and environmental dimensions aligned with sustainability. These situations are the responsibility of governments, citizens, and companies, which have fostered unsustainable production and consumption patterns since the planet's resources are not enough to satisfy them (Meadows et al. 2018).

In particular, companies have been under increasing pressure to implement good practices in their strategy and operations that generate non-financial returns in line with sustainable development. For instance, companies face increasing pressure to align with ESG criteria due to national and international commercial requirements. (Chiou and Shu 2019). For example, companies are required to meet commercial demands on a national and international scale, which are now more closely aligned with ESG criteria (Global Reporting Initiative 2020; Hahn 2013; International Standards Organization 2010, 2016; Sartor et al. 2016); the growing demonstrations by activist groups, NGOs and communities for environmental protection or human rights; the increasingly informed consumers, who demand better practices and cleaner production from companies and who may also be more willing to pay premium prices for sustainable products (Bianchi et al. 2019; Wongpitch et al. 2016); the availability of raw materials, which in some cases are scarcer and therefore more expensive to obtain; not to mention the increasingly stringent and demanding public policies and regulations that governments impose on companies to mitigate the adverse impacts of their operation (Haji et al. 2023). In addition, some financial institutions offer financial products such as bonds and loans to companies that demonstrate better ESG performance (Driessen 2021).

In addition, since the 1990s, various international entities have been promoting initiatives to encourage and promote best practices in the business world. Examples of this include public and private efforts, such as the World Business Council on Sustainable Development in 1995 (WBCSD 2024), The European Business Network for Corporate Sustainability and Responsibility established in 1995 (Idowu 2013), the Global Reporting Initiative in 1997 (Global Reporting Initiative 2020), the United Nations Global Compact in 1999 (UN Global Compact 2021), the ISO26000 about Corporate Social Responsibility (2010), the OECD guidelines for multinational enterprises on responsible business conduct (2011), the UN Sustainable Development Goals established in 2015 (2021) among others.

In one way or another, these pressures and incentives have caused some companies to respond with more responsible practices toward their stakeholders. For example, they are ratifying their commitment to social causes, expressing their categorical rejection of the violation of human rights, increasing investment in more efficient technologies with the use of resources and energy, improving transparency and labor relations mechanisms within companies as well as in the supply chain and mitigating externalities in the communities where they operate. This set of practices relates to what is known as corporate social responsibility (Dahlsrud 2008).

Since the 1970s, researchers around the world (Carroll 1999; Drucker 1984; Moskowitz 1972; Steiner 1971) took on the task of conducting empirical research to identify the type of relationship between corporate social responsibility (CSR) and the benefits for companies of carrying out such actions or what is known in the literature as the *CSR business case* (Carroll and Shabana 2010; Davis 1973; Kurucz et al. 2008; Salzmann et al. 2005). As for whether these practices generate a positive impact on financial performance, the results have been inconclusive, as positive, negative, neutral, and mixed relationships have been found (Margolis et al. 2007; Miras Rodriguez et al. 2014; Orlitzky et al. 2003).

Other authors have theorized and found evidence to support that CSR practices can generate benefits such as increased customer loyalty (Pérez and Del Bosque 2015), better attraction and retention of human talent (Bhattacharya et al. 2008), or reduced operating costs (Zafar and Farooq 2014). Conversely, research continues to investigate how firms obtain tangible or intangible benefits and why such benefits are sometimes not realized (Kong et al. 2020). This is a relevant issue for entrepreneurs, business people, and managers because their organizations practically have to carry out CSR actions to respond to the multiple and growing pressures; however, for them, it is not entirely clear if, in addition to complying with the requirements or demands of their stakeholders, their organizations can benefit from investing limited resources and capabilities in CSR actions.

It is at this point that Burke and Logsdon's (1996) research is relevant because it suggests that research should not only investigate the correlation between CSR and benefits for companies but should focus on reviewing the implementation of CSR policies and programs that can create long-term benefits within the organization, that is, to carry out CSR actions in a strategic way. According to these authors, corporate social responsibility (policy, program, or process) is strategic when it provides the company with substantial business-related benefits, particularly by supporting core business activities and thereby contributing to its effectiveness in fulfilling its mission.

Strategic CSR research in Latin America represents important opportunity areas (Jaén et al. 2018); thus, there is a need to study whether the strategy of companies is aligned with the implementation of social responsibility practices so that companies can obtain benefits in the long term in this context since most research has been conducted in Europe and the United States and very few in developing economic contexts (D'Amato Herrera 2013; Fatima and Elbanna 2022). Based on the above, this research aims to determine the extent to which Colombian companies that carry out CSR actions do so strategically and to determine company profiles based on their practices associated with the five strategic CSR dimensions suggested by Burke and Logsdon (1996).

2 Literature Review

There is evidence that social responsibility had its beginnings in the nineteenth century (Husted 2015); however, it was in the mid-twentieth century that Bowen (1953) formally defined CSR as "…the obligation of businessmen to pursue those policies, to make those decisions, or to follow those lines of action which are desirable in terms of objectives and values of our society." (p. 6). Since then, CSR has been defined and redefined by academia, multilateral organizations, and practitioners (Dahlsrud 2008; Rodriguez-Gomez et al. 2020; Sarkar and Searcy 2016), which has hindered the advancement of knowledge in this field. However, CSR can be considered one of several existing tools that allows a company's management to respond to the consequences that may be generated by the impacts of its actions when developing its corporate purpose in its environment.

2.1 Strategic CSR (SCSR)

The idea that companies can harness social and environmental issues to create value is not new and has been discussed elsewhere (Camarena-Martínez et al. 2016; Drucker 1984; Schumpeter 2011). Nevertheless, it is a debate that is still relevant in the literature (Čater and Golob 2020; Ślęzak 2020). In general terms, Strategic CSR (SCSR) implies that companies can obtain long-term tangible or intangible benefits while positively impacting society or the environment while matching consumer demand for responsible actions (Camarena-Martínez and Wendlandt-Amézaga 2017; Xiang et al. 2024). According to Kuokkanen and Sun (2023), CSR is strategic when a company systematically invests in responsible behavior that suits consumer behavior. For their part, McWilliams and Siegel (2011, p. 1480) define SCSR as: "responsible activity that allows a firm to achieve a sustainable competitive advantage, regardless of motive." Among the reasons why companies may adopt CSR actions are societal expectations, globalization, constant flows of information, and external pressures to adopt corporate sustainability (Ooi et al. 2017). Furthermore, strategic corporate social responsibility (CSR) has been associated with increased innovative, responsible initiatives (Yousfi and Loukil 2021). Therefore, managers can implement different CSR actions; however, the challenge lies in designing them so the organization can reap its benefits.

Implementing SCSR can provide firms with opportunities for growth, innovation, and competitive advantage (McWilliams and Siegel 2011; Werner 2009). This approach ensures businesses minimize their operations' negative impacts or externalities rather than just engaging in isolated philanthropic actions (Milliman et al. 2008). Additionally, by employing a strategic vision of CSR, a company might improve engagement with its primary stakeholders, thus improving competitivity (de Sousa de Melo et al. 2022). For their part, Vishwanathan et al. (2020) argue that CSR, when strategic, provides financial benefits for firms through four mechanisms:

enhancing firm reputation, increasing stakeholder reciprocation, mitigating risk, and strengthening innovation capacities. Thus, SCSR should combine all these mechanisms, where companies obtain financial results, minimize negative externalities, and engage with their stakeholders while transparently exercising accountability.

As for the reasons why companies may adopt this strategic approach to CSR, Milliman, Ferguson, and Sylvester (2008) argue for four main reasons why companies may engage in strategic CSR: (1) when the organization seeks to solve a business problem that cannot be solved within traditional business boundaries, it must therefore address the underlying social issue that is the primary source of that problem; (2) when an organization creates an opportunity based on a social problem or trend (e.g., marketing green products to consumers in response to environmental issues such as global warming); (3) when an organization seeks to protect itself from a threat posed by a social problem (e.g., scarcity of natural resources) and; (4) when an organization seeks to have a more significant business and social impact from its existing CSR programs.

Now, when discussing the SCSR implementation, there are several visions. For example, Vitolla et al. (2017), following a philosophy of social management, suggest that the most critical strategic processes related to CSR integration are the processes of strategic formulation, execution, and the generation of bottom-up innovations. On the other hand, Marques-Mendes and Joao Santos (2016) mention that the stages of the integration of SRHR in organizations consist of five levels of maturity and correspond to roles or attitudes in companies: (1) negation with a skeptical role about the benefits of CSR, (2) observance with a compliance role,(3) efficiency with an attitude of risk reduction and support, (4) strategic innovation, with a role of disseminator and, (5) transformational with a role of promoter. Additionally, Xiang et al. (2024) suggest three fundamental aspects to consider in strategic CSR implementation: (1) social value orientation, which emphasizes implementing activities resulting from social value; (2) strategic embeddedness, referring to the degree of alignment among firm core activities and the social value it produces and; (3) corporate responsible competitiveness, defined as the market feedback resulting from the social value provided by the company.

For the purposes of this work, we will consider the approach by Burke and Logsdon (1996) and as outlined by these authors CSR is strategic "...when it brings benefits to company-related business, especially when it supports core activities and contributes to the effectiveness of the company to achieve its mission" (p. 496). Based on the above definition, the authors proposed five dimensions that capture strategic behaviors for companies to benefit from CSR. These dimensions are fundamental for the company's success and valuable for relating CSR policies, programs, and processes to creating value within the company. This is the central axis of the manager's decision-making. The strategic dimensions of CSR are centrality, specificity, proactivity, voluntarism, and visibility.

2.2 Centrality

Centrality refers to the closeness between CSR actions and the company's mission, goals, and objectives. Actions with high centrality tend to be a priority within the company and produce long-term benefits. Galbreath (2009) argues that centrality is usually a critical issue in the organization because this variable provides direction and feedback to the company by revealing whether the application of an action or program is consistent with a firm current strategic planning. Moreover, evidence suggests that a sustainability-oriented mission statement may influence customers purchasing decisions (Ding et al. 2023). Thus, programs with high centrality should be prioritized within the company to obtain benefits (Burke and Logsdon 1996). Likewise, a central issue, which is part of the company's mission, is typically operationalized as plans, projects, and programs so that it is possible to have goals and measurement indicators. Husted et al. (2008) argue that a company could internalize its CSR actions or, on the contrary, outsource them through support to foundations or other means of corporate philanthropy; however, they found that companies with CSR themes aligned to a greater extent with their mission and strategic planning tend to internalize these actions to a greater extent. Likewise, another way in which CSR actions are central to the company's mission is by aligning it with one or more of the United Nations Sustainable Development Goals (Mathur and Poddar 2023). In addition, material CSR, that is, CSR activities related to a firm's core activities, is considered a relevant factor for investors (Pratoomsuwan and Chiaravutthi 2023).

2.3 Specificity

This dimension refers to a company's ability to internalize the benefits of CSR actions, such as competitive advantage, long-term economic sustainability, minimizing additional regulations, and obtaining a first-mover advantage in setting high ethical standards (Fioravante 2024). The benefits can be financial (Havlinova and Kukacka 2023), while others may include enhancing firm reputation, increasing stakeholder reciprocation, mitigating firm risk, and strengthening innovation capacity (Vishwanathan et al. 2020). For their part, Heslin and Ochoa (2008) propose seven strategic principles of CSR, which could be interpreted as the benefits that companies can obtain from exercising it: cultivate needed talent, develop new markets, protect labor markets, reduce environmental footprint, profit from by-products, involve customers and fostering a green supply chain. Other studies point out that the usefulness of the SCSR is to gain competitive advantages (McWilliams and Siegel 2011; Sousa-Filho et al. 2010) as well as product and process innovation (Bocquet et al. 2019; Maury 2022; Yousfi and Loukil 2021). Thus, measuring, evaluating, and internalizing these impacts in the company is essential to improve decision-making (Kowszyk and Vanclay 2020; Porter et al. 2012). In addition to generating benefits, specificity implies that the company can internalize or capture them for itself so that other firms do not take advantage of them (Bowman and Ambrosini 2000; Lepak et al. 2007).

2.4 Proactivity

The proactivity dimension refers to a company's ability to plan the development of CSR activities by observing the behavior and evolution of global trends (economic, technological, social, environmental, legal, etc.). This dimension is related to long-term planning, where organizations can innovate by anticipating tendencies, such as trend analysis and scenario planning techniques (Fauré et al. 2017). Proactivity is relevant because organizations must first manage their responsibilities strategically in their planning processes so that they execute them accordingly (George et al. 2024). Additionally, acting in this way can prevent negative impacts on the reputation of companies in times of crisis (Rim and Ferguson 2020); likewise, when CSR is reactive, this can be troublesome for companies and may need the disbursement of significant resources. Also, it is worth noting that business stakeholders increasingly expect companies to take a proactive approach to environmental and societal issues (Mochales and Blanch 2022).

2.5 Visibility

This dimension refers to a company's ability to gain recognition from stakeholders—internal and external—for its socially responsible initiatives, although it can generate positive or negative appreciation. The commitments and actions companies make regarding CSR can be seen by companies' stakeholders as a mechanism to forge strategic alliances and link efforts aligned with corporate values, producing favorable results for both parties (Camilleri 2022). Companies typically communicate their good practices through sustainability reports; however, visibility also implies communication within the company, which is very important to align all organization members to CSR initiatives. Evidence also suggests that the more strategically oriented CSR initiatives are, the more favorably the media perceive them (Fessmann et al. 2023). Additionally, Sanchez-López and Bejarano (2022) found that companies' CSR efforts may not have a relevant impact on returns to investors if efforts to communicate and disseminate these practices are deficient. The research of Arora et al. (2022) identified six manners in which companies can communicate CSR efforts to their stakeholders: disclosures in annual CSR reports, CSR policy published on companies websites, award recognition, visibility by expenditure on sustainable practices, sustainability reporting, international disclosures such as the Global Reporting Initiative.

2.6 Voluntarism

Finally, voluntarism refers to the extent of discretionary decision-making by a company regarding CSR activities and the absence of externally imposed compliance requirements (Arevalo and Aravind 2017). Various international sustainability frameworks consider CSR and its communication a voluntary commitment beyond compliance with standards (Global Reporting Initiative 2020; International Standards Organization 2010). Voluntary CSR requires a lot of commitment from managers to build links with stakeholders that allow them to improve the company's results and gain a reputation and good name while making the company competitive and sustainable (Husted and Allen 2000).

However, as previously mentioned, several pressures force companies to adopt this type of practice, whether they want to or not. These pressures can range from mandatory regulations and public policies (Othman et al. 2011) to consumer demands and certifications required by the focal companies of the supply chains to their suppliers and distributors (Acosta et al. 2019). Likewise, stakeholder groups have been related as a source of pressure in determining why companies act responsibly (Belay et al. 2023) The voluntarism dimension implies that companies carry out CSR actions voluntarily without pressures or situations that force them to do so as an imposition. If a company waits for a law or regulation to force it to act sustainably, it can lead to significant costs such as fines or penalties or large investments in equipment.

In some developing countries, businesses practice voluntary CSR because regulations for mandatory social responsibility practices are lacking or outdated (Putri et al. 2023). However, in Colombia, organizations have shown interest in participating in voluntary CSR initiatives such as the United Nations Global Compact, which currently has more than 590 member organizations (Linkedin 2024). Likewise, following the international B Corporations movement, the Colombian government, through the issuance of Law 1901 of 2018, created commercial companies of benefit and collective interest—BIC, companies that, in addition to this benefit, act in the interest of the community and the environment, a rule of voluntary adoption, which allows companies that accept this quality to obtain some benefits as long as they comply with the premise that their business model has a triple purpose: create economic, social and environmental value.

2.7 Previous Empiric Strategic CSR Dimensions Studies

Concerning the empirical research that has studied the relationship between the strategic dimensions of social responsibility, we can cite the work of Husted and Allen (2007), who, in the light of the resource-based view approach, applied Burke and Logsdon's (1996) model regarding strategic corporate social responsibility as a source of value for companies in the context of Spanish companies. Also, Husted and Allen (2009) evaluated the relationship between SCSR and value creation in 111 multinational firms operating in Mexico, finding that the dimensions of centrality and visibility positively explained value creation while voluntarism negatively. On the other hand, Novita (2012) applied the five-dimension model to companies in Indonesia and Malaysia and found a positive relationship between strategic social responsibility, return on assets (ROA), and value creation.

For their part, Bocquet et al. (2013) explored the relationship between CSR actions and innovation, finding that those with a strategic profile (measured by the Burke and Logsdon model) tend to innovate more compared to those who do reactive CSR. Camarena and Wendlandt (2017), in their research on the relationship between strategic dimensions and the creation of shared value, with information from 50 Mexican companies, found through multiple linear regression that the specificity dimension is positively related to the strategic value of companies while voluntarism is negatively related, coinciding to some extent with the results of Husted and Allen (2009). In the context of Palestinian companies, Sleibi and Abdelkarim's (2021) findings suggest that strategic dimensions relate in different ways to strategic value creation: a strong positive correlation with proactivity in the banking sector, a moderate negative correlation with voluntarism in the industrial sector, and a strong positive correlation with visibility in the investment and services sector. Moreover, Lu and Abeysekera (2021) used the five-dimensional model to create strategic CSR disclosure ratings, and they found that the market has a positive reaction to SCSR disclosures, but the market fails to perceive whether such disclosures are credible.

More recent studies, such as Anlesinya and Abugre's (2022) provided evidence that activities involving strategic CSR planning, implementation, and positioning contributed considerably to value creation in multinational subsidiaries. Xue et al. (2023) found that voluntary CSR disclosure has a stronger negative effect on firm financial performance than mandatory CSR disclosure, while firms with voluntary CSR reporting attract more investors and have higher stock returns. Finally, Chwiłkowska-Kubala et al. (2023) found that the proactive approach of strategic CSR positively impacts organizational firm agility, as in the capability of awareness and seizing opportunities that benefit firms.

3 Methodology

For the development of this research, the data collection was conducted via surveys of companies under the supervision of Colombia's Superintendence of Companies in 2017. It is worth noting that the companies under the supervision of the Superintendence of Companies are commercial companies that belong to the real sector of the economy under the provisions of articles 83, 84, 85, and 86 of Law 222 of 1995, 2° of Decree 1080 of 1996 and Decree 1074 of 2015 in Colombia.

In determining the universe of companies, the Superintendency of Companies takes into account the following criteria (Superintendency of Companies 2014): (1) type of commercial organization: corporations, limited companies, simplified corporations, general partnerships, limited partnerships, limited partnerships, branches of foreign companies and sole proprietorships; (2) size, measured as the number of assets or income; (3) historical information reported in previous years; (4) grounds for surveillance; (5) control status; (6) supervisory policy[1]; (7) additional criteria such as technological capacity, operational capacity, macroeconomic variables.

For this study, we used a database of 5,136 observations corresponding to the information reported by the same number of companies. The information on social responsibility corresponds to 29 questions available on this subject in the business practices survey. The answers to these questions were coded in values of 1 or 0 and classified into five strategic dimensions. Table 1 relates the instrument's items to the related CSR literature.

3.1 Procedure

The exploratory study of CSR data involved descriptive statistics, contingency tables, and box plots to determine the presence of outliers. Likewise, missing data verification was carried out, and the CSR variables were typified to avoid problems derived from the scale. Through principal component analysis, a variable reduction technique was applied to the 29 corporate social responsibility variables according to the strategic dimension to which they belong to determine the transformed variables that best capture the information by strategic dimension. Subsequently, with fewer variables, the k means clustering technique determined groups of similar companies regarding their strategic CSR dimensions. The avobe techniques were conducted using IBM SPSS Statistics software version 25. This methodology has been used in prior studies to identify profiles according to CSR and sustainability actions (Jamali et al. 2009; Zeng et al. 2008).

4 Results

4.1 Principal Component Analysis (PCA)

As shown in Table 1, it is evident that four out of five dimensions have five or more variables, so it is necessary to use the principal component technique, a statistical technique of data reduction, to obtain new variables that capture as much information as possible from the original variables of each dimension. The dimensional structure of the strategic social responsibility dimensions instrument was identified through principal component analysis and varimax rotation using IBM SPSS version 25. The minimum factor loading criteria was set to 0.50 (Pett et al. 2003). Todos los componentes extraídos mostraron eigen values myores a 1. The results of the GWP

[1] These are the sectors determined by the Superintendent of Societies for special monitoring.

Table 1 Measured variables by Strategic CSR Dimension

Dimension	Items	Source
Centrality	*Which of the following are related to your company's mission?* Respect for human rights in all the company's activities	Mathur and Poddar (2023)
	The use of mechanisms to prevent bribery and corruption	
	Respect, care, and consideration for customers and consumers	Heslin and Ochoa (2008)
	Improvement of employees living conditions	
	Reducing the environmental impact of operations	
Specificity	*What benefits does your company attain by doing CSR?* Increased sales	Porter and Kramer (2006)
	Reduction of costs and/or expenses	
	Ease of access to funding sources	
	Tax benefits	
	Fulfillment of contractual obligations	
	Improved relations with government and authorities	
	Improved employee relations	Bhattacharya et al. (2008)
	Improved image and reputation	Rim and Ferguson (2020)
	Improved relations with neighbors	Camilleri (2022)
	Improvement of supply relationships (suppliers and distributors)	Acosta et al. (2019)
Proactivity	Formal identification of stakeholders	Camilleri (2022)
Visibility	*Does your firm report sustainable practices?*	Fessmann et al. (2023), Sanchez-Lopez and Bejarano (2022)
	Social responsibility report	
	Social balance report	
	Sustainability report	
	Promotion and evaluation of responsible practices among groups of suppliers and/or distributors	Acosta et al. (2019), Heslin and Ochoa (2008)
	Support to improve the productivity and efficiency of your suppliers and/or distributors	Acosta et al. (2019)
	Product and service impact analysis and life cycle analysis and extended responsibility	Haji et al. (2023)
Voluntarism	*Why does your company engage in CSR activities?* Linking to social and community projects	Ndhlovu (2011)
	Moral obligation	Haji et al. (2023)
	Operational and/or commercial customs	Acosta et al. (2019)

(continued)

Table 1 (continued)

Dimension	Items	Source
	Contribution to foundations and donations in general	Husted et al. (2008)
	Requirements of international markets	Chiou and Shu (2019), Othman et al. (2011)

Source own elaboration

concerning the related factors, items, and factor loadings are summarized in Table 2. The factors identified are described as follows:

Factor 1. The centrality dimension is configured of five items that relate companies' mission and strategic objectives with CSR issues. For this dimension, the items were grouped into a single component, which explains 80.45% of the variance. Cronbach's alpha yielded a value of 0.94, considered acceptable (Cronbach 1951; Nunnally and Bernstein 1994).

Factors 2 and 3. Two components were extracted for the specificity dimension factor that explained 61.57% of the variance and obtained a Cronbach's alpha coefficient of 0.88, which is acceptable. The first component groups items related to the appropriation of economic value for the company, such as increasing sales, reducing costs, accessing sources of financing, obtaining tax benefits, and complying with contracts. The second component grouped items related to improving the relationship with stakeholders.

Factor 4. Proactivity. Data reduction was not used for this dimension since only one item exists. This question reflects companies' progress in identifying their stakeholders, which is done in a premeditated and anticipated manner.

Factors 5 and 6. The items of the visibility dimension were grouped into two components, which explain 59.60% of the variance. Also, an item was eliminated, so Cronbach's alpha finally resulted in 0.73. The first component characterizes the items that measure companies' recognition from their stakeholders (excluding suppliers and distributors) when presenting and disseminating their sustainability reports. On the other hand, the second component shows the recognition of the members of the supply chain (suppliers and distributors) of the company when it carries out CSR actions, including actions to promote, evaluate, and support sustainable practices in the chain, as well as life cycle analysis and extended responsibility actions.

Factor 7. Finally, the items of the voluntarism dimension were grouped into a single component with an eigenvalue of 2.59, which explained 51.75% of the variance and presented a Cronbach's alpha reliability index of 0.76. This component includes companies' motivations to carry out corporate social responsibility actions. Most companies contribute to foundations and donations and are engaged in social projects without a legal mandate; they do so because of their institutional values or their vocation for service.

The five strategic dimensions of corporate social responsibility defined by Burke and Logsdon (1996) were expressed in the following seven variables (see Fig. 1), where, except for proactivity, the others resulted from applying the PCA.

Table 2 Strategic CSR dimensions' constructs and loadings

Items	1	2	3	4	5	6	7
Centrality							
Respect for human rights in all the company's activities	0.92						
The use of mechanisms to prevent bribery and corruption	0.92						
Respect, care, and consideration for customers and consumers	0.90						
Improvement of employees living conditions	0.90						
Reducing the environmental impact of operations	0.85						
Specificity (firm)							
Increased sales		0.80					
Reduction of costs and/or expenses		0.78					
Ease of access to funding sources		0.70					
Tax benefits		0.67					
Fulfillment of contractual obligations		0.52					
Specificity (stakeholders)							
Improved relations with government and authorities			0.86				
Improved employee relations			0.82				
Improved image and reputation			0.81				
Improved relations with neighbors			0.75				
Improvement of supply relationships (suppliers and distributors)			0.68				
Proactivity							
Formal identification of stakeholders				NA			
Visibility (reporting)							
Does your firm report sustainable practices?					0.84		
Social responsibility report					0.77		
Social balance report					0.68		
Sustainability report					0.68		
Visibilidad (supply chain)							

(continued)

Table 2 (continued)

Items	1	2	3	4	5	6	7
Promotion and evaluation of responsible practices among groups of suppliers and/or distributors						0.80	
Support to improve the productivity and efficiency of your suppliers and/or distributors						0.79	
Product and service impact analysis and life cycle analysis and extended responsibility						0.75	
Voluntarism							
Linking to social and community projects							0.81
Moral obligation							0.76
Operational and/or commercial customs							0.70
Contribution to foundations and donations in general							0.69
Requirements of international markets							0.61

Note The extraction method used was principal components analysis with varimax rotation. *Source* own elaboration

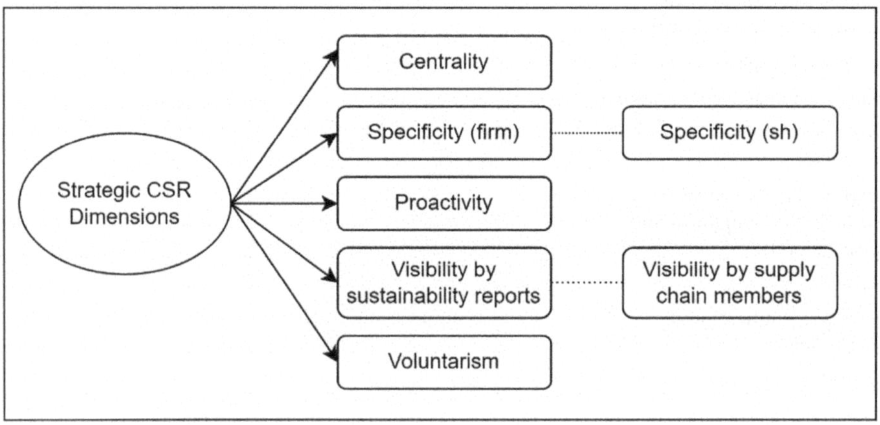

Fig. 1 Strategic CSR dimensions extended model. *Source* own elaboration based on Burke and Logsdon (1996)

These findings suggest that, to some extent, Colombian companies apply the five strategic dimensions of CSR defined by Burke and Logsdon (1996), and even in the case of specificity and visibility, it was found that these were subdivided into two components, thus expanding the original model. In the first case, specificity contains financial benefits for the company and positively impacts stakeholder relations. On the other hand, the visibility dimension is subdivided into those generated by sustainability reports and obtained by responsible management in the company's supply chain, i.e., suppliers and distributors.

4.2 *Correlation Between Variables*

Next, Pearson's bivariate correlation test was applied to determine the relationship between the strategic dimensions of CSR (see Table 3).

According to Zikmund et al. (2013) discriminant validity is established when a scale that should not correlate too highly with a measure of different dimensions does not correlate very highly. It is observed that the dimension of centrality correlates significantly with the dimensions of specificity for stakeholders (0.799) and voluntarism (0.842). This could be because centrality implies a certain way to communicate the company's mission and objectives with stakeholders. However, for the other dimensions, the correlation is not high enough (<0.70) to have discriminant validity problems.

4.3 *Cluster Análisis Results*

Next, the K-means cluster analysis was applied to the six components and the proactivity variable to determine the groups of companies with similar CSR strategies. Two clusters were obtained based on the seven variables. This number of clusters was determined from the silhouette measurement of cohesion and separation. It showed that the quality is good[2] for two clusters, which allows us to continue with the analysis of K means because the data fit the structure of the clusters. The two clusters generated after 50 iterations are presented in Table 4.

Based on the results in Table 4, two clusters of companies that share similar behaviors concerning the dimensions of SCSR were identified and described as follows.

Cluster 1 (weak SCSR): This cluster grouped 1,914 companies that, on average, neglect to obtain recognition from their stakeholders when they present sustainability reports (visibility_reports) and anticipation of economic and technological trends

[2] The bad, fair, and good results of the cluster quality chart were based on the work of Kaufman and Rousseeuw (2009) with respect to the interpretation of clusters.

Table 3 Correlations between study variables

	Centrality	Specificity _sh	Specificity_firm	Proactivity	Visibility_reports	Visibilidad_sc	Voluntarism
Centrality	1						
Specificity _sh	,799**	1					
Specificity_firm	,228**	,000	1				
Proactivity	,174**	,168**	,064**	1			
Visibility_reports	,115**	,132**	,012	,231**	1		
Visibilidad_sc	,619**	,582**	,292**	,164**	,000	1	
Voluntarism	,842**	,735**	,352**	,164**	,124**	,606**	1

Note **. The correlation is significant at the 0.01 level (two-sided). *. The correlation is significant at the 0.05 level (two-sided). *Source* own elaboration

Table 4 Centers of the final clusters

SCSR Dimension	Cluster	
	2	1
Centrality	**0,70**	−1,2
Specificity _sh	**0,62**	−1,0
Specificity_firm	0,17	**−0,3**
Proactivity	0,15	**−0,2**
Visibility_reports	0,09	**−0,2**
Visibility_sc	0,47	−0,8
Voluntarism	**0,63**	−1,1

Note Source Own elaboration

(proactivity). They have shortcomings in the dimensions of centrality, specificity_sh, specificity_firm, visibility_sc, and voluntarism.

Cluster 2 (strong SCSR): A total of 3,222 companies were convened in this cluster, emphasizing the closeness of their CSR programs to their mission and objectives (centrality), the scope of decision-making without the need for a legal mandate (voluntarism) and the ability to internalize the benefits of a CSR program with their stakeholders (specificity_sh). They also make significant efforts in communicating CSR in their supply chains (visibility_sc), internalizing the benefits of a CSR program, mainly related to financial aspects (specificity_firm), anticipating economic and technological trends (proactivity), and recognizing their stakeholders through reporting (visibility_reports).

Next, Fig. 2 presents the proportion of companies in each cluster according to their economic sector and size.

Regarding the economic sectors, the strong SCSR cluster includes a considerable proportion of companies in the commerce and manufacturing sector, while cluster 1 concentrates mainly on companies in the service sector, which is usually not as pressured to adopt CSR practices as the manufacturing and construction sectors.

In terms of size (see Fig. 3), both conglomerates mainly concentrate on large and medium-sized companies, while cluster 1 has a more significant number of small companies, which do not show interest in CSR practices. This result responds to the research objective of identifying similar behavior patterns in Colombian companies relating to the dimensions of SCSR, such as cluster 2 with strong SCSR and cluster 1 with weak SCSR.

5 Discussion

The most relevant results of this study lie in the expansion of the original model proposed by Burke and Logsdon (1996), where they proposed five strategic dimensions of CSR. The results of the principal component analysis also made it possible

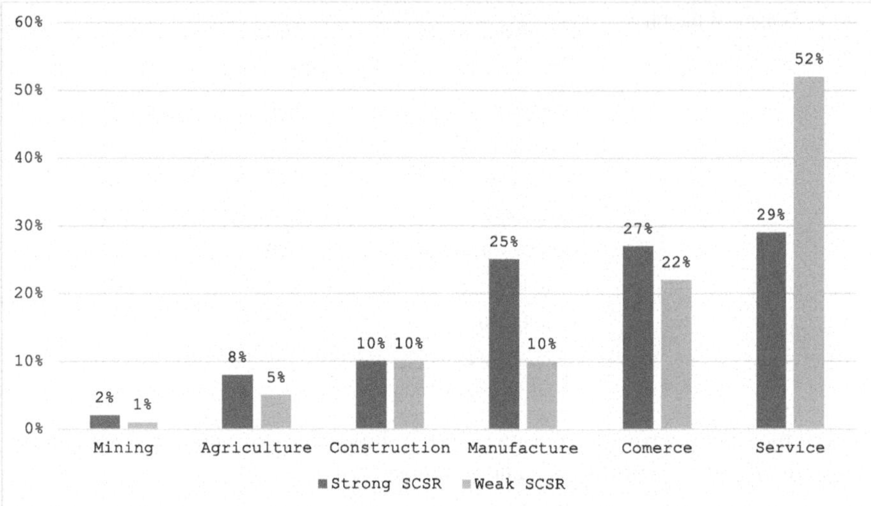

Fig. 2 Proportion of firms by cluster type (n = 5,136). Source. Own elaboration

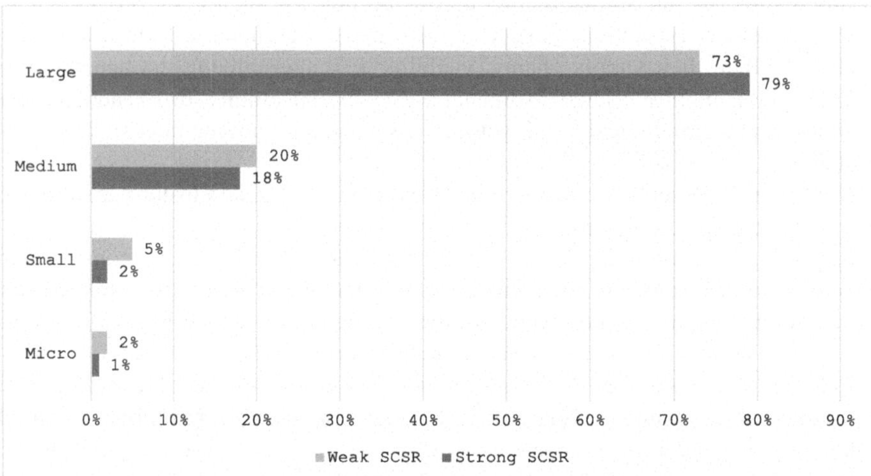

Fig. 3 Proportion of firms by size (n = 5,136). *Note* Firm classification corresponds to annual revenue from ordinary activities (Decree 1074 of 2015, Colombian Presidency 2023) Source. Own elaboration

to measure a sub-dimension called stakeholder specificity, which aligns with the SCSR approach suggested by Camilleri (2022), where to generate closer relationships with these groups, companies can obtain strategic benefits or with the shared value approach, which according to Porter and Kramer (2011) it is an evolution of traditional CSR and generates win–win relationships between business and society.

In addition, the findings of the PCA made it possible to measure the visibility focused mainly on the supply chain members in companies, which is also a contribution to the original model where only the visibility of CSR actions to mainstream stakeholders is considered. The visibility of CSR practices in the supply chain can be considered a sustainable supply chain management strategy where the focal companies share their corporate values, promote sustainability-related actions, and manage and evaluate these elements in the members of their chain (Formentini and Taticchi 2016).

On the other hand, the cluster analysis results align with the findings of Bocquet et al. (2013) in Luxembourg companies, in which two types of company profiles were obtained regarding the levels of SCSR they practice. Their results showed a cluster that they called strategic CSR and another called reactive CSR; in our case, they were called strong SCSR and weak SCSR, where the companies of the first group are primarily composed of large companies that operate in the services, trade, and manufacturing sectors; while those of the latter are also made up of large companies, notably in the service sector, followed by the trade sector. One reason these differences may exist is the demands that important clients, e.g., multinational corporations, may place on companies (Othman et al. 2011), as well as internalizing CSR instead of outsourcing it (Husted et al. 2008). Also, the size of the company is a factor that could explain why only large companies had more representation in the strong SCSR cluster since smaller companies are usually more focused on the operation and do not have as many resources available to plan and formalize their CSR aspects, which can lead them to practice traditional CSR (Afrin 2013). However, the data obtained limits the ability to infer, and it is not possible to clarify the reasons for the differences.

On the other hand, Colombian companies in the strong CSR cluster obtained high values in the dimensions of centrality and voluntarism, which can be compared to the findings of Husted and Allen in Spanish (2007) and Mexican (2009) companies and with those reported by Camarena-Martínez and Wendlandt-Amézaga (2017) where the companies consulted reported high levels in these dimensions. Moreover, the results reported by Sleibi and Abdelkarin (2021) show a preference for companies in the banking sector for the visibility dimension and those in the industrial sector for centrality, which is related to our results in the number of companies in the services and manufacturing sectors in the strong SCSR group.

6 Conclusion

The objective of this study was to determine the extent to which Colombian companies that carry out CSR actions do so strategically and to determine company profiles based on their practices associated with the five strategic CSR dimensions suggested by Burke and Logsdon (1996). Our results made it possible to identify two types of profiles of Colombian companies concerning the five strategic dimensions of CSR and expand the model with two new sub-dimensions.

Burke and Logsdon's original model comprised five dimensions: centrality, specificity, proactivity, and voluntarism. The results of the PCA allowed us to expand the specificity and visibility dimensions with two additional factors. The specificity dimension originally referred to the benefits companies can internalize from their CSR activities. However, the additional factor obtained alludes to the benefits a company's stakeholders obtain from engaging in responsible activities. As for the visibility dimension, the added factor refers to the visibility of CSR activities concerning a company's supply chain members.

Regarding the cluster analyses, the first cluster identified was called *strong SCSR,* which contains companies that have their strategy aligned with CSR because they emphasize the closeness of their CSR programs to their mission and objectives (centrality), the scope of decision-making without the need for a legal mandate (voluntarism), and the ability to internalize the benefits of a CSR program with their stakeholders (specificity_sh). They also present significant efforts in recognizing suppliers and distributors (visibility_sc) the ability to internalize the benefits of a CSR program, mainly related to financial aspects (specificity_firm), in anticipating economic and technological trends (proactivity), and in recognizing their stakeholders through reporting (visibility_reports). On the other hand, the second cluster denominated the *weak SCSR*, grouped companies that showed disinterest in CSR because they neglected to obtain recognition from their stakeholders when they submitted sustainability reports (visibility_reports) and anticipated economic and technological trends (proactivity). In addition, this group is characterized by shortcomings in the dimensions of centrality, stakeholder specificity, firm specificity, visibility in the supply chain, and voluntarism.

As for the implications of our study, the results show that in our sample of Colombian companies, there is still a significant proportion of firms that engage in CSR activities that do not necessarily do it bearing in mind a strategic mindset, probably because of lax legislation or because firms are seeking short term benefits such as reputation and tax deductions. However, this represents an opportunity for managers to embed sustainability issues in their companies' strategy and core activities, which might provide medium- and long-term benefits. As for implications for academia, our findings helped in broadening Burke and Logsdon's SCSR model and considering firms' stakeholders in the specificity dimension, which can be aligned with more recent currents of thought in CSR, such as shared value and inclusive business, where the company benefit is not the only target but also societal value. As for the visibility of supply chain members, this could pave the way for new research in sustainable supply chain and value chain governance fields.

Within the study's limitations, this exploratory research had a descriptive scope and focused only on companies surveyed by the Superintendence of Companies of Colombia, so it is impossible to generalize this behavior beyond the sample. Likewise, the data collected do not allow us to establish relationships between the dimensions of SCSR and other variables of interest, such as performance or the creation of business and social value; thus, future research could apply the extended model of strategic dimensions of CSR resulting from this research and apply it in different contexts to make comparisons, as well as determine which strategic dimensions have the significant impact on dependent variables. In addition, studies could be conducted to explore whether there are significant differences between company profiles and investigate which mediating or moderating variables can explain them. The findings of this study show that CSR is important for business operations and strategy, reflecting an ongoing cultural change in Colombia as it evolves from traditional CSR (philanthropic) management to strategic management, increasingly integrating with its stakeholders and giving importance to their perception of their actions in society; however, the job is far from done.

References

Acosta P, Acquier A, Gond J-P (2019) Revisiting Politics in Political CSR: how coercive and deliberative dynamics operate through institutional work in a Colombian company. Organ Stud 42(7):1111–1134. https://doi.org/10.1177/0170840619867725

Afrin S (2013) Traditional vs strategic corporate social responsibility: in pursuit of supporting sustainable development. J Econ Sustain Dev 4(20):153–157

Alimonti G, Mariani L (2023) Is the number of global natural disasters increasing? Environ Hazards, 1–17.https://doi.org/10.1080/17477891.2023.2239807

Anlesinya A, Abugre JB (2022) Strategic CSR practices, strategic orientation and business value creation among multinational subsidiaries in Ghana. Soc Bus Rev 17(2):257–279. https://doi.org/10.1108/SBR-02-2021-0021

Arevalo J, Aravind D (2017) Strategic outcomes in voluntary CSR: reporting economic and reputational benefits in principles-based initiatives. J Bu Ethics 144(1):201–217. https://doi.org/10.0.3.239/s10551-015-2860-5

Arora S, Kaur J, Arora G (2022) Visibility dimension of strategic corporate social responsibility: level-based measurement of selected Indian companies. Int J Soc Ecol Sustain Dev (IJSESD) 13(6):1–25. https://doi.org/10.4018/IJSESD.301250

Belay HA, Hailu FK, Sinshaw GT (2023) Linking internal stakeholders' pressure and Corporate Social Responsibility (CSR) practices: the moderating role of organizational culture. Cogent Bus Manag 10(2):2229099. https://doi.org/10.1080/23311975.2023.2229099

Bhattacharya CB, Sen S, Korschun D (2008) Using corporate social responsibility to win the war for talent. MIT Sloan Manag Rev 49(2):36–44

Bianchi E, Bruno JM, Sarabia-Sanchez FJ (2019) The impact of perceived CSR on corporate reputation and purchase intention. Eur J Manag Bus Econ 28(3):206–221. https://doi.org/10.1108/EJMBE-12-2017-0068

Bocquet R, Le Bas C, Mothe C, Poussing N (2013) Are firms with different CSR profiles equally innovative?: empirical analysis with survey data. Eur Manag J 31(6):642–654. https://doi.org/10.1016/j.emj.2012.07.001

Bocquet R, Le Bas C, Mothe C, Poussing N (2019) Strategic CSR for innovation in SMEs: Does diversity matter? Long Range Plan 52(6):101913. https://doi.org/10.1016/j.lrp.2019.101913

Bowen HR (1953) Social responsibility of the businessman. Harper

Bowman C, Ambrosini V (2000) Value creation versus value capture: towards a coherent definition of value in strategy. Br J Manag 11(1):1–15. https://doi.org/10.1111/1467-8551.00147

Burke L, Logsdon JM (1996) How corporate social responsibility pays off. Long Range Plan 29(4):495–502. https://doi.org/10.1016/0024-6301(96)00041-6

Calvin K, Dasgupta D, Krinner G, Mukherji A, Thorne PW, Trisos C, Romero J, Aldunce P, Barrett K, Blanco G, Cheung WWL, Connors S, Denton F, Diongue-Niang A, Dodman D, Garschagen M, Geden O, Hayward B, Jones C, … Ha M. (2023). IPCC, 2023: climate change 2023: synthesis report. contribution of working Groups I, II and III to the sixth assessment report of the intergovernmental panel on climate change [Core Writing Team, Lee H , Romero J (eds.)]. IPCC, Geneva, Switzerland. (Arias P, Bustamante M, Elgizouli I, Flato G, Howden M, Méndez-Vallejo C, Pereira JJ, Pichs-Madruga R, Rose SK, Saheb Y, Sánchez Rodríguez R, Ürge-Vorsatz D, Xiao C, Yassaa N, Romero J, Kim J, Haites EF, Jung Y, Stavins R, … Péan C, Eds). https://doi.org/10.59327/IPCC/AR6-9789291691647

Camarena-Martínez JL, Ochoa-Silva B, Wendlandt-Amezaga TR (2016) Exploring the origins of creating shared value in the CSR literature. Int J Bus Manag 11(12):97–109. https://doi.org/10.5539/ijbm.v11n12p97

Camarena-Martínez JL, Wendlandt-Amézaga TR (2017) Strategic CSR dimensions and value creation in socially responsible Mexican firms. Int Rev Manag Bus Res 6(2):749–769

Camilleri MA (2022) Strategic attributions of corporate social responsibility and environmental management: the business case for doing well by doing good! Sustain Dev 30(3):409–422. https://doi.org/10.1002/sd.2256

Carroll AB (1999) Corporate social responsibility evolution of a definitional construct. Bus Soc 38(3):268–295. https://doi.org/10.1177/000765039903800303

Carroll AB, Shabana KM (2010) The business case for corporate social responsibility: a review of concepts, research and practice. Int J Manag Rev 12(1):85–105. https://doi.org/10.1111/j.1468-2370.2009.00275.x

Čater B, Golob U (2020) Strategic CSR: mapping the state-of-the-art. In: Žabkar V, Redek T (Eds), Challenges on the path toward sustainability in Europe, pp 57–80. Emerald Publishing Limited. https://doi.org/10.1108/978-1-80043-972-620201005

Cheng L, Abraham J, Trenberth KE, Fasullo J, Boyer T, Locarnini R, Zhang B, Yu F, Wan L, Chen X, Song X, Liu Y, Mann ME, Reseghetti F, Simoncelli S, Gouretski V, Chen G, Mishonov A, Reagan J, Zhu J (2021) Upper ocean temperatures hit record high in 2020. Adv Atmos Sci 38(4):523–530. https://doi.org/10.1007/s00376-021-0447-x

Chiou C-L, Shu P-G (2019) How does foreign pressure affect a firm's corporate social performance? Evidence from listed firms in Taiwan. J Multinational Financ Manag 51:1–22. https://doi.org/10.1016/j.mulfin.2019.04.006

Chwiłkowska-Kubala A, Cyfert S, Malewska K, Mierzejewska K, Szumowski W, Prause G (2023) What drives organizational agility in energy sector companies? The role of strategic CSR initiatives and the dimensions of proactive CSR. Sustainable Futures 6:100133. https://doi.org/10.1016/j.sftr.2023.100133

Cronbach LJ (1951) Coefficient alpha and the internal structure of tests. Psychometrika 16(3):297–334

Dahlsrud A (2008) How corporate social responsibility is defined: an analysis of 37 definitions. Corp Soc Responsib Environ Manag 15(1):1–13. https://doi.org/10.1002/csr.132

D'Amato Herrera GM (2013) Academic trends in the study of corporate social responsibility and development issues in Latin America, 2000–2010. Cuadernos De Administración (Universidad Del Valle) 29(49):85–94

Davis K (1973) The case for and against business assumption of social responsibilities. Acad Manag J 16(2):312–322. https://doi.org/10.2307/255331

de Sousa de Melo MF, Souza Pião R de C, Campos-Silva WL, Palheta Nery D (2022) Corporate social responsibility and competitiveness: a study of Brazilian multinationals. Production/Produção, 32

Ding M-C, Lii YS, Lai KF (2023) A great start? Sustainability mission statements affect consumer behavior. Environment, development and sustainability. https://doi.org/10.1007/s10668-023-04247-0

Driessen M (2021) Sustainable finance: an overview of ESG in the financial markets. In: Busch D, Ferrarini G, Grünewald S (Eds), Sustainable finance in Europe: corporate governance, financial stability and financial markets, pp 329–350. Springer International Publishing. https://doi.org/10.1007/978-3-030-71834-3_10

Drucker PF (1984) Converting social problems into business opportunities: the new meaning of corporate social-responsibility. Calif Manage Rev 26(2):53–63. https://doi.org/10.2307/41165066

Fatima T, Elbanna S (2022). Corporate social responsibility (CSR) implementation: a review and a research agenda towards an integrative framework. J Bus Ethics.https://doi.org/10.1007/s10551-022-05047-8

Fauré E, Arushanyan Y, Ekener E, Miliutenko S, Finnveden G (2017) Methods for assessing future scenarios from a sustainability perspective. Europ J Futures Res 5(1):17. https://doi.org/10.1007/s40309-017-0121-9

Fessmann AN, Bruyaka O, Hirshman A, Dawley D (2023) Does strategic CSR pay off? A study of CSR events, CSR Portfolio and media favorability. Acad Manag Proc 2023(1):11641

Fioravante R (2024) Beyond the business case for responsible artificial intelligence: strategic CSR in light of digital washing and the moral human argument. Sustainability 16(3):1232. https://doi.org/10.3390/su16031232

Formentini M, Taticchi P (2016) Corporate sustainability approaches and governance mechanisms in sustainable supply chain management. J Cleaner Prod 112:1920–1933. ELSEVIER SCI LTD. https://doi.org/10.1016/j.jclepro.2014.12.072

Galbreath J (2009) Building corporate social responsibility into strategy. Eur Bus Rev 21(2):109–127. https://doi.org/10.1108/09555340910940123

George B, Worth MJ, Pandey S, Pandey SK (2024) Strategic management of social responsibilities: a mixed methods study of US universities. Public Money Manag 44(1):15–25. https://doi.org/10.1080/09540962.2023.2197253

Global Reporting Initiative (2020) GRI Standards. https://www.globalreporting.org/how-to-use-the-gri-standards/resource-center/

Hahn R (2013) ISO 26000 and the standardization of strategic management processes for sustainability and corporate social responsibility. Bus Strateg Environ 22(7):442–455. https://doi.org/10.1002/bse.1751

Haji AA, Coram P, Troshani I (2023) Consequences of CSR reporting regulations worldwide: a review and research agenda. Account Aud Account J 36(1):177–208. https://doi.org/10.1108/AAAJ-05-2020-4571

Havlinova A, Kukacka J (2023) Corporate social responsibility and stock prices after the financial crisis: the role of strategic CSR activities. J Bus Ethics 182(1):223–242. https://doi.org/10.1007/s10551-021-04935-9

Heslin PA, Ochoa JD (2008) Understanding and developing strategic corporate social responsibility. Organ Dyn 37(2):125–144. https://doi.org/10.1016/j.orgdyn.2008.02.002

Husted BW (2015) Corporate social responsibility practice from 1800–1914: past initiatives and current debates. Bus Ethics Q 25(1):125–141. https://doi.org/10.1017/beq.2014.1

Husted BW, Allen DB (2000) Is it ethical to use ethics as strategy? In: Business challenging business ethics: new instruments for coping with diversity in international business, pp 21–31. Springer

Husted BW, Allen DB (2007) Strategic corporate social responsibility and value creation among large firms: lessons from the Spanish experience. Long Range Plan 40(6):594–610. https://doi.org/10.1016/j.lrp.2007.07.001

Husted BW, Allen DB (2009) Strategic corporate social responsibility and value creation: a study of multinational enterprises in Mexico. MIR: Manag Int Rev 49(6):781–799. https://doi.org/10.1007/s11575-009-0016-5

Husted BW, Allen DB, Rivera JE (2008) Governance choice for strategic corporate social responsibility: evidence from central America. Bus Soc 49(2):201–215. https://doi.org/10.1177/0007650308315504

Idowu SO (2013) CSR Europe. In: Idowu SO, Capaldi N, Zu L, Das Gupta A (Eds), Encyclopedia of corporate social responsibility, pp 699–701. Springer, Berlin, Heidelberg. https://doi.org/10.1007/978-3-642-28036-8_194

International Standards Organization (2010) ISO 26000—Social responsibility. Standards. https://www.iso.org/iso-26000-social-responsibility.html

International Standards Organization (2016) ISO 37001:2016 Anti-bribery management systems—Requirements with guidance for use. https://www.iso.org/standard/65034.html#:~:text=ISO 37001%3A2016 is applicable,commitments applicable to its activities

Jaén MH, Auletta N, Bruni Celli J, Pocaterra M (2018) Bibliometric analysis of indexed research on corporate social responsibility in Latin America (2000–2017). Academia Revista Latinoamericana De Administración 31(1):105–135. https://doi.org/10.1108/ARLA-06-2017-0190

Jamali D, Sidani Y, El-Asmar K (2009) A three country comparative analysis of managerial CSR perspectives: Insights from Lebanon, Syria and Jordan. J Bus Ethics 85(2):173–192. https://doi.org/10.1007/sl0551-008-9755-7

Kaufman L, Rousseeuw PJ (2009) Finding groups in data: an introduction to cluster analysis. John Wiley & Sons

Kong Y, Antwi-Adjei A, Bawuah J (2020) A systematic review of the business case for corporate social responsibility and firm performance. Corp Soc Responsib Environ Manag 27(2):444–454. https://doi.org/10.1002/csr.1838

Kowszyk Y, Vanclay F (2020) The possibilities and limitations regarding the use of impact evaluation in corporate social responsibility programs in Latin America. Corporate Govern Int J Bus Soc 21(2):279–293

Kuokkanen H, Sun W (2023) Willingness to pay for corporate social responsibility (CSR): does strategic CSR management matter? J Hospitality & Tourism Res 48(4):657–670. https://doi.org/10.1177/10963480231182990

Kurucz E, Colbert B, Wheeler D (2008) The business case for corporate social responsibility. The Oxford Handbook of Corporate Social Responsibility, 83–112

Lepak DP, Smith KG, Taylor MS (2007) Value creation and value capture: a multilevel perspective. Acad Manag Rev 32(1):180–194. https://doi.org/10.5465/AMR.2007.23464011

Linkedin (2024) Pacto Global Colombia. https://www.linkedin.com/company/pacto-global-colombia/about/

Lu Y, Abeysekera I (2021) Do investors and analysts value strategic corporate social responsibility disclosures? Evidence from China. J Int Financ Manag Acc 32(2):147–181. https://doi.org/10.1111/jifm.12126

Margolis JD, Elfenbein HA, Walsh JP (2007) Does it pay to be good? A meta-analysis and redirection of research on the relationship between corporate social and financial performance. In Ann Arbor (Vol 1001)

Marques-Mendes A, Santos MJ (2016) Strategic CSR: an integrative model for analysis. Social Responsib J 12(2):363–381. https://doi.org/10.1108/SRJ-04-2015-0055

Mathur A, Poddar A (2023) The impact of strategic CSR on small holder farmers: a study of agri-input firms in India. In: Narula SA, Raj SP (Eds), Sustainable food value chain development: perspectives from developing and emerging economies, pp 205–217. Springer Nature Singapore. https://doi.org/10.1007/978-981-19-6454-1_10

Maury B (2022) Strategic CSR and firm performance: the role of prospector and growth strategies. J Econ Bus 118:106031. https://doi.org/10.1016/j.jeconbus.2021.106031

McWilliams A, Siegel DS (2011) Creating and capturing value: Strategic corporate social responsibility, resource-based theory, and sustainable competitive advantage. J Manag 37(5):1480–1495. https://doi.org/10.1177/0149206310385696

Meadows DH, Meadows DL, Randers J, Behrens WW (2018) The limits to growth. In: Green planet blues, pp 25–29. Routledge

Milliman J, Ferguson J, Sylvester K (2008) Implementation of Michael porter's strategic corporate social responsibility model. J Global Bus Issues Conf Edn 2:29–33

Miras Rodriguez M del M, Carrasco Gallego A, Escobar Perez B (2014) Corporate social responsibility and financial performance: a meta-analysis. Spanish J Finance Account-REVISTA ESPANOLA DE FINANCIACION Y CONTABILIDA 43(2):193–215

Mochales G, Blanch J (2022) Unlocking the potential of CSR: an explanatory model to determine the strategic character of CSR activities. J Bus Res 140:310–323. https://doi.org/10.1016/j.jbusres.2021.11.002

Moskowitz M (1972) Choosing socially responsible stocks. Bus Soc Rev 1:71–75

Ndhlovu TP (2011) Corporate social responsibility and corporate social investment: the South African case. J Afr Bus 12(1):72–92. https://doi.org/10.1080/15228916.2011.555264

Novita N (2012) Strategic corporate social responsibility and firm performance. Int Conf Bus Entrepreneurship Manag 2012:290–299

Nunnally JC, Bernstein IH (1994) Psychometric theory (3rd ed.). McGraw Hill

OECD (2011) OECD guidelines for multinational enterprises. http://www.oecd.org/daf/inv/mne/48004323.pdf

Ooi SK, Amran A, Yeap JAL (2017) Defining and measuring strategic CSR: a formative construct. Global Bus Manag Res 9:250–265

Orlitzky M, Schmidt FL, Rynes SL (2003) Corporate social and financial performance: a meta-analysis. Organ Stud 24(3):403–441

Othman S, Darus F, Arshad R (2011) The influence of coercive isomorphism on corporate social responsibility reporting and reputation. Social Respon J 7(1):119–135. https://doi.org/10.1108/17471111111114585

Oxfam International (2024) Extreme inequality and essential services. https://www.oxfam.org/en/what-we-do/issues/extreme-inequality-and-essential-services

Pérez A, Del Bosque IR (2015) Corporate social responsibility and customer loyalty: exploring the role of identification, satisfaction and type of company. J Serv Market

Pett MA, Lackey NR, Sullivan JJ (2003) making sense of factor analysis: the use of factor analysis for instrument development in health care research. SAGE Publications.https://doi.org/10.4135/9781412984898

Porter ME, Hills G, Pfitzer M, Patscheke S, Hawkins E (2012) Measuring shared value: how to unlock value by linking social and business results. FSG. Creative commons attribution

Porter ME, Kramer MR (2006) Strategy and society: the link between competitive advantage and corporate social responsibility. Harv Bus Rev 84(12):78–92. https://hbr.org/2006/12/strategy-and-society-the-link-between-competitive-advantage-and-corporate-social-responsibility

Porter ME, Kramer MR (2011) Creating shared value. Harv Bus Rev 89(1/2):62–77

Pratoomsuwan T, Chiaravutthi Y (2023) The interaction of corporate social responsibility (CSR) materiality and explicit assessment on willingness to invest: perception of financial analysts. Asian J Account Res 8(2):122–132. https://doi.org/10.1108/AJAR-01-2022-0012

Putri CNT, Masse FAU, Kesuma SA (2023) The implementation of mandatory or voluntary for corporate social responsibility: a systematic literature review. UTSAHA (J Entrepreneurship) 2(3):1–14. https://journal.jfpublisher.com/index.php/joe/article/view/337

Rim H, Ferguson MAT (2020) Proactive versus reactive CSR in a crisis: an impression management perspective. Int J Bus Commun 57(4):545–568. https://doi.org/10.1177/2329488417719835

Rodriguez-Gomez S, Arco-Castro ML, Lopez-Perez MV, Rodríguez-Ariza L (2020) Where does CSR come from and where does it go? A review of the state of the art. Administr Sci 10(3):60. https://doi.org/10.3390/admsci10030060

Salzmann O, Ionescu-Somers A, Steger U (2005) The business case for corporate sustainability: literature review and research options. Eur Manag J 23(1):27–36. https://doi.org/10.1016/j.emj.2004.12.007

Sanchez-Lopez JC, Bejarano JMG (2022) ¿Reconoce el mercado la responsabilidad social corporativa? Revista de Administração de Empresas, 62

Sarkar S, Searcy C (2016) Zeitgeist or chameleon? A quantitative analysis of CSR definitions. J Clean Prod 135:1423–1435. https://doi.org/10.1016/j.jclepro.2016.06.157

Sartor M, Orzes G, Di Mauro C, Ebrahimpour M, Nassimbeni G (2016) The SA8000 social certification standard: Literature review and theory-based research agenda. Int J Prod Econ 175:164–181. https://doi.org/10.1016/j.ijpe.2016.02.018

Schumpeter R (2011) Oh, Mr Porter–The new big idea from business's greatest living guru seems a bit undercooked. The Economist. http://www.economist.com/node/18330445
Sleibi R, Abdelkarim N (2021) Strategic corporate social responsibility and value creation: empirical study in Palestine 12(4):25–37. https://doi.org/10.5430/rwe.v12n4p25
Ślęzak M (2020) From traditional to strategic CSR: systematic literature review. J Corporate Respons Leadership 7(1):39–53
de Sousa-Filho JM, Wanderley LSO, Gómez CP, Farache F, de Sousa Filho JM, Wanderley LSO, Gómez CP, Farache F, de Sousa-Filho JM, Wanderley LSO, Gómez CP, Farache F (2010) Strategic corporate social responsibility management for competitive advantage. BAR-Brazilian Administr Rev 7(3):294–309. https://doi.org/10.1590/S1807-76922010000300006
Steiner GA (1971) Business and society. Random House Inc.
The Guardian (2023) Death Valley approaches global heat record as US reels from extreme weather. https://www.theguardian.com/world/2023/jul/17/death-valley-record-earth-temperature-california-extreme-weather
UN Global Compact (2021) Act globally engage locally. Colombia. https://www.unglobalcompact.org/engage-locally/latin-america/colombia
United Nations Development Program (2021) Sustainable Development Goals. https://www.undp.org/content/undp/en/home/sustainable-development-goals.html
Vishwanathan P, van Oosterhout H (J), Heugens PPMAR, Duran P, van Essen M (2020) Strategic CSR: a concept building meta-analysis. J Manag Studies 57(2):314–350.https://doi.org/10.1111/joms.12514
Vitolla F, Rubino M, Garzoni A (2017) The integration of CSR into strategic management: a dynamic approach based on social management philosophy. Corporate Govern Int J Bus Soc 17(1):89–116. https://doi.org/10.1108/CG-03-2016-0064
WBCSD (2024) World business council for sustainable development. https://www.wbcsd.org
Werner WJ (2009) Corporate social responsibility initiatives addressing social exclusion in Bangladesh. J Health Populat Nutrition, 545–562.https://doi.org/10.3329/jhpn.v27i4.3401
Wongpitch S, Minakan N, Powpaka S, Laohavichien T (2016) Effect of corporate social responsibility motives on purchase intention model: An extension. Kasetsart J Soc Sci 37(1):30–37. https://doi.org/10.1016/j.kjss.2016.01.010
Xiang G, He Z, Chen L (2024) How to implement strategic CSR? A mechanism oriented to corporate responsible competitiveness. J Adv Manag Sci 12(1):1–7. https://doi.org/10.18178/joams.12.1.1-7
Xue S, Chang Q, Xu J (2023) The effect of voluntary and mandatory corporate social responsibility disclosure on firm profitability: evidence from China. Pac Basin Financ J 77:101919. https://doi.org/10.1016/j.pacfin.2022.101919
Yousfi O, Loukil N (2021) Why should firms achieve strategic CSR? In: Ayhan F (Ed), Local governance and regional development: Current Perspectives. https://hal.science/hal-03145064/
Zafar F, Farooq M (2014) Corporate social responsibility in small and medium enterprises to achieve organizational benefits. Int J Sci Eng Res 5(1):2276–2283
Zeng SX, Liu HC, Tam CM, Shao YK (2008) Cluster analysis for studying industrial sustainability: an empirical study in Shanghai. J Clean Prod 16(10):1090–1097. https://doi.org/10.1016/j.jclepro.2007.06.004

Integrating Social Responsibility in Corporate Strategies to Promote Sustainability

Anubhav Roy and Priyanka Singh

Abstract Over the last few decades, one can observe a humongous number of business organizations undertaking several initiatives to benefit society. These initiatives can vary from something as simple as avoiding business malpractices to doing charity or helping the world combat global ecological issues. These companies undertake these activities for several reasons, such as a genuine willingness to aid people, to fulfil the compulsory CSR requirements, or, to build a good customer rapport in the eyes of the prospective buyers and the public in general. Another major concern about these activities is whether are they sustainable or not and if yes, how are they getting integrated with business operations to promote sustainability. Therefore, this research paper aims to find out the rationale behind the same, how it is being done, how it can be done, and what are the potential benefits to all the stakeholders and also to society. For the same, the research has been conducted by using secondary data in an extensive manner that was obtained from journals, blogs, research papers, articles, etc. Through the data gathered, the study delves into the multiple dimensions of CSR and its potential impact on sustainable development. Some of the key themes discussed ahead include risk management, enhancing corporate reputation and stakeholder engagement, etc. The study also dives into the opportunities and challenges attached with the integration of social responsibility with business functions. Ultimately, the findings depict the critical role of businesses in advancing sustainability goals and highlighting their positive impact.

Keywords Charity · Willingness · CSR · Customer rapport · Potential benefits · Corporate reputation

---

A. Roy (✉)
IMT Faridabad, MDU, Faridabad, India
e-mail: royanubhav123456@gmail.com

P. Singh
Department of Commerce, Maharaja Surajmal Institute, Janakpuri, New Delhi, India
e-mail: priyanka@msijanakpuri.com

P. Singh et al. (eds.), *Integrating Environmental Sustainability and Corporate Social Responsibility: A Move Beyond Profit*, World Sustainability Series,
https://doi.org/10.1007/978-3-031-81469-3_7

1 Introduction

Carrie Underwood has rightly stated that successful people have a social responsibility to make the world a better place and not just take from it. Indeed, society provides all the requirements of the people and that necessitates their moral obligation to give something in return to this society. In a time when social issues and environmental concerns are becoming more and more pressing, corporations must look beyond traditional profit-driven motives. With an emphasis on promoting an integrated strategy, this study explores the crucial connection between business activity, social responsibility, and sustainability. A rising number of companies are realizing that their operations impact society and the environment. Moreover, getting financially successful at the cost of social progress and ecological integrity is not a wise decision. The amalgamation of business activity and social responsibility is significant from the standpoint of India, as the country is experiencing swift economic growth and urbanization. Indian firms face numerous social concerns, from environmental degradation to financial disparity, forcing them to reevaluate their position within the societal framework. Concurrently, there has been a boom in the worldwide conversation about sustainability, which has forced companies to implement strategies that support more expansive environmental and social goals. Recent studies show that integrating CSR into core business strategies not only enhances brand reputation but also improves financial performance (Gupta 2023).

Using examples from India and worldwide, this study examines effective ways of incorporating social responsibility into corporate strategies. The article aims to comprehensively grasp the complex relationship between corporate conduct, social responsibility, and sustainable development by combining theoretical frameworks, empirical evidence, and case studies. For instance, Indian companies like Tata and Infosys have pioneered adopting sustainable practices, showing that responsible business can lead to long-term profitability (Kumar and Mehta 2024). Similarly, global corporations such as Unilever and Patagonia have set benchmarks for integrating sustainability into their business models, demonstrating that ethical practices can drive economic success (Rao and Singh 2024).

Furthermore, research indicates that companies that actively engage in CSR activities tend to have more motivated employees and higher customer loyalty, which are critical factors for sustaining business in competitive markets (Sharma and Patel 2023). By aligning business goals with environmental and social responsibilities, companies can impact the society positively along with long-term success (Das and Roy 2023). As we navigate an intricate era of corporate evolution, this study endeavors to find pathways for businesses to thrive economically and contribute substantively to the well-being of societies and the preservation of our planet.

The urgent need to address critical environmental and social issues spurred this research. Conventional profit-driven business models are getting harder and harder to maintain. The purpose is to investigate how incorporating social responsibility into a business strategy can benefit the environment and local communities. By showcasing effective case studies, it guides companies to match their operations with more general societal objectives. The ultimate objective is to encourage businesses to implement sustainable practices that can guarantee long-term prosperity for their organizations and the communities they serve.

To structure the research paper effectively, the subsequent chapters will explore various critical aspects of integrating social responsibility into corporate strategies. The literature review will synthesize existing theoretical frameworks and research findings on CSR integration. Objectives of the study will outline the specific aims of this research. The research methodology chapter will detail the methods used to gather and analyze data. The conceptual framework will present the theories guiding the study. Data analysis and interpretation will examine the collected data. The findings chapter will summarize the primary outcomes, while the conclusions will discuss their broader implications. Recommendations will offer actionable strategies for CSR integration. Finally, the paper will address the limitations of this research study and suggest areas for research as prospects.

2 Literature Review

Integrating social responsibility into corporate strategies is increasingly recognized as a vital approach to promoting sustainability. This literature review synthesizes findings from recent research, highlighting key theoretical frameworks, practical implementations, challenges, and benefits associated with this integration.

Several studies underscore the importance of several theoretical frameworks in guiding the integration of social responsibility into corporate strategies. Freeman's Stakeholder Theory emphasizes the requirement for businesses to consider the interests of all the stakeholders, including workers, consumers, suppliers, and society in general, to attain long-term success (Freeman 1984). The approach highlights the interconnectedness between a business and its broader environment. Similarly, Elkington's (1997) Triple Bottom Line (TBL) framework instructs businesses to measure their success by financial performance and also by their impact on society and the environment. This holistic perspective ensures a balance between economic growth, environmental stewardship, and social equity. Recent research by Banerjee et al. (2023) reinforces the relevance of TBL in the context of emerging markets, showing that firms adopting this framework can achieve better overall performance. Porter and Kramer's (2011) Creating Shared Value (CSV) concept further support the idea that business success and social progress are interdependent, proposing that companies can create economic value through means that also enhances societal value by addressing societal needs and the problems. Kumar and Mehta (2024) provide contemporary evidence that CSV initiatives lead to innovative business models that drive both profitability and social impact.

Practical implementations of these frameworks are varied and widespread. Sustainable supply chain management is a common approach, where companies incorporate ethical sourcing, reduce waste, and minimize environmental impacts throughout their supply chains. Carter and Rogers (2008) illustrate how such sustainable practices can enhance both environmental and economic performance. Recent findings by Rao and Singh (2024) indicate that companies in the Global South are increasingly adopting sustainable supply chain practices to improve their competitive advantage and sustainability credentials. Another significant approach is green innovation, where companies invest in green technologies and innovative practices to develop products and processes that minimize environmental harm. Schiederig et al. (2012) highlight the critical role of green innovation in achieving sustainability goals. Recent advancements by Gupta (2023) show that companies that prioritize green innovation see significant gains in market share and customer loyalty. Effective corporate governance structures also play a crucial role, as they are essential for embedding sustainability into business strategies. Eccles et al. (2014) emphasize the importance of transparency, accountability, and long-term planning in sustainable corporate governance. Das and Roy (2023) further argue that robust governance mechanisms are pivotal in aligning corporate activities with broader sustainability goals.

However, the integration of social responsibility into corporate strategies is not without its challenges. One significant challenge is the cost and resource allocation required for implementing sustainable practices. Epstein and Buhovac (2014) note that the substantial initial investment needed can be a significant barrier for many companies. Additionally, there is a lack of standardized metrics for measuring and reporting CSR activities, which makes it difficult for companies to demonstrate their impact and progress. Searcy (2012) discusses the challenges associated with developing comprehensive performance measurement systems for corporate sustainability. Recent research by Sharma and Patel (2023) highlights ongoing difficulties in establishing universally accepted CSR metrics. Furthermore, internal cultural resistance and a lack of understanding of sustainability can hinder the integration of CSR into corporate strategies. Lozano (2015) underscores the importance of fostering a corporate culture that supports sustainability initiatives. New insights by Rao and Singh (2024) suggest that leadership commitment and employee engagement are crucial in overcoming internal barriers to CSR integration.

Despite these challenges, the advantages of combining social responsibility into corporate activities are substantial. Corporations that actively engage in CSR activities often enjoy enhanced brand reputations, leading to increased customer loyalty and trust. Fatma et al. (2015) explore how CSR activities can build a company's image and brand equity. Additionally, sustainable practices can also lead to improved operational efficiencies, for example: reducing waste and consumption of various sources of energy, that can further lower the costs in the long run. King and Lenox (2001) demonstrate the positive relationship between lean production and environmental performance. Recent studies by Gupta (2023) confirm that companies investing in sustainable operations achieve significant cost savings and enhanced efficiency. Moreover, integrating CSR helps companies manage risks related to environmental regulations, social unrest, and market volatility. Orlitzky et al. (2003) provide evidence of the positive impact of corporate social and financial performance. Recent case studies by Kumar and Mehta (2024) illustrate that proactive CSR strategies can mitigate regulatory and market risks effectively. Several case studies illustrate the successful integration of CSR into corporate strategies. For instance, Unilever's Sustainable Living Plan integrates sustainability into its core business model, focusing to improve the health and well-being, along with reducing environmental impact, and upgrading livelihoods (Unilever n.d.). Similarly, Patagonia has embedded sustainability into its mission, emphasizing environmental responsibility through initiatives like the Worn Wear program and sustainable sourcing (Chouinard et al. 2011). New research by Das and Roy (2023) showcases additional successful examples from various industries, highlighting the scalability and adaptability of CSR strategies across different business contexts.

3 Objectives of the Study

The motives of this research paper are to identify several potential benefits that businesses get by integrating social responsibility into corporate activities, such as improved brand reputation and operational efficiency. It also seeks to tackle the challenges associated with these practices, such as cost and resource allocation and the lack of standardized metrics for CSR reporting. Additionally, the study aims to explore the opportunities that businesses can leverage for sustainability, such as risk management and market advantages.

Finally, it intends to analyze the trend of corporate social responsibility (CSR) activities at both national and global levels, providing insights into the evolving landscape of CSR initiatives.

4 Research Methodology

Research Design adopted for this research paper is Descriptive and Analytical. The data has been sourced from several secondary sources such as webpages, journals, research papers, blogs, etc. Also, the data has been collected by using Derived/ Compiled data and survey data. Lastly, this research includes both, quantitative and qualitative data.

5 Conceptual Framework

5.1 *What is Corporate Social Responsibility and Its Types?*

According to the theory of corporate social responsibility, or CSR, businesses should operate according to principles and values that advance society and the environment. Beyond just following the law, CSR involves businesses making a conscious decision to follow morally justified, sustainable towards the society, and accountable business practices to benefit society, employees, stakeholders, and consumers. The following are the main areas of CSR:

Environmental responsibility: Taking part in CSR initiatives to reduce climate change and preserve natural resources. CSR encourages sustainability by employing eco-friendly techniques including reducing waste, using renewable resources, and using less energy.

Ethical Responsibility

CSR initiatives usually center on human rights and social impact issues, such as making sure that workers receive fair wages, have safe working conditions, and are treated well by suppliers and employees. They also encourage responsibility from the inside out. A few instances of ethical CSR include supply chain transparency, the elimination of workplace discrimination, and adherence to fair labor practices.

Philanthropic responsibility: Donating cash, materials, or time to worthy causes and organizations, such as national and local disaster relief efforts, educational initiatives, and charities. Businesses that engage in charitable corporate social responsibility (CSR) give back to the communities where they operate by funding skill-development programs, volunteering, sponsoring local events, and donating to local non-profits.

Economic responsibility: Businesses set policies and procedures to ensure that their decisions align with their core values, even when there are more profitable or cost-effective options. Economic CSR also entails activities to boost the development and prosperity of the communities in which a business operates (MacGrath and Jonker 2023).

5.2 *Historical Context/Background of Corporate Social Responsibility*

CSR has evolved significantly over the past century. In the early 1900s, industrialists like Andrew Carnegie and John D. Rockefeller began philanthropic efforts, hinting at the idea that businesses have responsibilities beyond profit (Carroll 1999). However, it wasn't until Howard R. Bowen's (1953) work, "Social Responsibilities of the Businessman," that CSR gained formal recognition (Carroll 2008).

The 1960s and 1970s saw a surge in public awareness due to the civil rights movement, environmentalism, and consumer advocacy. This led to the creation of regulatory bodies like the Environmental Protection Agency (EPA) in 1970, emphasizing the need for environmental responsibility in business (Vogel 2005).

By the 1980s and 1990s, CSR began integrating into corporate strategies. Companies started adopting voluntary codes of conduct and reporting standards. The Global Reporting Initiative (GRI), introduced in 1997, provided guidelines for businesses to disclose their social and environmental impacts (Kolk 2008).

In the twenty-first century, CSR has continued to evolve with a focus on sustainability and corporate governance. The United Nations' Sustainable Development Goals (SDGs), introduced in 2015, have reinforced the importance of aligning business strategies with global sustainability objectives (Montiel and Delgado-Ceballos 2014). Recent studies show a trend of integrating CSR into core business operations driven by stakeholder expectations and regulatory pressures (Lee 2021; Jamali and Karam 2018).

5.3 Potential Benefits of CSR

Enhanced Brand Reputation and Customer Loyalty

One of the most significant benefits of CSR is enhancing a company's brand reputation. Engaging in socially responsible activities helps build a positive image, which in turn increases customer loyalty and trust. Fatma et al. (2015) found that CSR initiatives majorly contribute to building a corporation's image along with its brand equity by fostering trust among consumers. When consumers perceive a corporation to have social responsibility, they would like to support its products and services more, which will eventually lead to increased customer loyalty.

Operational Efficiency and Cost Savings

Implementing CSR practices often leads to improved efficiency in the operations. Sustainable initiatives like waste reduction and conservation of energy, can lead to major cost savings for companies. King and Lenox (2001) demonstrated that firms adopting lean production methods, which emphasize waste reduction and efficiency, often experience enhanced environmental performance and cost savings. These operational efficiencies not only benefit the company financially but also reduce its environmental footprint.

Risk Management and Regulatory Compliance

CSR initiatives can play an important role in risk management and regulatory compliance. Companies that address social and environmental issues proactively are positioned to anticipate and respond to regulatory changes and potential legal challenges. Orlitzky et al. (2003) conducted a meta-analysis showing that companies with robust

CSR practices tend to have better financial performance due, in part, to their effective risk management strategies. By adhering to higher standards of social and environmental responsibility, companies can mitigate risks associated with regulatory non-compliance and reputational damage.

Attracting and Retaining of Talent

An impactful CSR strategy is able to enhance a corporation's ability of attracting and retaining the best talent. Human resources are excessively seeking superiors whose values align along with theirs, and they prefer to work for companies that have a strong commitment towards social and environmental responsibility. Research by Greening and Turban (2000) indicates that companies with strong CSR practices are more attractive to prospective employees and are more likely to retain their workforce. This not only reduces recruitment and training costs but also fosters a motivated and engaged workforce.

Accessing the Capital

Companies with an impactful CSR strategy usually find it easy in accessing the capital. Investors are progressively taking into consideration aspects like environmental, social, and governance (ESG) criteria while taking investment decisions. Eccles et al. (2014) found that companies with high sustainability performance have a higher probability of attracting investment from socially responsible investors. This access to capital can support further innovation and growth, reinforcing the company's competitive advantage.

Competitive Advantage and Market Differentiation

CSR can provide a significant competitive advantage by differentiating a company from its competitors. Corporations that combine CSR into their business activities are able to create unique selling points that resonate with consumers, investors, and other stakeholders. Porter and Kramer (2011) have argued that creating a shared value through addressing such issues can drive innovation and open up new market opportunities. This differentiation helps companies stand out in the marketplace and build long-term sustainable success.

Long-Term Sustainability and Growth

Ultimately, the integration of CSR into corporate strategies promotes long-term sustainability and growth. By addressing the needs and expectations of various stakeholders, companies can build resilient and sustainable business models. Lozano (2015) highlights that a holistic approach to sustainability, which includes social responsibility, ensures that companies are better equipped to navigate the challenges and opportunities of the future.

5.4 *How Does the Integration of Social Responsibility in Corporate Strategies Promote Sustainability?*

1. **Environmental Stewardship**: CSR initiatives often focus on reducing a company's ecological footprint. This may involve implementing sustainable practices, reducing energy consumption, and adopting eco-friendly technologies to contribute to environmental conservation.
2. **Social Responsibility**: Companies engage in CSR by supporting local communities through various social programs. This could include initiatives like education and healthcare projects, infrastructure development, and job creation, promoting social well-being and economic development.
3. **Ethical Business Practices**: CSR pushes companies to maintain moral principles in their operations. This covers ethical labor standards, conscientious material sourcing, and open communication. Companies promote ethical behavior as a means of establishing trust with stakeholders.
4. **Philanthropy and Charity**: Many CSR programs involve financial contributions to charitable organizations and community projects. This direct involvement in philanthropy supports various social causes, addressing issues such as poverty, education, and healthcare.
5. **Stakeholder Engagement**: CSR highlights how crucial it is to consider the interests of all the stakeholders involved, such as clients, staff, vendors, general public, etc. Interacting with these groups promotes goodwill and guarantees that a wider variety of viewpoints are taken into account when making business decisions.
6. **Long-Term Value Creation**: Sustainable practices promoted through CSR contribute to long-term business success. By investing in sustainable technologies and practices, companies can enhance efficiency, reduce costs, and create value for both the business and society.
7. **Risk Management**: CSR strategies often involve identifying and mitigating potential risks associated with social and environmental issues. Proactive risk management helps companies avoid negative impacts on their reputation and financial performance.
8. **Compliance to the regulations**: Adhering to Corporate Social Responsibility principles will ensure compliance with the ever-changing environmental and social regulations. Such a proactive approach helps corporations stay ahead of the legal requirements. It also positions them as responsible corporate citizens.
9. **Employee Engagement and Well-being**: CSR initiatives positively impact employees by fostering a feeling of objective and pride in their workplace. This further leads to increased worker satisfaction, enhanced productivity, employee retention, etc. contributing to a sustainable corporate culture.
10. **Innovation for Sustainability**: Embracing CSR often drives innovation toward more sustainable business practices. Companies may like investing in R&D to develop eco-friendly products or find innovative solutions to reduce environmental impact.

5.5 *Link Between CSR & SDGs*

On September 25, 2015, in Paris, 193 nations, including India, endorsed and signed the Sustainable Development Goals (SDGs). The aim of this convention is to achieve 17 out of the 169 SDGs by 2030. These goals adopt a practical approach to foster sustainable development across the social, economic, and environmental dimensions of the planet. The SDGs are designed to be flexible, allowing nations and organizations to create their own action plans to meet these objectives, setting a roadmap for the next twelve years. India committed to the SDGs by integrating them into its national development plan and establishing a monitoring dashboard within Niti Aayog.

An examination of the eleven CSR areas outlined in Schedule VII of section 135 of the amended Companies Act of 2013 shows their alignment with multiple SDGs. For example, CSR area #1, which aims to eliminate poverty, hunger, and malnutrition, and to promote preventive healthcare and sanitation, corresponds with SDG goals #1 (No Poverty), #2 (Zero Hunger), #3 (Good Health & Well-Being), and #6 (Clean Water & Sanitation).

CSR area #2, focused on promoting education and vocational skills, including special education for women, the elderly, and people with disabilities, as well as livelihood enhancement, aligns with SDGs #1 (No Poverty), #2 (Zero Hunger), #4 (Quality Education), and #8 (Decent Work and Economic Growth).

CSR area #3, which is dedicated to empowering women, promoting gender equality, providing housing for women, orphans, and the elderly, and establishing daycare centers for children, also aims to reduce inequalities faced by socially and economically disadvantaged groups. This area aligns with SDGs #1 (No Poverty), #5 (Gender Equality), and #10 (Reduced Inequalities).

5.6 *How Corporations Can Undertake CSR Activities?*

Following are the ways how corporations/businesses can go for sustainable activities through CSR:

Volunteering in the community: One of the best things that companies can do is to invest in the community. Volunteering is an activity that will help establish a meaningful connection with the local community, regardless of industry or specialty.

In addition to fostering the development of critical soft skills and strengthening team dynamics, volunteering will provide staff members with an additional sense of purpose in their work.

Here Are Some Examples

- Local campaigns to pick up litter and recycle
- assisting with neighborhood events in the community
- planting shrubs, plants, trees, or flowers

- delivering food and meals or serving as a volunteer at food or clothing banks

Joining forces with a nonprofit or charity: Working with a nearby charity is another CSR initiative that will broaden a company's reach and have a very positive effect on the world.

Developing a connection with a non-profit or charity that is associated with the company or sector will guarantee that they are continuously striving to create a better future.

Employees will be encouraged to do good deeds by participating in a variety of fundraising initiatives through a charity partnership. To make mealtimes a little bit easier, McCain, a frozen food company in the UK, has partnered with the Family Fund, a charity that enhances the lives of children who are disabled or very ill.

Adopting appropriate sustainability actions: It's widely acknowledged that our planet is facing numerous environmental challenges, such as deforestation and global warming, and it is imperative for society to collaborate in finding solutions. In this context, one of the most impactful actions a contemporary business can take is to invest in environmental initiatives or sustainability measures. Here are some potential environmental CSR initiatives:

- Using social media and blogs to spread inspiring words and environmentally friendly advice.
- Making a policy for workplace sustainability that describes initiatives related to paper waste, energy consumption, composting, and recycling and offers incentives for adopting more environmentally friendly forms of transportation.
- Introducing schemes like the "Cycle to Work Scheme."
- Reducing carbon footprint by using more environmentally friendly suppliers and materials.

Updating the regulations in the workplace: Investing in CSR projects will enhance corporate culture. Consequently, businesses will improve employee well-being, which is good for business. Reworking workplace policies should involve everyone in the organization so that they can decide what programs will best support the health and wellness of their workers. Several crucial CSR areas, including fitness and well-being, internal values, social and team bonding, diversity and inclusion, and benefits and incentives, should be considered when developing workplace policies. Here are some CSR initiatives or pursuits to think about:

- Enhanced parental leaves and leave for religious or cultural reasons.
- Frequent social gatherings and team-building exercises.
- Putting in new furnishings, office designs, and breakout spaces to create a more airy and motivating work environment.
- Organizing projects and events that will have a major positive impact on workers' health and happiness. (United Heroes 2022)

5.7 *Some Real-Life Examples of How Some Companies Integrated CSR into Their Practices*

1. **TOMs**: TOMs, a Certified B Corporation, adheres to high standards for environmental conservation, social responsibility, and accountability. The company has woven CSR into its brand's core. From 2018 to 2021, TOMs significantly boosted its score from 96.3 to 121.5, well above the average business score of 50.9. TOMs implements a diverse CSR strategy, organizing its efforts into three distinct categories:

   - Purpose: By assisting local nonprofits, TOMs seeks to make the world more equitable. As a result, TOMs gives away a third of its earnings to charitable groups that support mental health, expand access to opportunities, and put an end to gun violence.
   - Planet: TOMs aims to enhance its enterprise's ecological sustainability through various approaches, particularly by utilizing reusable and recycled materials. By 2025, TOMs wants all of its packaging to be made entirely of sustainably grown cotton, up from the current 80% made of recycled materials.
   - People: Apart from contributing to charitable organizations, TOMs seeks to enhance its internal operations through the promotion of its anti-racist stance, the organization of volunteer events on Giving Tuesday, and collaboration with suppliers who align with its principles.

2. **PayPal**: Through the PayPal Giving Fund, PayPal's payment processing services leverage technology to increase donations to charities. The four steps below are how the PayPal Giving Fund assists nonprofits:

   - Nonprofit organizations can set up a PayPal account to take advantage of several benefits. Registered charities enjoy reduced processing fees and gain access to donation opportunities on PayPal and its partner sites, such as GoFundMe, eBay, and Humble Bundle.
   - Donations are made by supporters using eBay, PayPal, or another PayPal partner. Donations to nonprofit organizations can be made even more impactful by the fact that many PayPal partners will even pay the processing fees.
   - Donors receive a receipt from PayPal after their donation is received. PayPal will use this data to generate monthly donation reports for participating charities.
   - The charity's PayPal account is credited with the funds. Charities can use PayPal to receive grants and significant gifts in addition to gifts from individual donors.

3. **Toast**: Toast, a restaurant point-of-sale software provider, established the Toast.org foundation. Toast's corporate social responsibility program has a strong connection to the food industry, much like its products do. Using its foundation,

Toast seeks to tackle hunger and facilitate the availability of nutritious food. Additionally, Toast invites its patrons and staff to join it in this endeavor. Furthermore, Toast offers features in its software that help restaurants monitor and cut down on food waste.

4. **Unilever**: Unilever owns numerous renowned brands, including Ben & Jerry's, Dove, and Knorr, in both the ice cream and personal care sectors. As a substantial conglomerate with many successful businesses, Unilever has the resources to make a considerable impact with its CSR initiatives. The Unilever Compass social good program focuses on the following key areas for CSR efforts.
    - Changes in climate
    - Environmental preservation and reforestation
    - Minimizing waste
    - Uplifting the living standards of persons worldwide
    - Human Rights
    - Corporate accountability and transparency

"Future of Work" is a distinctive CSR initiative by Unilever. In response to advancements in automation and artificial intelligence, the skill sets required for workers to remain competitive in the job market are swiftly changing. In partnership with UNICEF, Unilever's "Future of Work" program seeks to assist 10 million young people, aged 15 to 24, in gaining the necessary skills for employment. Additionally, the program focuses on reskilling or upskilling existing employees across its businesses to better prepare them for the digital era. (Weigner A.)

5.8 Regulations Based on CSR in INDIA

Under section 135 of the Companies Act, 2013 ("Act"), specific companies are required to allocate a prescribed amount for corporate social responsibility (CSR) activities. "Corporate Social Responsibility," as outlined by the Act, includes, but is not limited to, the following:

- Initiatives or plans related to the activities listed in Schedule VII of the Act.
- Projects or programs overseen by the board of directors to ensure the CSR Committee adheres to the stated CSR Policy, provided the policy covers the topics outlined in Schedule VII of the Act.

The CSR regulations apply to any company that satisfied any of the following criteria in the most recent fiscal year:

- A net worth exceeding 500 crores INR.
- Revenue surpassing 1,000 crores INR.
- Net profit exceeding 5 crores INR.

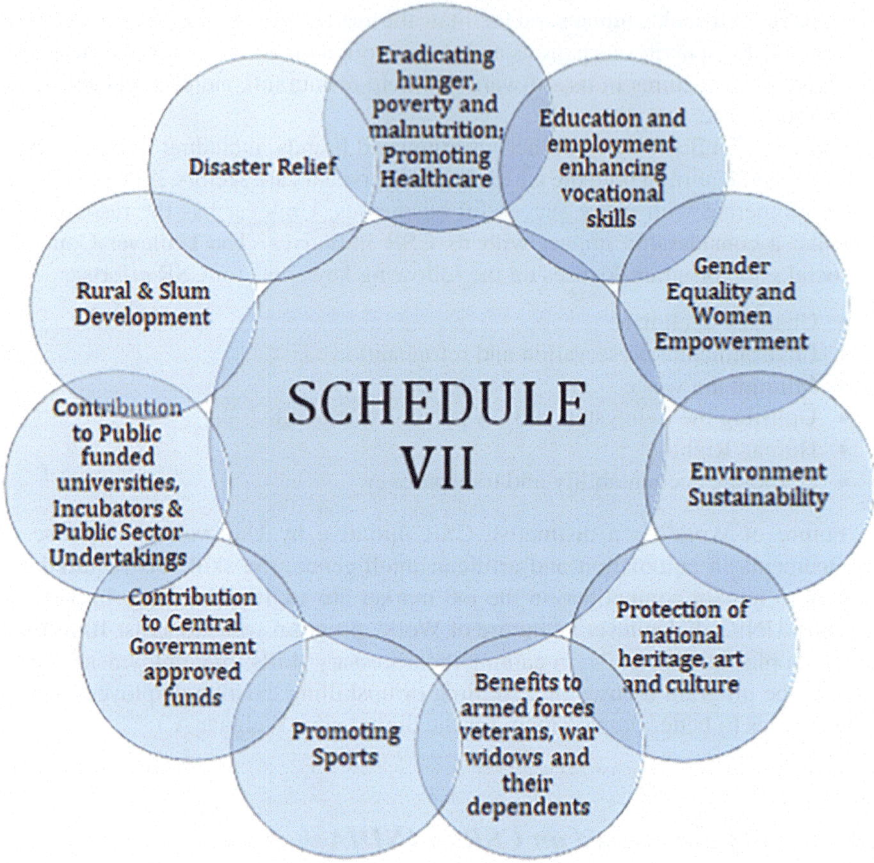

Fig. 1 List of activities under Schedule VII. *Source* CSR Services for Companies Applicable Under the Companies Act, 2013 (compliancecalendar.in)

Any company subject to the CSR provisions must have a Board of Directors that pledges to uphold the company's CSR policy and allocate at least 2% of its average net profits from the three most recent fiscal years to CSR initiatives each fiscal year. If the three financial years before incorporation haven't passed, the company must allocate 2% of the average net profits from the available years to CSR initiatives. (Acharya 2023) (Fig. 1).

5.9 *Some Popular CSR Campaigns in INDIA*

- **ITC**: For a considerable amount of time, ITC has carried out important CSR work in India. As of 2019, ITC's 2,334 Supplementary Learning Centers served

2,52,329 children in its educational program. Additionally, the support of 1183 self-help groups has empowered roughly 15,378 women members.

- **Reliance**: Reliance has taken proactive steps to ensure the safety of villagers in the Balangir district by introducing a life insurance program. Additionally, the company launched the "Education for All" initiative, which focuses on guaranteeing that every Indian has access to quality education. Their efforts extend to safeguarding the educational rights of underprivileged children, women, and individuals with disabilities, and they also contribute to disaster relief funds.
- **P&G**: A key part of Procter & Gamble's CSR strategy is the "Shiksha—Padhega India, Badhega India" initiative. Through this program, approximately 280,000 disadvantaged children have benefited, gaining their educational rights. Moreover, P&G has constructed and funded over 140 schools across India.
- **DLF**: The Swapana Sarthak Informal School was established to educate the children of construction workers. Under the DLF LIFE banner, the company has also set up centers offering vocational training to promote employability. DLF Inspire was also created to uplift the lives of children from low-income backgrounds.
- **CCD**: Cafe Coffee Day, a well-known coffee shop, employs 50 individuals with speech and hearing impairments across its locations. This initiative is part of their CSR program and a commitment to equal employment opportunities. They refer to these employees as the "Silent Brew-masters." Additionally, CCD partners with Enable India, a nonprofit organization, to empower disabled individuals. Their primary CSR project is at the SVGH Vocational Training College in Chikmagalur, Karnataka.
- **Tata Steel**: For over a century, Tata Steel has been dedicated to social progress in Jharkhand. The agenda of the CSR committee revolves around enhancing the quality of life in the communities it serves. They also introduced the "Lifeline Express" in 1991, a hospital train that provides medical supplies to impoverished areas across India. Furthermore, Tata Steel has conducted various campaigns to spread awareness about AIDS.
- **Amul**: Over the years, Amul has actively promoted rural development and health through the Tribhuvandas Foundation. They also initiated the comprehensive Swarnjayanti Gram Swarozgar Yojana in the Kheda district, encouraging self-employment. Additionally, they oversee blood donation drives, tree-planting campaigns, and rural sanitation projects. Amul also provides scholarships through the Amul Scholar program (Chandra 2019).

5.10 Challenges in CSR Implementation

The absence of a clear framework for implementing and assessing corporate social responsibility (CSR) initiatives poses a significant challenge for businesses. Unlike other business functions like marketing or finance, corporate social responsibility lacks a standardized framework that businesses can easily adopt. Consequently, businesses often need to develop their own metrics and strategies for CSR, which can be both costly and time-consuming. Additionally, assessing the impact of CSR initiatives can be complex due to their long-lasting effects and the difficulty in quantifying them.

A related issue is the lack of transparency and accountability in CSR practices. For successful implementation of CSR, companies must be forthcoming about their environmental and societal impact, including the steps they take to mitigate any negative effects. However, many companies fail to disclose comprehensive information regarding their CSR efforts, leading to skepticism and uncertainty regarding their commitment to environmental and social responsibility. To overcome these challenges, businesses can establish robust CSR management systems that provide comprehensive data about their initiatives and outcomes. Engaging with stakeholders, such as employees, customers, and investors, for feedback on CSR initiatives can also be beneficial. Another difficulty lies in balancing the need for sustainable, long-term CSR initiatives with the immediate financial concerns. In the dynamic business environment today, companies must continue to generate profits and create value for stakeholders. As a result, some businesses may be reluctant to allocate funds for CSR projects that may not yield immediate financial returns (Karmakar 2022).

6 Data Analysis and Interpretation

According to the India Philanthropy Report 2023 by Dasra and Bain & Co., CSR has increased at a rate of 15% annually over the last seven years, propelled by the 2% mandate. In fiscal year (FY) 2021, it accounted for 23% of all private giving, up from approximately 12% in FY 2015. This suggests that an increasing number of businesses are launching CSR projects, which ultimately help the economy alone. (HT Correspondent 2023) Additionally, the graph attached below (Fig. 2) shows that from the FY 2014–15, the CSR spending has nearly doubled in FY 2018–19. This again is an indicator of increased CSR compliance. Certainly, based on the upward trend, it can be easily estimated that by the current financial year (2023–24), CSR spending must have been increased to an even higher level. (India CSR 2019).

Figure 3 shows the top 10 companies who had exceeded their prescribed CSR spending in the FY 2020–21. As can be seen, some of them had manyfold actual spending too.

6.1 Trends of CSR in India

- Approximately 65% of companies have surpassed their annual mandated budget for corporate social responsibility.
- During the fiscal year, 23% of all CSR funds were allocated to public sector projects.

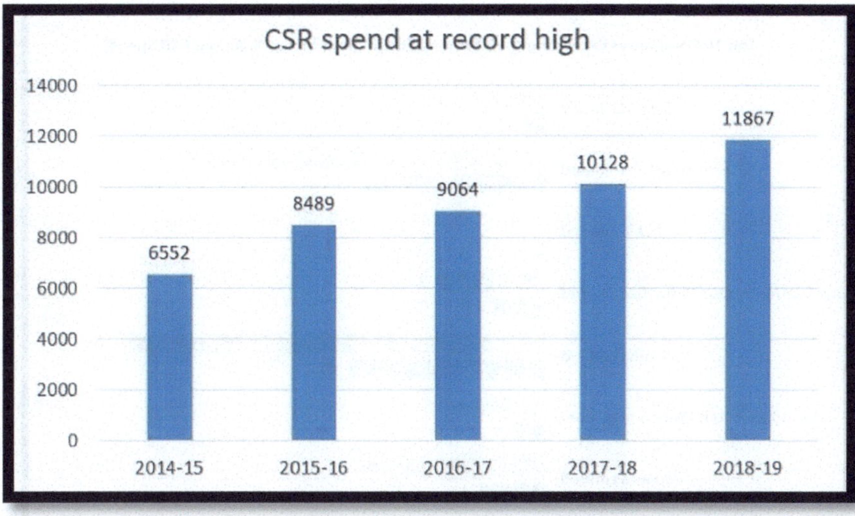

Fig. 2 Graph of CSR spending in India from 2014–15 to 2018–19. *Source* https://indiacsr.in/as-indias-csr-spending-touchs-rs-12000-cr-investigators-stalk-ghost-beneficiaries/

- Reliance Industries, Tata Consultancy Services, HDFC Bank, and ONGC are among the top 5 Indian companies based on actual CSR expenditures.
- Of these, over 40% have their headquarters in Maharashtra, with 10% each in Karnataka and Delhi.
- These five companies collectively account for more than 25% of all CSR expenditures. Maharashtra, Gujarat, and Odisha received nearly one-fifth of the total CSR funds in India.
- Seventy percent of India's CSR budget is allocated to the four primary theme areas of poverty alleviation, healthcare & WASH, education & skills, and rural development.
- Implementing agencies have completed about half of all CSR projects.
- Around 8% of CSR initiatives are conducted in collaboration with governmental bodies. (CSR BOX. Abridged Version (November 2021) India CSR Outlook Report, CSR Analysis of large 301 Listed companies (FY 2020–21).)

6.2 Notable Global Leaders in the Corporate Social Responsibility

1. LG H&H Co., Ltd. (Republic of Korea): Tops the list with an impressive S&P Global ESG Score.
2. Klabin S.A. (Brazil): Known for its commitment to sustainability in the container & Packaging industry.

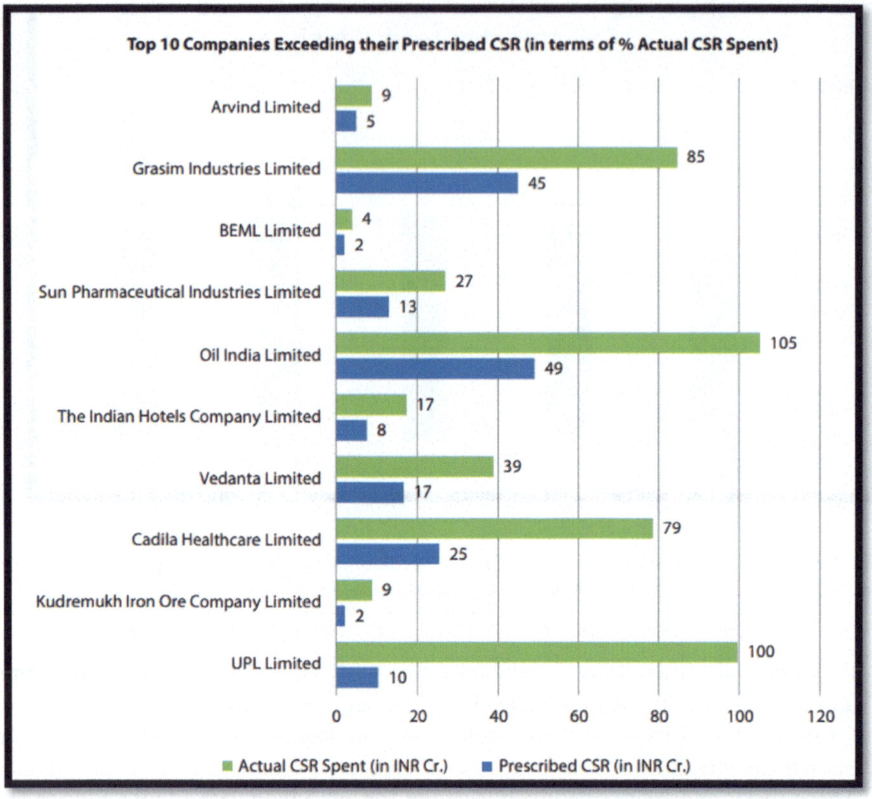

Fig. 3 Top 10 Companies exceeding their prescribed CSR (2021 data). *Source* https://csrbox.org/media/CSR-in-India-2021-CSR-Report-by-CSRBOX.pdf

3. Indra Sistemas, S.A. (Spain): A leader in providing IT services with a strong ESG focus.
4. Coca-Cola HBC AG (Switzerland): An industry mover in the beverage sector.
5. Deutsche Telekom AG (Germany): A telecommunications giant with a stellar ESG track record.
6. Reckitt Benckiser Group plc (United Kingdom): Known for its contributions in the Household Products category.
7. PTT Global Chemical Public Company Limited (Thailand): A standout in the Chemicals industry.
8. Galp Energia, SGPS, S.A. (Portugal): A leader in Oil & Gas Upstream & Integrated.
9. Moncler S.p.A. (Italy): Recognized for its sustainable practices in Textiles, Apparel & Luxury Goods.

10. Terna—Rete Elettrica Nazionale Societa per Azioni (Italy): An electric utilities company with a strong ESG profile. (S&P Global, Sustainability Yearbook 2023)

6.3 *Quantitative and Qualitative Trends in CSR*

Quantitative Trends Analysis

- **Big Data and Analytics**: Big data and analytics have revolutionized the study of CSR trends. Companies and researchers now utilize vast datasets from social media, financial reports, and third-party ratings to identify patterns and predict future trends (Li et al. 2023). For example, a study analyzing 10,000 CSR reports from 200 multinational corporations found a 35% increase in environmental initiatives from 2018 to 2022.
- **Statistical Techniques**: Advanced statistical methods, such as regression analysis and time-series analysis, are employed to understand the trajectory of CSR initiatives and their impact over time. A recent analysis using time-series data from 150 Fortune 500 companies showed that those with strong CSR practices had a 20% higher stock performance compared to those with weaker CSR activities (Hair et al. 2023).

Qualitative Trends Analysis

- **Content Analysis**: This method involves analyzing the content of CSR reports, press releases, and other corporate communications to identify recurring themes and narratives. For instance, a thematic analysis of 500 CSR reports found that terms related to "sustainability" and "community engagement" increased by 40% between 2019 and 2023 (Miles et al. 2020).
- **Case Studies**: In-depth case studies of leading companies provide insights into best practices and innovative approaches in CSR. A case study on Unilever's Sustainable Living Plan revealed a 50% reduction in their environmental footprint and a 75% increase in social investment from 2015 to 2023 (Bansal and Song 2023).

6.4 *Some Figures that Indicate the Benefits of CSR Integration*

- **Alignment with Corporate Strategy**: Recent studies show that companies increasingly integrate CSR into their core strategies, aligning social and environmental goals with business objectives to achieve sustainable competitive advantage. According to a survey of 250 CEOs, 85% reported that CSR activities are now a central part of their strategic planning (Lee 2021).

- **Performance Metrics**: The development and use of comprehensive performance metrics, such as the Balanced Scorecard, help companies measure the impact of CSR activities on overall business performance. Companies that implemented these metrics reported a 30% improvement in stakeholder trust and a 25% increase in long-term profitability (Bansal and Song 2023).
- **Supply Chain Management**: Companies are incorporating CSR principles into their supply chain management, ensuring ethical sourcing, reducing environmental impact, and improving labor conditions. For example, a study found that companies with sustainable supply chains saw a 15% reduction in operational costs and a 20% increase in brand loyalty (Montiel and Delgado-Ceballos 2014).
- **Corporate Governance**: Enhanced governance structures, including CSR committees and dedicated sustainability officers, ensure that CSR initiatives are effectively managed and integrated across all levels of the organization. A report showed that firms with dedicated CSR committees had 25% higher CSR performance scores compared to those without such structures (Pang, Zhang, and Tan 2022).

6.5 Top Regions of the World in Terms of CSR

- Europe: European businesses actively participate in social responsibility and sustainable practices, as evidenced by their CSR rating of 55.
- South America: With a CSR rating of 55, South American businesses likewise receive high marks.
- With a CSR rating of 54, Southeast Asia demonstrates a dedication to ethical business practices.
- North America: North American companies have a CSR rating of 52, despite the region having fewer countries overall.
- South Asia: With a score of 53, businesses there demonstrate a high commitment towards corporate social responsibility.
- Asia: With South and Southeast Asia included, the region as a whole continues to have a CSR rating of 48.
- Pacific: The CSR rating of businesses in the Pacific area is 51.
- Caribbean: Caribbean businesses actively engage in CSR initiatives, scoring a 50.
- Middle East: The CSR rating of Middle Eastern businesses is 49.
- Africa: African businesses have a CSR rating of 51 and support social impact and sustainable development. (CSR HUB)

7 Findings

Findings from the analysis of Corporate Social Responsibility (CSR) trends reveal significant shifts in both quantitative and qualitative approaches. Big data and analytics have been pivotal, showing a notable 35% increase in environmental initiatives among multinational corporations from 2018 to 2022. Statistical methods underscored a 20% higher stock performance for firms with robust CSR practices. Qualitatively, thematic analyses highlighted a 40% rise in terms like "sustainability" and "community engagement" in CSR reports. Case studies further illustrated impactful strategies, such as Unilever's achieving a 50% reduction in environmental footprint and a 75% increase in social investment by 2023. Strategic integration of CSR into business strategies has become prevalent, with 85% of CEOs emphasizing its central role, while metrics like the Balanced Scorecard correlated CSR efforts with a 30% boost in stakeholder trust and 25% higher long-term profitability. Operational integration through sustainable supply chains has yielded a 15% cut in operational costs and a 20% rise in brand loyalty. Enhanced corporate governance structures have similarly enhanced CSR performance by 25% compared to peers without dedicated CSR committees (Bansal and Song 2023; Hair et al. 2023; Lee 2021; Li et al. 2023; Miles et al. 2020; Montiel and Delgado-Ceballos 2014; Pang et al. 2022).

Additionally, based on the conceptual framework and Data Analysis and Interpretation done in the above section, it is found that there are a large number of benefits that can be availed by the owners, customers, economy, and all the other stakeholders. This makes it extremely crucial to combine CSR initiatives with corporate activities. Despite the numerous benefits, the CSR sector has been facing varying challenges such as balancing the need for long-term growth with social welfare, the lack of transparency, and also the absence of well-defined for carrying out and evaluating. Under the conceptual framework, numerous opportunities have been found for businesses that can avail towards sustainability such as joining forces with non-profit organizations or charities, adopting appropriate sustainability actions, volunteering in the community, and updating the regulations in the workplace. For trends at the National and Global levels, it has been found that in India, the majority of companies are exceeding the prescribed limits of CSR, which indicates an increased sense of becoming socially responsible. Additionally, companies like LG h&H Co. Ltd., Klabin S.A, Indra Sistemas, etc. are found to be global leaders in CSR. Regions like Europe, North America, South America, South Asia, etc. have been found to have one of the highest CSR scores. While in India, Karnataka, Maharashtra, and Delhi have been found one of the flourishing regions with CSR initiatives.

8 Conclusion

Gone are the days when companies believed that their only objective was to earn profit and increase their wealth. Over the years, the importance of being socially responsible has increased gradually. It has become so integral that even the Companies Act of India necessitates a fraction of the revenue generated to be used in CSR if they fulfill certain criteria.

Following the same, a large number of corporations have integrated their social responsibility activities with their corporate activities, some of them have been working to provide healthcare services, some are working to provide education to impoverished children, and some are emphasizing reducing the impact on the environment, while, some of them are working on inclusion and diversity. It is found that CSR activities not only help the underprivileged people, but also the environment, economy, and society in general as well. Despite the numerous pros, the CSR sector has been facing varying challenges such as balancing the need for long-term growth with social welfare, the lack of transparency, etc., which can be overcome by the use of blockchain, cloud computing, etc. as discussed under the recommendations given below.

Additionally, there isn't a fixed path for undertaking CSR activities since there are numerous opportunities for businesses that they can avail towards sustainability such as joining forces with non-profit organizations or charities, adopting appropriate sustainability actions, volunteering in the community, and updating the regulations in the workplace.

Both, at the National and Global levels, the CSR sector is prospering. Hence, the main point of the given text is that integrating social responsibility into corporate strategies promotes sustainability. This integration involves various aspects such as environmental stewardship, social responsibility, and ethical business practices, which contribute to environmental conservation, social well-being, and economic development. Therefore, it can be concluded that given the environmental and societal issues faced by the world, integrating social responsibility with corporate activities is the best way to promote sustainability and also aid towards the fulfillment of the Sustainable Development Goals.

9 Recommendations

Initiatives that businesses can undertake to promote sustainability while integrating CSR with their operations include:

1. **Inclusion of Digital Media**: By offering courses and training programs in digital skills, businesses connect underprivileged populations to the internet and essential digital tools, enhancing employability. Additionally, these internal programs assist employees in developing their digital literacy and skill sets.

2. **Diversity Management in the Workplace**: Diversity management promotes an inclusive and equitable work environment, attracting and retaining qualified employees. Digital platforms help train staff about diversity and raise awareness, offering skill-building materials to foster collaboration and reduce bias and discrimination.
3. **Application of Blockchain**: Blockchain technology increases the accountability, transparency, and efficiency of social and environmental impact reporting. Startups use blockchain to track commodity and raw material movements, identifying risks like forced labor and environmental degradation. Businesses issue virtual tokens to replace carbon credits, offsets, incentives, and emissions trading.
4. **Use of Cloud Computing**: Cloud-based CSR initiatives allow scalable, cost-effective, and flexible data management, communication, and collaboration. They ensure continuity in CSR activities by facilitating collaboration through video conferencing and document sharing from any location. Additionally, cloud computing helps reduce energy use and carbon emissions related to business travel.
5. **Marketing for Causes**: Companies engage in cause marketing to attract consumers who align with their beliefs or highlight the advantages of their products. They utilize social media to promote their services and raise funds or awareness for causes, generating donations and support.
6. **Conscious Supply Chains**: Supply chain management and ethical sourcing are critical for CSR goals due to the carbon emissions generated during the transport of goods. Startups leverage technologies like IoT, blockchain, and tags to enhance supply chain management, ensuring transparency and accountability.
7. **Reporting Solutions for CSR**: CSR reporting evaluates and shares a company's social and environmental performance with stakeholders. These metrics aid governments, legislators, and stakeholders in creating guidelines and suggesting policy changes. Innovations in CSR reporting improve transparency, accountability, and stakeholder engagement. Businesses may use integrated reporting to provide a comprehensive picture of their performance by combining financial and non-financial data (Startus-Insights 2023).

References

Acharya M (2023) Corporate social responsibility under section 135 of companies Act 2013. https://cleartax.in/s/corporate-social-responsibility

Banerjee R, Sen A, Khanna P (2023) Corporate social responsibility in emerging markets: case studies from India. J Bus Ethics 176(2):345–362. https://doi.org/10.1007/s10551-022-05187-4

Bansal P, Song H-C (2023) Data and methods in sustainability management research: current trends and future directions. Sustain Manag Rev 2(1):1–23. https://doi.org/10.1007/s12325-022-00594-1

Bowen HR (1953) Social responsibilities of the Businessman. Harper & Row

Carroll AB (1999) Corporate social responsibility: evolution of a definitional construct. Bus Soc 38(3):268–295

Carroll AB (2008) A history of corporate social responsibility: concepts and practices. In: Crane A, Matten D, McWilliams A, Moon J, Siegel DS (eds) The Oxford Handbook of Corporate Social Responsibility. Oxford University Press, pp 19–46

Carter CR, Rogers DS (2008) A framework of sustainable supply chain management: moving toward new theory. Int J Phys Distrib Logist Manag 38(5):360–387. https://doi.org/10.1108/09600030810882816

Chandra A (2019) List of CSR activities in India by Indian Companies. https://www.avinashchandra.com/list-of-csr-activities-in-india

Chouinard Y, Ellison J, Ridgeway R (2011) The sustainable economy. Harv Bus Rev 89(10):52–62. https://hbr.org/2011/10/the-sustainable-economy

CSR BOX. Abridged Version (2021) India CSR Outlook Report, CSR Analysis of large 301 Listed companies (FY 2020–21). https://csrbox.org/media/CSR-in-India-2021-CSR-Report-by-CSRBOX.pdf#:~:text=Reliance%20Industries%2C%20Tata%20Consultancy%20Services%2C%20HDFC%20Bank%2C%20and,CSR%20spent%20is%20covered%20by%20these%205%20companies.

CSR HUB. https://www.csrhub.com/CSR_ratings_by_region_and_country/

CSR Services for Companies Applicable Under the Companies Act, 2013 (compliancecalendar.in), https://www.compliancecalendar.in/csr-services

Das K, Roy S (2023) Sustainability and corporate governance: the role of social responsibility in emerging economies. Sustain Dev 31(1):78–94. https://doi.org/10.1002/sd.2297

Eccles RG, Ioannou I, Serafeim G (2014) The impact of corporate sustainability on organizational processes and performance. Manage Sci 60(11):2835–2857. https://doi.org/10.1287/mnsc.2014.1984

Elkington J (1997) Cannibals with forks: the triple bottom line of 21st century business. Oxford: Capstone. https://www.wiley.com/en-us/Cannibals+with+Forks%3A+The+Triple+Bottom+Line+of+21st+Century+Business-p-9781841120843

Epstein MJ, Buhovac AR (2014) Making sustainability work: best practices in managing and measuring corporate social, environmental, and economic impacts. Berrett-Koehler Publishers. https://www.bkconnection.com/books/title/making-sustainability-work

Fatma M, Rahman Z, Khan I (2015) Building company reputation and brand equity through CSR: the mediating role of trust. Int J Bank Market 33(6):840–856. https://doi.org/10.1108/IJBM-11-2014-0166

Freeman RE (1984) Strategic management: a stakeholder approach. Boston: Pitman. https://www.cambridge.org/core/books/strategic-management/53E9BB0D2789D13F9707D0E2D6E6D7F1

Greening DW, Turban DB (2000) Corporate social performance as a competitive advantage in attracting a quality workforce. Bus Soc 39(3):254–280. https://doi.org/10.1177/000765030003900302

Gupta A (2023) The impact of corporate social responsibility on financial performance: evidence from India. Asian Bus Manag 22(3):293–309. https://doi.org/10.1057/s41291-022-00177-3

Hair JF, Black WC, Babin BJ, Anderson RE (2023) Multivariate data analysis (8th ed.). Cengage Learning. https://www.cengage.com/c/multivariate-data-analysis-8e-hair/

HT Correspondent (2023) Unlocking the power of CSR: impactful initiatives and future outlook. https://www.hindustantimdes.com/india-news/unlocking-the-power-of-csr-impactful-initiatives-and-future-outlook-101681226328588.html

India CSR (2019) As India's CSR spending touchs Rs 12,000 cr, investigators stalk ghost beneficiaries. https://indiacsr.in/as-indias-csr-spending-touchs-rs-12000-cr-investigators-stalk-ghost-beneficiaries/

Jamali D, Karam C (2018) Corporate social responsibility in developing countries as an emerging field of study. Int J Manag Rev 20(1):32–61

Karmakar S (2022) Corporate social responsibility (CSR): challenges and resolutions. Cyberswift. https://www.cyberswift.com/blog/csr-challenges-and-resolutions/

King A, Lenox M (2001) Lean and green? An empirical examination of the relationship between lean production and environmental performance. Prod Oper Manag 10(3):244–256. https://doi.org/10.1111/j.1937-5956.2001.tb00373.x
Kolk A (2008) Sustainability, accountability and corporate governance: exploring multinationals' reporting practices. Bus Strateg Environ 17(1):1–15
Kumar S, Mehta R (2024) Integrating sustainability into corporate strategy: Insights from Indian companies. Int J Corporate Strategy 15(1):102–118. https://doi.org/10.1504/IJCS.2024.100112
Lee MDP (2021) A review of the theories of corporate social responsibility: its evolutionary path and the road ahead. Int J Manag Rev 23(1):20–40. https://doi.org/10.1111/ijmr.12214
Li X, Liu J, Zhu Y (2023) Big data analytics in corporate social responsibility: opportunities and challenges. J Bus Res 156:113405. https://doi.org/10.1016/j.jbusres.2023.113405
Lozano R (2015) A holistic perspective on corporate sustainability drivers. Corp Soc Responsib Environ Manag 22(1):32–44. https://doi.org/10.1002/csr.1325
MacGrath A, Jonker A (2023) What is corporate social responsibility (CSR)? IBM. https://www.ibm.com/topics/corporate-social-responsibility
Miles MB, Huberman AM, Saldaña J (2020) Qualitative data analysis: a methods sourcebook (4th ed.). SAGE Publications. https://us.sagepub.com/en-us/nam/qualitative-data-analysis/book246128
Montiel I, Delgado-Ceballos J (2014) Defining and measuring corporate sustainability: are we there yet? Organ Environ 27(2):113–139
Orlitzky M, Schmidt FL, Rynes SL (2003) Corporate social and financial performance: a meta-analysis. Organ Stud 24(3):403–441. https://doi.org/10.1177/0170840603024003910
Pang Z, Zhang Y, Tan L (2022) Enhancing data quality in CSR research: best practices and methodological advancements. Bus Ethics Q 32(4):501–524. https://doi.org/10.1017/beq.2022.10
Porter ME, Kramer MR (2011) Creating shared value. Harv Bus Rev 89(1/2):62–77. https://hbr.org/2011/01/the-big-idea-creating-shared-value
Rao S, Singh N (2024) Corporate social responsibility and environmental sustainability: trends and practices in the global south. Bus Strateg Environ 33(2):556–570. https://doi.org/10.1002/bse.3124
S&P Global, The Sustainability Yearbook—2023 Rankings, https://www.spglobal.com/esg/csa/yearbook/2023/ranking/
Schiederig T, Tietze F, Herstatt C (2012) Green innovation in technology and innovation management—an exploratory literature review. R&D Manag 42(2):180–192. https://doi.org/10.1111/j.1467-9310.2011.00672.x
Searcy C (2012) Corporate sustainability performance measurement systems: a review and research agenda. J Bus Ethics 107(3):239–253. https://doi.org/10.1007/s10551-011-1038-z
Sharma V, Patel D (2023) Social responsibility and corporate ethics: balancing profit and principles. J Corporate Respon 20(3):123–136. https://doi.org/10.1007/s40991-023-00312-x
Startus-Insights (2023) Discover the top 10 corporate social responsibility trends in 2023. https://www.startus-insights.com/innovators-guide/corporate-social-responsibility-trends/
Unilever (n.d.) Sustainable living plan. Retrieved from https://www.unilever.com/sustainable-living
United Heroes (2022) 4 Solid examples of corporate social responsibility activities. https://www.united-heroes.com/blog/4-solid-examples-of-corporate-social-responsibility-activities#:~:text=Here%20are%20some%20environment%20CSR%20activities%20you%20could,Work%20Scheme%E2%80%99%20and%20%E2%80%98ride%20share%E2%80%99%20rotas%20More%20items
Vogel D (2005) The market for virtue: the potential and limits of corporate social responsibility. Brookings Institution Press
Weinger A, Companies committing social good: 7 real-world CSR examples, DoubletheDonation. https://doublethedonation.com/corporate-social-responsibility-examples/#:~:text=Companies%20Committing%20Social%20Good%3A%207%20Real-World%20CSR%20Examples,6%206.%20Unilever%20...%207%207.%20Bombas%20

The Role of Government and International Institutions in Fostering Social Responsibility and Environmental Harmony: Experience of Russia and BRICS Countries

Anna V. Belitskaya and Yuliya S. Kharitonova

Abstract Russia has a long history of social entrepreneurship, which differs drastically from the "third sector" of European countries, and the concept of green economy is pretty new to it despite the fact of having developed ecology law. Analyzing and comparing the experiences of Russia and other BRICS countries in legal regulation of sustainable development can help to work out the unified understanding of the concept and reveal best practices in fostering it.

Keywords Sustainable development · Green economy · Social responsibility · Environmental harmony · ESG · State regulation · New Development Bank · Impact finance

1 Introduction

Sustainable development including environmental harmony and social responsibility has become a trendy concept recently, gathering states, businesses and people around the idea of the responsibility to future generations. Unfortunately, the concept is not yet that mature and natural, so that every person on Earth understands it and relies to it. This is why the role of governments and international institutions is so vast in promoting this idea by means of different incentives and restrictions.

BRICS countries nowadays have a goal to foster social responsibility and environmental harmony to achieve sustainable development. Governments and international institutions, including BRICS New Development Bank, put effort into impact finance and encouraging their businesses and societies to do the same. In comparison with the Western World the BRICS countries have less experience in implementing ESG

A. V. Belitskaya · Y. S. Kharitonova (✉)
Business Law Department, Lomonosov Moscow State University, Law Faculty, Moscow, Russia
e-mail: sovet2009@rambler.ru

P. Singh et al. (eds.), *Integrating Environmental Sustainability and Corporate Social Responsibility: A Move Beyond Profit*, World Sustainability Series,
https://doi.org/10.1007/978-3-031-81469-3_8

concept in their corporate businesses or developing the "third sector" that aims to support social and publicly important projects. At the same time, BRICS countries have a long history or building solidarity economy, where the state works together with the society to make life in the country better: reduce poverty, improve the housing conditions, ensure that there is enough food to eat and water to drink for the population.

In the twenty-first century with all its pollution, pandemic and global warming issues the global trend is to care not only about social aspect of people's well-being but also about environmental protection of current and future generations which highlights the sustainability-based social value function of the state or the alternative of ESG concept implemented in corporate life of businesses in the Western world. The concept of green economy in BRICS members is currently not as developed as in the Western countries, although protecting ecology as part of the support for sustainable development is deemed to be the important objective of BRICS policy. The reason for it is the priority of economy before ecology, present time before future, as the luxury of caring about future generations is relevant only if there is no need of survival to the existing one. In our opinion, when most of the population cannot provide for their basic needs and the businesses cannot handle this role either, it is important that the governments take care of such needs by guaranteeing minimal social and ecological standards should be met.

2 Social and Environmental Functions of the State in Russia and BRICS Countries

2.1 The Concept of Welfare (Social) State in Russia and BRICS Countries

The concept of welfare state, in which governmental institutions are responsible for the protection and promotion of the economic and social well-being of citizens,[1] including high-quality education, health care, old-age support, reduction of poverty and unemployment,[2] is nowadays depicted in some constitutions of the world. The BRICS countries serve a bright example of such welfare states with the paternalistic approach in comparison to minimalist states of the Western countries, which share the belief that citizens should help themselves with healthcare, education and pensions instead of delegating this role to the government.

It should be emphasized that in England, the term "welfare state" appeared in the 1940s. In American English, it has stronger associations with paternalism. This to some extent comes into conflict with the idea of capitalism as a system.

[1] Welfare state. Written and fact-checked byThe Editors of Encyclopaedia Britannica. Updated: Feb. 22, 2024 https://www.britannica.com/money/welfare-state.

[2] Saez, E. (2022). THE SOCIAL STATE. Fiscal, 23.

Capitalist economy is known as it is an ideology presented by Adam Smith as an economic system controlled by the private sector in which the government does not play an excessive role in business activities. The impact of the capitalist economic system on social relations is reflected in the fact that there is a meaningful social gap between rich and poor people. At the same time, the ESG agenda in Western countries focuses primarily on climate and environmental issues, while the economic situation in the Global South and other countries like Russia looks different. In particular, in Russia, under sanctions pressure, the focus is on social and governance development. The authorities convince society and business that it is necessary to strengthen the development and support of those areas of people's economic activity that lead to the strengthening of the country's economic wellfare for the benefit of all citizens.

For instance, according to Article 7 of the Constitution of the Russian Federation, Russia is a social state that aims to create the necessary conditions for the development of people's potential and their well-being, including the realization of their right to have a house. The state protects the health and labor of people, pays special attention to family, children, disabled and old people social protection. According to Article 42 of the Constitution of the Russian Federation, Russia guarantees environmental protection of people as well.[3] Currently the Russian government carries out several national projects to improve the quality of life of Russian people. They are devoted to such spheres as healthcare, education, demography, culture, safe high-quality roads, housing and urban environment, ecology, science and universities, small and medium entrepreneurship, labor productivity, data and digital economy, etc.

There are certain criteria, that show the extent of the state involvement in carrying social responsibilities, and the most important of them are providing for education, health care and old-age support. Russia guarantees free education and health care for its citizens. Although there was an attempt to develop the institute of private pension funds in addition to the existing State pension fund, Russia has a decent state pension system. It is fair to say, that Russia has a paternal outlook on the relations between the state and the society, where the government takes care of its people. State regulation of the social-economic development of the country plays a significant role in Russia, unlike in Europe, where the institute of civil society is strong enough to provide balanced self-regulation. Contemporary model of state regulation in Russia is definitely influenced by the inheritance of the Soviet Union, although the market economy came in place of the planned economy more than two decades ago. At the current stage of development, it is hard for the state to give up its social and ecological responsibilities in hope that the society itself will take care of them. People depend on the state and the state tries to be dependable.

The Brazilian welfare regime is described as controversial.[4] Social policies in Brazil were heavily dependent on economic development projects of different political regimes.[5] The process of building a welfare state model in Brazil, from the elitist social insurance model to the challenges of developing a universal institutional model

[3] Constitution of the Russian Federation http://archive.government.ru/eng/gov/base/54.html.

[4] Lavinas et al. (2017).

[5] Lobato (2016).

of social security began in 1988. Latin American countries had to face the challenges of achieving social integration and economic redistribution, as well as strengthening democratic institutions. Governance models were created to respond to the emerging demands of decentralization and social partnership. According to Article 194, 195 of the Constitution of Brazil, social welfare comprises an integrated whole of actions initiated by the government and by society, with the purpose of ensuring the rights of people to health, social security and assistance. Social welfare shall be financed by all of society, either directly or indirectly. According to Article 225, all have the right to an ecologically balanced environment, which is an asset of common use and essential to a healthy quality of life, and both the government and the community's.[6]

The educational system in Brazil is nationally regulated but strongly decentralized, both in funding and in provision, into state and municipal levels. Primary education is compulsory and free, while secondary is free but not mandatory.[7] The experts state that public university education favors the elite, which leads to the conclusion, that resources could be better spent for the elementary education, benefiting poorer individuals.[8] Public health care in Brazil is provided for free through Unified Health System.[9] With regard to the pension system, both employers and employees in Brazil pay social security contributions.[10] For almost a decade, the debate on social security in the region has centered around risk diversification, the macroeconomic effects of the systems, and the participation of the private sector in their administration. Now, however, many analysts are beginning to turn their attention to the issue of coverage.

While analyzing Indian concept of welfare, it is important to take into consideration the relationship between its colonial past and democratic present.[11] The state's capacity to redistribute wealth and alleviate poverty has been questioned along with the robustness of its institutions and the negotiation of boundaries between state and civil society.[12] As Atul Kohli writes, "over time, the state in India has evolved from a reluctantly pro-capitalist state with a socialist ideology to an enthusiastically pro-capitalist state with some commitment to inclusive growth. This shift has significant implications for the possibility of redistributive development in India."[13] India has adopted the ideal of welfare state and is fully committed to its realization.[14] According to Article 38 of the Constitution of India, the state aims to secure a social order for the promotion of welfare of the people, and according to Article 48A the protection and improvement of environment is also the responsibility of the

[6] Constitution of the Federative Republic of Brazil https://www.globalhealthrights.org/wp-content/uploads/2013/09/Brazil-constitution-English.pdf.

[7] Draibe (2007).

[8] Almeida (2019).

[9] Roman (2023).

[10] chrome-extension://efaidnbmnnnibpcajpcglclefindmkaj/https://www.oecd.org/els/public-pensions/PAG2021-country-profile-Brazil.pdf.

[11] Kumar (2005).

[12] Verma (2018).

[13] Kohl (2009).

[14] Robson (1954).

state.[15] The Indian state plays an essential role in promoting economic growth and in reducing inequalities and poverty.[16] Education continues to be free in India at public schools, although private education is more common at the secondary level.[17] India ensures that every individual has access to the necessary healthcare services to live a healthy life.[18] Although India, like most other developing countries, does not have a universal social security system to protect the elderly against economic deprivation, it has a long tradition of pension and other forms of formal old age income support system.[19]

The most unique example of the role of the government in social and environmental spheres is depicted in the Constitution of the People's Republic of China. In Article 1 of the Constitution China is described as a socialist state under the people's democratic dictatorship led by the working class and based on the alliance of workers and peasants. Article 1 states that the socialist system is the basic system of the People's Republic of China. According to Article 6, the basis of the socialist economic system of the People's Republic of China is socialist public ownership, which supersedes the system of exploitation of man by man and applies the principle of "from each according to his ability, to each according to his work".[20] The most important to our research is the Article 7, which is devoted to the concept of state-owned economy, namely, the socialist economy. According to Article 7 the state ensures the consolidation and growth of the state-owned economy. Moreover, the state undertakes the development of socialist education and works to raise the scientific and cultural level of the whole nation (Article 19), develops medical and health services, promotes modern medicine and traditional Chinese medicine (Article 21), promotes family planning so that population growth may fit the plans for economic and social development (Article 25). According to Article 26, the state also protects and improves the environment in which people live and the ecological environment. It prevents and controls pollution and other public hazards.

The Constitution of the Republic of South Africa in its preamble states that the Republic aims to improve the quality of life of all citizens and free the potential of each person. It also contains the articles, devoted to the state social and environmental responsibilities.[21] South Africa has one of the world's most expansive social grant system: 47% of the population relies on a monthly payment,[22] which includes care dependency grant, child support grant, disability grant, foster child grant, grant in aid,

---

[15] The Constitution of India. Article 38. https://cdnbbsr.s3waas.gov.in/s380537a945c7aaa788ccfcdf1b99b5d8f/uploads/2023/05/2023050195.pdf.

[16] Kohl (2012).

[17] Education System in India https://www.scholaro.com/db/Countries/India/Education-System.

[18] Understanding India's Healthcare System https://www.internationalinsurance.com/health/systems/india.php.

[19] Goswam (2001).

[20] http://www.npc.gov.cn/zgrdw/englishnpc/Constitution/2007-11/15/content_1372963.htm.

[21] The Constitution of the Republic of South Africaю Article 27. https://www.wipo.int/edocs/lexdocs/laws/en/za/za107en.pdf.

[22] https://theconversation.com/47-of-south-africans-rely-on-social-grants-study-reveals-how-they-use-them-to-generate-more-income-203691.

war veterans grant. South Africa also pays attention to old age pension, social relief of distress, etc.,[23] it is widely praised for the coverage, generosity and efficiency of its social protection system.[24]

Summing it up, it is important to state, that the government plays a key role in Russia, as in other BRICS countries, in guaranteeing the minimal social and environmental protection of the citizens. The society of the developing countries itself is too weak to carry such responsibilities on its own based on self-regulation regime. Although the state can delegate the implementation of some of the state functions in social and environmental spheres to the society, the government will be still there to monitor and provide support. The people made the state responsible for the society welfare by adopting their Constitutions.

2.2 *Delegating Social and Environmental State Functions to the Society in Russia and BRICS Countries*

The idea of a minimal state, which gains the popularity in the Western word, is the idea of a state that protects people's property rights and does nothing else.[25] This concept has been widely spread in developed countries, where people have a good level of life and can invest their own money in education and healthcare at minimum. It is understandable that Environmental, Social and Governance (ESG) principles and impact investing approaches as well as the concept of "third sector" or "solidarity" economy started to develop in these countries.

ESG is a framework that helps stakeholders understand how an organization is managing risks and opportunities related to environmental, social, and governance criteria, and takes the holistic view that sustainability extends beyond just environmental issues.[26] ESG agenda mostly concerns investments by private companies. The "third sector" is an umbrella term that covers a range of different non-governmental, non-profit, values-driven organizations, belonging neither to the state, nor to the private sector. Most third sector organizations devote themselves either to a particular issue which needs solving (for example, climate change or unaffordable housing) or to a particular group in society (for example, dementia sufferers, women) who requires support and representation.[27] Basically, the "third sector" organizations are

[23] Social benefits https://www.gov.za/services/social-benefits.

[24] Vidojević J. In pursuit of public good: the need for a welfare revolution in South Africa on 30/07/2023 https://www.globalgovernmentforum.com/in-pursuit-of-public-good-the-need-for-a-welfare-revolution-in-south-africa/.

[25] Wendt (2016).

[26] Peterdy K. What is ESG? https://corporatefinanceinstitute.com/resources/esg/esg-environmental-social-governance/.

[27] What is the third sector and what does it do? http://toolkit.northernbridge.ac.uk/engagingwithpolicymakers/engagingwiththethirdsector/whatisthethirdsectorandwhatdoesitdo/.

the example of the self-regulation form of society that willingly take on itself some social and environmental responsibilities.

International experience shows, that the legislation in various countries reflects different legal structures aimed to implement the concept of sustainable development and social responsibilities of society into law enforcement practice. There are norms in relation to social or green bonds, corporate ESG bonds, stock exchanges declaring their intention to promote principles of sustainable development, socially oriented corporations as types of legal entities, specialized funds for the accumulation of social investments, crowdfunding, public–private partnerships in the social sphere, sustainable contracts, sustainability clauses, etc. In market economy governments usually use legal incentives and restrictions to delegate some of the social and environmental functions from the state to the society with preference given to legal incentives as the implementation of an indirect method of state regulation. Thus, governmental support for sustainable development in various countries includes, in particular, tax incentives, budget subsidies, state guarantees, preferential commercial loans, subsidies for lost income, etc. Apart from it, the governments create an infrastructure for delegating social and environmental state functions to the society, including the foundation of development institutions.

Russia has a long history of social entrepreneurship, which differs drastically from the "third sector" of other countries. Lots of Russian business people were philanthropists, supporting construction, culture and health care. According to the Skolkovo Business School, nowadays the contribution of companies makes up approximately half of the total spending on social projects in Russia.[28] Russian legislation since 2019 contains the term "social entrepreneurship", which is understood as entrepreneurial activity aimed at achieving socially beneficial goals, contributing to the solution of social problems of citizens and society.[29] According to the Federal Law No. 209-FZ dated 24.07.2007 "On the development of small and medium-sized businesses in the Russian Federation", social entrepreneurship is supported by the state financially and via other means, which shows that even in the process of delegation of its functions the state stays responsible for the final result. Special organizational and legal forms are necessary to increase the attractiveness of social enterprises as objects of financing and imply a ban or restriction on the distribution of the enterprise's profits, mandatory reporting on the achievement of the goals of the organization's mission, the ability to raise funds like in charitable organizations. This example shows that the state uses incentives as well as restrictions while delegating its social functions to the society.

All of this points to the importance of understanding the potential role of corporate social responsibility in companies is important for the design of regulation. At one level it overlaps with ideas of self-regulation and co-regulation, and at another it

[28] Устоять и выжить: что изменится в российской ESG-повестке в ближайшее время https://www.forbes.ru/forbeslife/458067-ustoat-i-vyzit-cto-izmenitsa-v-rossijskoj-esg-povestke-v-blizajsee-vrema.

[29] Federal Law of July 24, 2007 N 209-FZ "On the development of small and medium-sized businesses in the Russian Federation" (Article 1, 24.1).

challenges the foundations of modern liberal market economics. The personal and moral responsibility of directors and employees must also be considered alongside the role of the corporation in meeting the legitimate expectations of its investors and shareholders.

Corporate social responsibility can certainly be seen as a good thing. Nevertheless, governments must assume a shared responsibility to ensure that failures in behavior—both market and non-market—are managed appropriately, taking into account each of the three pillars of sustainable development: economic, social and environmental.

The cooperation in between the state and the businesses is deemed very important in Russia. According to E.P. Gubin, the state and business are located in the same system of coordinates, and their opposition can only lead to the collapse of the economic system, on the one hand, and the destabilization of the state, on the other.[30] According to A.G. Bykov, the state, defined by the Constitution of the Russian Federation as social, has the right and the obligation to assign to entrepreneurial activity social functions and the solution of social problems.[31]

2.3 *The Solidarity Economy and Principle of Economic Solidarity in Russia and BRICS Countries*

Delegating social and environmental state functions to the society is the part of the concept of solidarity economy, which grows from the idea of social justice. Solidarity economy means that not only the state but also private entities are actively interested in supporting the social, environmental or other types of state policies. The concept of solidarity economy is close to the concept of social economy, which is widespread in the Western world, but they are not the same. The Western countries have strong businesses whose role in the economy is as prominent as the states' role. The Western business has created the "third sector", which includes the organizations, whose activities are mostly non-for-profit. These organizations are mostly independent and do not cooperate with the government as they have their own strong agenda. The solidarity economy on the other hand aims to build collaboration in between of the state and business by finding common goals and objectives for them. In our opinion, this is a different paradigm of development that seeks to change the whole system of traditional economy and put it through flexible transformation. In most BRICS countries the state traditionally has played the key role in fostering the economic development and the Western concept of the primacy of the maximization of profit is new and foreign to the BRICS members. This is why the solidarity economy concept that puts people and planet at its core first and where people can influence

[30] Gubin (2015).

[31] Bykov (2004).

all dimensions of human life including economic, social and environmental[32] has gained its popularity in Brazil, China, India, South Africa and Russia, where the concept of solidarity is depicted in its Constitution.

In 2020 the Constitution of the Russian Federation was amended with Article 75.1, which states that the Russian Federation creates the conditions for the sustainable economic growth of the country and improvement of the well-being of citizens, for mutual trust of the state and the society. The protection of dignity of citizens and respect for working people are guaranteed, the balance of the rights and responsibilities of a citizen, social partnership, economic, political and social solidarity are ensured. By declaring a focus on ensuring economic solidarity, the state introduces a new principle and new constitutional value. Using the word "solidarity" in relation to the economic, political and social sphere in the context of the formation of mutual trust between the state and society, the Russian legislator expects the creation of a new vector of interaction between them on the basis of partnership, integration, and co-regulation. Solidary interaction presupposes a relationship where the expectations and actions of the interacting parties coincide and mutually help in achieving goals, and therefore solidarity should concern not only the goals themselves, but also the methods leading to their achievement. The balance of private and public interests in implementing the principle of economic solidarity is manifested in the fact that the state stimulates and encourages business entities when they take part in the implementation of state policy, in particular through various incentives and measures of state support, as well as through the creation of legal regimes of stimulating nature, ensuring stability of business conditions of investors and entrepreneurs. As the Constitutional Court of the Russian Federation noted, in order to develop a state economic policy that meets the interests and needs of society, the most favorable conditions for a market economy are created both through direct regulatory government influence and through stimulating free economic activity based on the principles of self-organization, a balance of private and public interests, corporate interaction and cooperation.

As far as ESG principles are concerned, Russia is at the initial stage of its implementation. The ecology law in Russia is well developed, mostly regulated by the Federal Law "On Environmental Protection" No.7-FZ dated 10.01.2002. Special measures devoted to ESG are also implemented in Russian legislation. Thus, Government Decree No. 709 dated 05.06.2023 on the rules for subsidizing the Clean Air program allows enterprises of the fuel and energy complex in 12 cities participating in the federal project on quotas for pollutant emissions to receive preferential loans at a rate of 3% to reduce emissions into the atmosphere.[33] It is also important to mention national project "Ecology" (2019), decarbonization strategy (2021) and other decisions that influence the ESG agenda in Russia, for example, the Government Decree

---

[32] Global Vision for a Social Solidarity Economy: Convergences and Differences in Concepts, Definitions and Frameworks. 2015. chrome-extension://efaidnbmnnnibpcajpcglclefindmkaj/https://www.ess-europe.eu/sites/default/files/publications/files/ripess_global_vision_-_en.pdf.

[33] ESG-повестка в России: основные итоги 2023 года https://journal.ecostandard.ru/esg/ustoychivoe-razvitie/esg-povestka-v-rossii-osnovnye-itogi-2023-goda-/.

No. 1912-p dated 14.07.2021 "On approval of the goals and main directions of sustainable (including green) development of the Russian Federation" and the Resolution No. 1587 dated 21.09.2021 "On approval of criteria for sustainable (including green) development projects in the Russian Federation and requirements for the verification system for financing instruments for sustainable development in the Russian Federation".

The solidarity in Brazil includes the concept of the cooperative movement[34] and sharing among small rural producers and consumers,[35] which helps to reduce the poverty of the population and shows the prioritization of social and sometimes environmental goals of businesses in the country. Although Brazil does not mention solidarity as a constitutional principle, the implementation of its social and economic policies shows the importance of this concept for investors, which participate in non-for-profit projects. In our opinion, Brazil has more experience in co-financing sustainability in the frame of public–private partnership than Russia. China mentions solidarity as part of its governmental policy, in which the state together with the people of China foster social and economic benefits and continuously improve people's well-being. The priority in financing in China belongs to resource-saving, environmentally friendly and labor-intensive industries,[36] which shows the importance of the concept of solidarity economy in the country. Mutual support and development of high scale projects, which are equally important for public and private sectors, can be seen as a trend in contemporary India as well. A lot of co-financing takes place in innovation and technology fields, where India has a huge potential of development.

We believe, that a true economic solidarity is achieved, when the state creates appropriate conditions for public–private partnerships and other types of collaboration in between of public and private sectors to grow and develop. The government and business should cooperate in co-financing sustainable development of national economies as well as high scale and high impact infrastructure projects.

3 Socially and Ecologically Responsible (Sustainable) Investment in Russia and Other BRICS Countries

3.1 The Concept of Sustainable Investment in Russia and Other BRICS Countries

To invest means to turn down current spending to gain more in the future. Most investments are aimed to gain profit but some countries legislation, including, Russian, also sees a social or other beneficial effect as a possible goal of investment. The analysis of investments in other BRICS countries also showed, that many foundations

---

[34] Morais et a.l (2017).

[35] Otsuki and Castro (2020).

[36] Wang (2021).

have tended to care more about social impact than financial return.[37] Ideally, the aim of investments should include achieving positive social impact as well as financial returns.[38] BRICS members are mentioned among the countries with top high impact projects in social and environmental spheres including South Africa Green Hydrogen Project, Ceara Costa Nordeste Offshore Wind Farm Project in Brazil, CSI Solar Haidong New Energy Whole Industry Chain Project in China, SMS Hyderabad Particle Characterization Laboratory Project in India. See below International private investment in the sustainable development: change in number of projects, 2021–2022 and 2015–2022.[39]

		2021–2022	2015–2022
Infrastructure Transport infrastructure, power generation and distribution (except renewables), telecommunication	7, 9, 11	+26%	+16%
Renewable energy Installations for renewable energy generation, all sources	13	+8%	+21%
WASH Provision of water and sanitation to industry and households	6	+20%	+13%
Agrifood systems Agricultural production and processes; fertilizers, pesticides and other chemicals; R&D; technology	2	+6%	-19%
Health and education Hospital facilities, school buildings and other infrastructure for service delivery	3, 4	+8%	+11%

Social and environmental investments are usually not very profitable, they prioritize other goals such as the overall impact on economy and social sphere. Social and environmental investment can be deemed as part of the Western social economy concept, where large businesses are responsible for the compensation for pollution and poverty problems and have to participate in reducing them. There are some activists in the Western countries that try to impose responsible investment agenda on corporations, build ESG principles in their daily business routine. We believe, that in BRICS countries this will not work taking into consideration the vast role that the state plays in economy and social sphere of BRICS members and that the people there mostly rely on the governments in terms of solving social and environmental issues such as reducing poverty, guaranteeing minimal social rights for education, health care and support of old age, that most of the population cannot afford on its own. While the state carries such responsibilities itself using taxes, people in BRICS countries, including Russia don't try to force business to participate in in fostering social responsibility and environmental harmony.

The concept of the investment that has a socially beneficial effect initially prevailed In Russian legislation as the inheritance of the Soviet Union. Thus, the Law of the

[37] Meng et al. (2022).

[38] Lo and Zhang (2023).

[39] World Investment Report chrome-extension://efaidnbmnnnibpcajpcglclefindmkaj/https://unctad.org/system/files/official-document/wir2023_en.pdf.

RSFSR № 1488-1 dated 26.06.1991 "On investment activity in the RSFSR" understands investments as cash, targeted bank deposits, shares and other securities, technologies, machinery, equipment, loans, and any other property or property rights, intellectual values invested in objects of business and other types of activity in order to generate profit (income) and achieve a positive social effect. This Law aims at the expected result from any and every investment. In the current Russian investment laws such characteristic as "socially positive effect" is changed the "positive effect" that can be beneficial not necessarily for the society but more for the investor himself. There is no specific legislation in Russia devoted to impact investment. Nether the less investors can use impact investing techniques in relation to different asset classes and with different return expectations. For example, although current legislation in Russia does not cover such instruments as green or social bonds, investment can still be attracted through general instruments such as investment funds and crowdfunding in accordance with the Federal Law "On Investment Funds" No. 156-FZ dated 29.11.2001 and the Federal Law "On attracting investments using investment platforms and on introducing amendments to certain legislative acts of the Russian Federation" No. 259-FZ dated 02.08.2019.

Currently, the implementation of the concept of sustainable development is mainly funded from the budget of the Russian Federation. The concept of public private partnership has only recently become a trend in Russia,[40] and most of the projects, which take place in the country, are related to the construction of transport and energy infrastructure, rather than green and social projects. Unlike the Western countries, which have their own legal schemes for building schools and hospitals in the frame of public–private partnership, Russia does not distinguish between infrastructure projects using the same concession and public–private partnership agreement schemes for all the objects. Currently using state contracts for construction of the social objects is more widespread in Russia as the private business entity under such contract does not bear the risks of loosing its investments. At the same time the Russian government realizes the importance of co-financing the high impact projects with the money of private investors, this is why the legislation of the Russian Federation gains more and more instruments to attract them. It is important to mention, that the state uses a lot of incentives to stimulate private investment in sustainable projects, which also shows its paternalistic role in social and environmental spheres.

Although China has adopted the concept of responsible investment from the Western World, it is believed that Chinese social enterprises generally lack the skills and capacity to continuously measure social impact.[41] Responsible investment in China is the continuation of the concept of solidarity economy. According to the State Council of the People's Republic of China, in this context, increasing people's sense of benefit, harnessing investment potential in the social sphere, maintaining

---

[40] Public–private partnerships in Russia takes various legal forms in accordance with such laws as, for example, the Federal Law "On Concession Agreements" dated 21.07.2005 No. 115-FZ, the Federal Law "On public–private partnership, municipal-private partnership in the Russian Federation and amendments to certain legislative acts of the Russian Federation" dated 13.07.2015 No. 224-FZ.

[41] Shen et al. (2016).

stable investment growth, cultivating new driving forces for economic development, promoting economic transformation and modernization, and achieving coordinated economic and social development is of particular importance.[42] Thus, there is an initiative in China for creating guidelines to foster the issuance of special bonds for industries in social fields (health, elderly care, education, culture and sports), which are characterized by low average returns and long return cycles, as well as developing provisions for incentives to promote the development of related industries. The National Development and Reform Commission of China has taken the initiative to actively promote these investments, the Securities Regulatory Commission of China and the People's Bank of China play a leading role in supporting qualified enterprises in relevant fields to issue corporate bonds, non-financial corporate debt financing instruments and asset securitization products, and explore the issuance of combined equity bond financing products to meet daily working capital needs. Great efforts are being made by government agencies to direct social capital to participate in the construction and operation of health care facilities, aged care facilities, educational institutions, cultural and sports facilities through the public–private partnership model. The Chinese authorities are trying to make full use of the guiding role of public funds, but where the current conditions allow, sectoral social capital investment funds can be established. Such funds are characterized by market-oriented activities and social capital as the basis. It is Important to mention, that responsibility for the implementation of above mentioned financial schemes are carried by the government of each province of China.[43]

Since the late twenty-first century, the South African political economy is structured around the "mineral-energy complex," historically supported and dominated by British imperial conglomerates. From the 1930's, marginalized Afrikaner elites promoted a nationalist project with the objective to develop a "volkskapitalism", i.e., a capitalism by and for Afrikaners.[44] Impact Investing South Africa is a cross sectoral initiative aimed at building the eco-system in order to accelerate the deployment of capital that optimises financial, social and environmental returns.[45] Impact investment in the country is mostly financed from foreign direct investment including the investment from international institutions. Foreign and international institutions investments is mostly the source of impact investment also in Brazil and India as in most developing countries.

Impact investment plays an important role in fostering sustainable growth and development. It allows businesses and non-profit organizations including institutes of development to finance social and environmental initiatives in addiction to budget funding. In developing countries impact investments are mostly financed from foreign

[42] 国务院办公厅关于进一步激发社会领域投资活力的意见(国办发〔2017〕21号)// Opinions of the General Office of the State Council on further stimulating the viability of social sector investment (Guobanfa [2017] No. 21).

[43] Guotai and Guotai (2023).

[44] Ducastel and Anseeuw (2020).

[45] https://www.impactinvestingsa.co.za/.

resources including international institutions, but Russia and China have their own ways of achieving solidarity economy through impact investing.

Below you can find trend comparison of BRICS countries' financial development from 1990 to 2020[46] to see how each country develops in terms of investment activity:

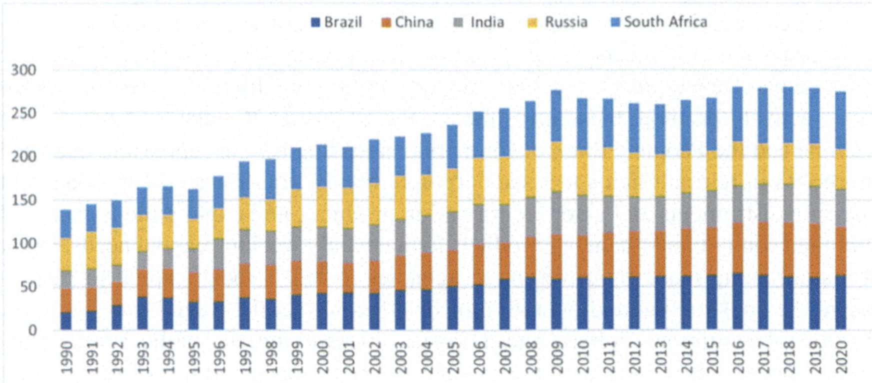

3.1.1 The Role of International Institutions in Fostering Responsible (Sustainable) Investment in Russia and Other BRICS Countries

Various institutes of development including banks of development of national or international origin play a significant role in fostering social responsibility and environmental harmony in BRICS countries, including Russia. BRICS has its own bank of development, called New Development Bank, founded in 2015 as an alternative to the World Bank with ambitious goals to become the central international institution of non-dollar based economy. The establishment of the Bank was one of the results of the political and economic dissatisfaction arising out of the growing disparity between the BRICS' share of the world economy and its representation in the institutions underpinning the global financial architecture.[47]

Below it is shown how BRICS and G7 countries' share of the world's total gross domestic product (GDP) in purchasing power parity (PPP) from 2000 to 2023[48]:

[46] Trend comparison of BRICS countries' financial development from 1990 to 2020 https://www.researchgate.net/figure/Trend-comparison-of-BRICS-countries-financial-development-from-1990-to-2020_fig3_361566740.

[47] Hofman and Srinivas (2022).

[48] BRICS and G7 countries' share of the world's GDP in PPP 2000–2023. Statista Research Department (2024) https://www.statista.com/statistics/1412425/gdp-ppp-share-world-gdp-g7-brics/#:~:text=The%20BRICS%20countries%20overtook%20the,held%20by%20the%20G7%20countries.

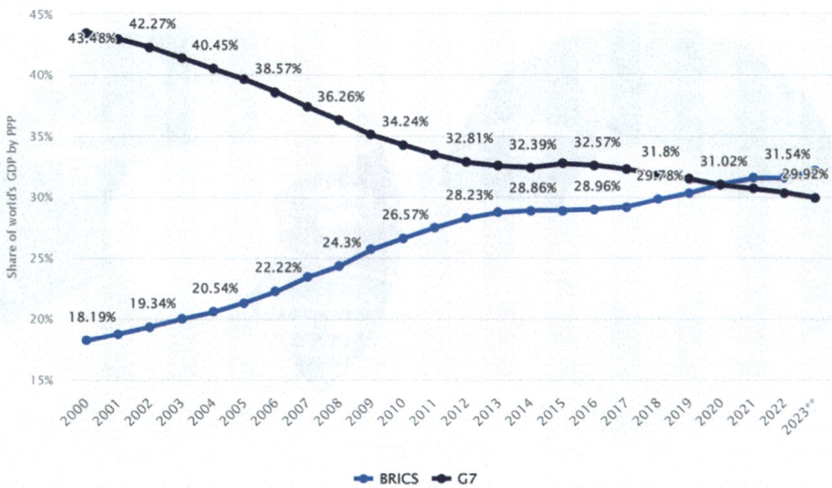

Unlike other integration or regional multilateral development banks, the New Development Bank aims to contribute in the world's sustainable future, so it does not support solely the projects in BRICS countries, as it has a much broader outlook on financing, which includes high impact operations all over the World. The New Development Bank seeks to ensure that all projects financed by it are implemented in a sustainable way and their ESG impacts are assessed, minimized and mitigated during project implementation.[49]

BRICS member states made their own contribution in the capital of the New Development Bank that now aims to finance the big and important projects in BRICS and other countries thus strengthening its influence and broadening its impact in the sustainable development of global and regional economies. The New Development Bank take into consideration the goals of BRICS members in their vision of infrastructure development, including large scale projects in housing units, bridges and roads to be constructed, irrigation facilities to be upgraded, drinking water supply to be improved. Thus it reduces overall poverty, protects social needs of the population of BRICS countries and their partners.

New Development Bank brings not only money, but also its unique expertise in the projects. It can participate in equity or debt schemes and gather collective investment to finance international projects as well as national projects of the countries, it which economies it aims to contribute. It can form different collaborations with other developments banks to fulfill its goals in the projects. It is important to mention, that the New Development Bank is much closer to the concept of the state rather than business as it does not aim at profit but aims at solving other social, infrastructure, ecological and other issues. It is important to mention that national bank of developments can even be seen as agents of governments to which the governments delegated some of the state functions.

---

[49] About NDB https://www.ndb.int/about-ndb/.

See the project breakdown by country and by area below[50]:

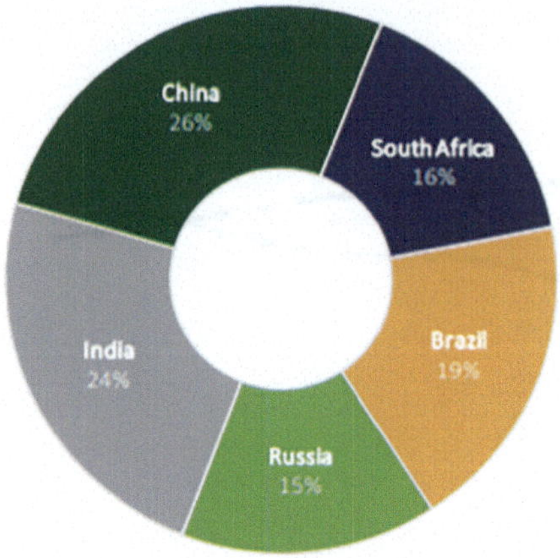

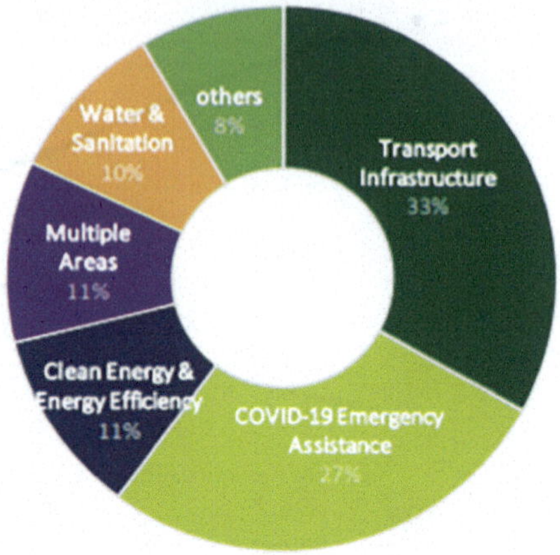

Each BRICS member has its own development bank, created under a law[51] or a special governmental act to support economic growth and to finance high impact projects. State corporation of development VEB.RF, Brazilian Development Bank, China Development Bank, Industrial Development Bank of India, Development Bank of South Afrcia and other national institutes of development in BRICS countries are mostly owned, controlled and financed by the states. Basically, they are state agents in fostering development of national economies and ensure the social and environmental responsibilities of the governments. National development banks provide responsible investment for national projects such as infrastructure, basic industries, energy and transportation, green economy, digital economy and innovation. These institutions are deemed to be methodological centers for the usage of financial instruments for sustainable development, they not only invest their own money in high impact projects but also attract and gather the collective investment from different sources, which can play a significant role in ensuring social and ecological welfare of society.

4 Conclusions

The state plays a significant role in guaranteeing the minimal social and environmental protection of the citizens in Russia and other BRICS countries in contrast of the Western world, where the concept of "minimal state" has become very popular. In Russia and other BRICS countries people depend on the state and the state tries to be dependable as per their Constitutions, where the state is declared to be responsible for the society welfare and environmental harmony. Unlike the Western countries, where

[50] Investor fact sheet https://www.ndb.int/.

[51] For example, Russian Development Bank VEB.RF is regulated by the Federal Law "On the State Development Corporation "VEB.RF"" No. 82-FZ dated May 17, 2007.

the institute of civil society is strong enough to provide balanced self-regulation, developing countries prefer to rely on the state in carrying the social and ecological functions.

Delegation of the social and environmental state functions to the society in Russia and BRICS countries follows the concept of solidarity economy, which is the alternative to the concept of social economy of the Western world, where the "third sector" organizations play an important role. The solidarity economy integrates the effort of the state and the whole society, including businesses and each and every person, which all play an important role in solving social and environmental issues. True economic solidarity is achieved, when the state creates conditions for mutually beneficial effective cooperation, including incentives and state support measures for social and green initiatives.

Impact investment plays an important role in fostering social responsibility and environmental harmony. When budget resources are not enough to finance solutions in social and ecological sphere, the involvement of businesses is of great help. Russia and China have their own ways of impact investment that implements the concept of solidarity economy in practice, in contrast Brazil, India and South Africa mostly fund impact investments from foreign resources including international institutions.

Development banks as the agents of the state that carry some of its functions in financial support of nationwide investment projects and play a significant role in gathering impact investment and organizing investment schemes for fostering social responsibility and environmental harmony in BRICS countries.

References

Almeida WMD (2019) About the end of free public higher education in Brazil. Cad Pesqui 49:10–27

Bykov AG (2004) On the content of the course in business law and the principles of its construction. Business law in a market economy. M.: New legal culture, 61

Draibe SM (2007) The Brazilian developmental welfare state: rise, decline and perspectives. Latin America: A new developmental welfare state model in the making? Palgrave Macmillan UK, London, pp 239–281

Ducastel A, Anseeuw W (2020) Impact Investing in South Africa. Historical Social Research/ historische Sozialforschung 45(3):53–73

Goswami R (2001) Indian pension system: problems and prognosis. Indian Institute of management, Bangalore, January, 262001

Gubin EP (2015) State and business in the context of legal reforms. J Russian Law 1:23–30

Guotai J, Guotai JS (2023). A study on the innovation of financing instruments in capital market for small and medium-sized enterprises. Multilevel Capital Markets Res 1:159 (国泰君安证券课题组, 国泰君安证券. 中小企业资本市场融资工具创新研究 //《多层次资本市场研究》. 2023. 1:159)

Hofman B, Srinivas PS (2022) New development Bank's role in the international financial architecture. National University of Singapore, East Asian Institute

Kohli A (2009) States and economic development. Brazilian J Political Econ 29:212–227

Kohli A (2012) State and redistributive development in India. Growth, inequality and social development in India: is inclusive growth possible?, 194–226

Kumar A (2005) The welfare state system in India. Welfare states and the future, 336–363

Lavinas L, Gentil D, Cobo B (2017) The controversial Brazilian welfare regime (No 2017–10). UNRISD Working Paper

Lo AW, Zhang R (2023) Quantifying the impact of impact investing. Manag Sci

Lobato LDVC (2016) Social policies and social welfare models: fragilities of the Brazilian case. Saúde Em Debate 40:87–97

Meng T, Newth J, Woods C (2022) Ethical sensemaking in impact investing: reasons and motives in the Chinese renewable energy sector. J Bus Ethics 179(4):1091–1117

Morais LP, Dash A, Bacic MJ (2017) Social and solidarity economics in India and Brazil. Soc Enterp J 13(1):95–112

On 30/07/2023. https://www.globalgovernmentforum.com/in-pursuit-of-public-good-the-need-for-a-welfare-revolution-in-south-africa/

Otsuki K, de Castro F (2020) Solidarity economy in Brazil: towards institutionalization of sharing and agroecological practices. Building More Sustainable and Resilient Society, Sharing Ecosystem Services, pp 159–178

Peterdy K, What is ESG?. https://corporatefinanceinstitute.com/resources/esg/esg-environmental-social-governance/

Robson WA (1954) India as a welfare state. Polit Quart 25(2):116–131. https://doi.org/10.1111/j.1467-923X.1954.tb00152.x

Roman A (2023) A closer look into Brazil's healthcare system: what can we learn?. Cureus 15(5)

Shen C, Xu Z, Yang Y (2016) A reduced pricing model for mezzanine financing based on options and support vector machines. Clust Comput 19:2169–2177

Verma V (2018) State in India: ideas, norms and politics. https://www.researchgate.net/publication/330093311_State_in_India

Vidojević J, In pursuit of public good: the need for a welfare revolution in South Africa

Wang Y (2021) Problems of sustainable development in international investment agreements: from external effects to internal effects. Wood's Rev Int Law 5:117 (王晔琼. 中国人民大学法学院//《武大国际法评论. 2021. 5. 117)

Wendt F (2016) Political authority and the minimal state. Soc Theory Practice, 97–122

Embracing Sustainability: Role of Corporate Social Responsibility in Promoting Green Growth Amidst a Competitive Business Environment in India

Jayashree Bhattacharjee, Ranjit Singh, and Nitish Kumar Singh

Abstract In today's ever-evolving business environment, the synergy between corporate achievement and societal accountability highlights the critical importance of integrating sustainability through Corporate Social Responsibility (CSR). This article explores CSR's pivotal role in cultivating green growth amid the intricacies of a highly competitive business environment. Starting with an examination of CSR's evolution and components, the narrative highlights the need for a paradigm shift in business strategies amidst global challenges like climate change and resource scarcity. Central to this transformation is the integration of sustainable practices, emphasizing eco-friendly measures and operational efficiency. The article delves into the multifaceted benefits of CSR, illustrating how companies can shape a positive corporate image by aligning with consumer preferences and ethical considerations. Crucial elements such as stakeholder engagement and collaboration are emphasized, showcasing successful CSR initiatives that foster sustainable development through cooperative efforts. Additionally, the article underscores CSR's role in navigating the evolving regulatory landscape, emphasizing the importance of proactive compliance for sustained business viability. In an era where investors increasingly consider Environmental, Social, and Governance (ESG) criteria, the article positions CSR as a magnet for socially responsible investments. It elucidates how a commitment to sustainable practices not only attracts investors but also fuels innovation, fostering market differentiation and enhancing a company's competitive edge. Ultimately, the article contends that incorporating CSR into corporate strategy is not

J. Bhattacharjee (✉)
Symbiosis School of Banking and Finance, Symbiosis International Deemed University, Lavale, Pune, India
e-mail: jayashree.bhattacharjee@ssbf.edu.in

R. Singh
School of Management Studies, Indian Institute of Information Technology, Allahabad, Prayagraj, India

N. K. Singh
School of Business and Management, CHRIST University, Bangalore, India

P. Singh et al. (eds.), *Integrating Environmental Sustainability and Corporate Social Responsibility: A Move Beyond Profit*, World Sustainability Series,
https://doi.org/10.1007/978-3-031-81469-3_9

merely a response to societal expectations but a strategic imperative for businesses seeking resilience, relevance, and prosperity amidst intense competition and global challenges.

Keywords Corporate Social Responsibility (CSR) · Sustainable business · Business environment · Global challenges · Environmental degradation · Competitive market

1 Introduction

India is emerging as one of the world's fastest growing economies, recently ranking as the fifth largest globally with a GDP of $3.75 trillion (Forbes India 2023). However, this growth trajectory raises concerns about its environmental impact, given the strain on natural resources such as land, water, minerals, and fossil fuels, leading to increased energy and commodity prices. Balancing economic development with environmental sustainability and social equity is crucial. The concept of "green growth" becomes pivotal here, emphasizing the need to reduce resource consumption over time while promoting job creation and social equity. Achieving sustainable development goals hinges on India's ability to embrace green growth practices.

Corporate Social Responsibility (CSR) is a dynamic and evolving concept that encapsulates a company's commitment to addressing and balancing its economic, social, and environmental impacts. The definition of CSR has evolved over time, reflecting changing societal expectations and perceptions of the role of businesses in the broader community. Initially, CSR was primarily seen through a philanthropic lens, where companies engaged in charitable activities and community initiatives as a way of giving back (Carroll 1999). This narrow understanding perceived CSR as a discretionary act, divorced from core business functions. However, as the global business landscape underwent profound changes, so did the definition of CSR. A more comprehensive perspective emerged, recognizing that CSR extends beyond charitable donations to encompass ethical business practices, environmental stewardship, and a commitment to societal well-being.

The evolution of CSR can be traced through different stages. Carroll's seminal CSR pyramid, introduced in 1979, outlined four layers: economic responsibilities, legal responsibilities, ethical responsibilities, and philanthropic responsibilities. This model helped establish a framework for understanding the multifaceted nature of CSR, emphasizing that businesses have responsibilities beyond mere profit-making (Carroll 1979). Over time, scholars expanded this pyramid to reflect a more holistic approach, recognizing that CSR is an integral part of a company's identity rather than a separate, discretionary activity (Garriga and Melé 2004).

According to CSRBOX's survey of major CSR funders (2023), 26% of companies prioritize Healthcare and WASH as their primary CSR focus. This is followed by Environment and Sustainability and Promotion of Education, each highlighted by

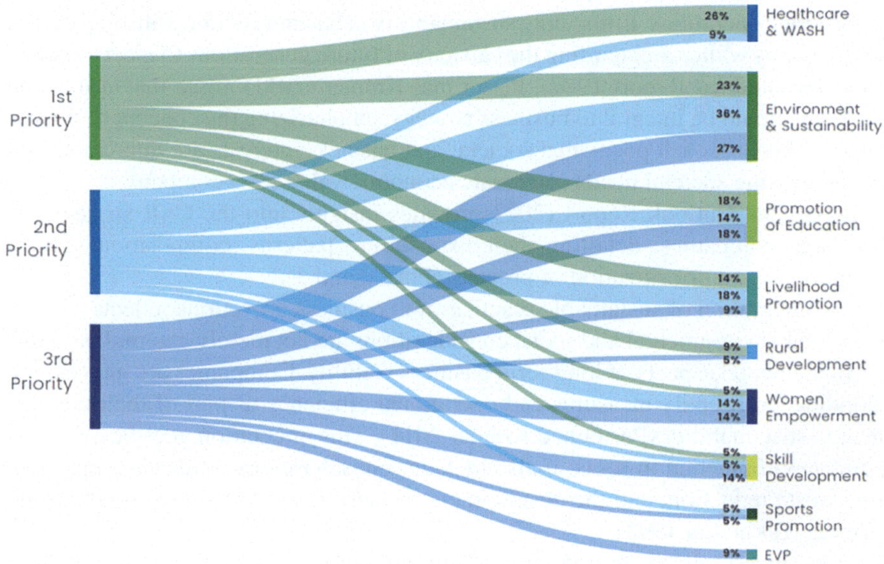

Fig. 1 CSR priority areas. *Source* CSRBOX's survey of major CSR funders (%age of companies participated)

23%. Notably, Environment and Sustainability also ranks as the second and third CSR priority for over one-third and one-fourth of companies, respectively (Fig. 1).

In a world facing pressing environmental challenges, businesses are increasingly expected to minimize their ecological footprint and contribute to environmental conservation. This involves adopting sustainable practices, reducing emissions, conserving resources, and, in some cases, investing in carbon offset programs (McWilliams and Siegel 2001). Ethical business practices constitute another vital component of CSR. This involves conducting business in a manner that aligns with moral and ethical standards. Fair labour practices, transparent supply chain management, and responsible marketing are integral to ensuring that a company's operations uphold ethical standards (Garriga and Melé 2004). Ethical considerations in CSR extend beyond legal compliance to encompass actions that go above and beyond what is legally required, reflecting a commitment to principled conduct.

Social responsibility is a cornerstone of corporates and involves engaging with and contributing positively to the communities in which companies operate. This component focuses on addressing societal issues such as education, healthcare, poverty alleviation, and community development. Engaging in social responsibility initiatives enhances a company's reputation and fosters a positive relationship with its stakeholders, including consumers, employees, and local communities (McWilliams and Siegel 2001).

There is a noticeable evolution in CSR towards prioritizing sustainable business practices, spurred by a heightened awareness of the environmental and social impacts of corporate operations. Embracing sustainability in business entails fulfilling current requirements while safeguarding the capacity of future generations to meet their own needs (Brundtland Report 1987). Porter and Kramer (2011) argue that sustainable practices should be integral to business strategy, emphasizing the concept of "shared value." This approach posits that societal and business interests are interconnected, and addressing societal needs can create economic value for a company.

India CSR Outlook Report (2023 provides insights into the CSR strategies of 301 large companies, detailing their expenditure patterns, composition of CSR committees, and prioritization of projects (Figs. 2 and 3).

The integration of sustainable practices into business operations reflects a recognition that environmental and social considerations are not merely external pressures but fundamental aspects of long-term business viability. Companies are increasingly adopting eco-friendly measures and operational efficiency as part of their commitment to sustainability (Porter and Kramer 2011). This integration goes beyond mitigating environmental impacts; it also includes considerations of resource optimization, waste reduction, and the development of innovative, sustainable products and services (Elkington 1998).

This paper discusses about business environment, its competition and challenges, how green initiative can make positive brand image, stake holder engagement and collaboration in green growth, regulatory standards, investors attraction and concluding remark. This paper is primarily a qualitative work and the methodology adopted in this paper is that of reviewing the existing literature available on this subject followed by the logical interpretation of those literature. The recent and relevant literature related to this has been chosen from the several reputed journals indexed in reputed databases. Thereafter, nomological network was established

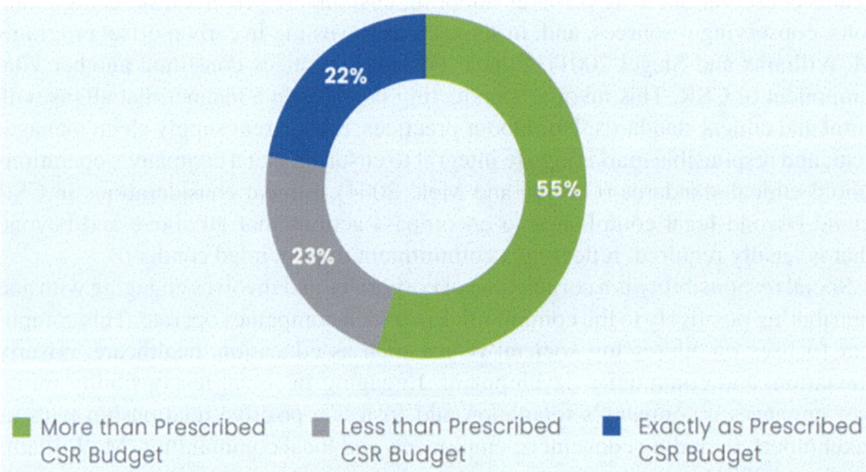

Fig. 2 Percentage of Companies Meeting their Respective CSR Compliance Requirements. *Source* CSRBOX Research Desk

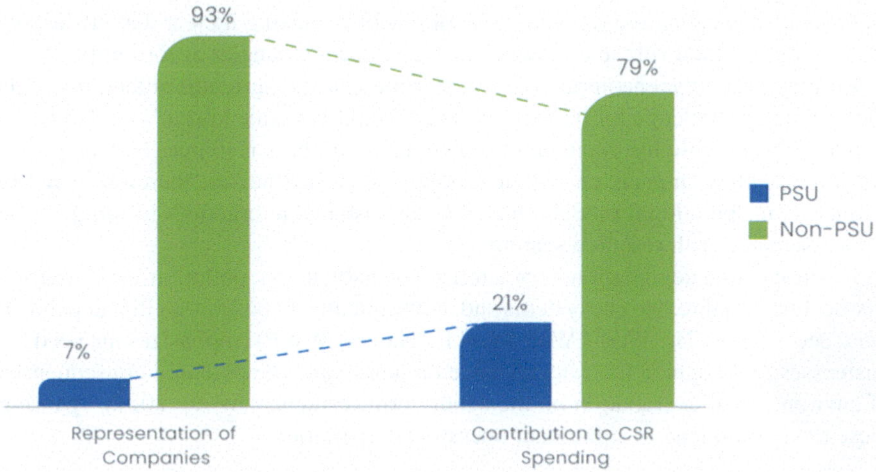

Fig. 3 Type of Companies and Contribution to CSR Spent. *Source* CSRBOX Research Desk

between the variables and constructs of interest. Various reports of several bodies have also been considered in this paper.

2 The Business Environment: A Landscape of Competition and Challenges

The business environment is undergoing rapid transformation, shaped by factors ranging from technological advancements to shifting consumer preferences. However, the emergence of global challenges, particularly climate change, resource scarcity, and environmental degradation, adds a layer of complexity that demands immediate attention (Porter and Heppelmann 2014). The interplay of these challenges poses risks not only to the environment but also to the long-term viability of businesses.

2.1 Emerging Global Challenges

With the increased industrialisation and technological development, certain new challenges such as climate change, resource scarcity, environmental degradation have evolved globally.

The impacts of climate change are no longer theoretical; they are tangible and pose substantial risks to businesses across sectors. Rising temperatures, extreme weather events, and regulatory changes are shaping a new reality. Brown (2005) underscores the economic repercussions of inaction, making it clear that businesses must recognize their role in climate change and adapt strategies accordingly.

Resource scarcity, encompassing water, minerals, and agricultural land, presents a multifaceted challenge for businesses. The World Economic Forum's Global Risks Report (2021) highlights the interconnected nature of these resources and the potential for cascading impacts on global supply chains. Businesses, therefore, need to rethink their operational models, focusing on sustainable practices to mitigate the risks associated with resource scarcity.

Environmental degradation, characterized by habitat loss, pollution, and biodiversity decline, is a threat to ecosystems and, consequently, to businesses that depend on these ecosystems. The World-Wide Fund for Nature (WWF) emphasizes the need for businesses to recognize their dependence on nature and the potential consequences of environmental degradation on their long-term resilience (WWF 2021). Ignoring these risks could lead to significant operational disruptions.

2.2 The Imperative for a Green Paradigm Shift

To address these challenges, a paradigm shift towards green initiatives is not just a strategic choice but a necessity for businesses. The adoption of sustainable practices aligns with broader global efforts to combat climate change and ensures a resilient and responsible business model. Urged by the Intergovernmental Panel on Climate Change (IPCC), transitioning industries to sustainable practices is imperative to mitigate global warming and its consequential impacts. (IPCC 2018).

One key aspect of the green paradigm shift involves the integration of renewable energy sources into business operations. The International Energy Agency (IEA) advocates for a rapid increase in renewable energy deployment to meet sustainability goals and reduce dependence on fossil fuels (IEA 2020). Businesses can contribute to this shift by investing in and transitioning to renewable energy, thereby reducing their carbon footprint and fostering environmental responsibility.

Adopting circular economy practices is essential in mitigating the impact of resource scarcity. The Ellen MacArthur Foundation (2017) promotes a circular economy model that prioritizes the reduction, reuse, and recycling of materials, minimizing waste and promoting sustainable resource use. Businesses can incorporate these principles into their supply chains, fostering resilience against resource shortages.

The integrity of supply chains is critical in the green paradigm. Businesses must embrace ethical supply chain management practices, ensuring fair labour conditions and responsible sourcing of materials. The Global Reporting Initiative (GRI) provides a framework for businesses to transparently communicate their environmental and social impacts, promoting accountability in the supply chain (GRI 2022).

Innovation for sustainability is a cornerstone of the green shift. The World Business Council for Sustainable Development (WBCSD) encourages businesses to innovate solutions that align with environmental goals (WBCSD 2018). Additionally, integrating corporate social responsibility (CSR) into business strategies, as outlined by the United Nations Global Compact (UNGC), fosters a holistic approach that considers social and environmental impacts alongside financial goals (UNGC n.d.).

The shift towards green initiatives is not isolated; it is influenced by consumer awareness and changing market dynamics. The Nielsen Global Sustainability Report (2015) highlights that consumer increasingly prefers products from companies committed to positive social and environmental impact. This growing consumer demand creates market incentives for businesses to align with sustainability principles.

3 Building a Positive Corporate Image Through Green Initiative

The imperative of adopting environmentally friendly practices permeates every facet of our lives. In the realm of business, meeting this demand necessitates a steadfast commitment to conducting eco-friendly operations throughout the entire business process. The viability of a company hinges on its ability to achieve competitive excellence. When a company integrates strategic initiatives that prioritize social responsibility, demonstrate a heightened sense of care, and actively engage with stakeholders, it not only meets ethical standards but also enhances the overall competitiveness of the organization (Jesús et al. 2016).

Corporate image, frequently interchangeable with reputation, stands out as a priceless intangible asset for businesses. It functions as a mirror reflecting a company's values, operational practices, and its overall position in the perception of stakeholders, encompassing consumers, investors, and employees. In a market marked by intense competition and a plethora of choices, a favorable corporate image emerges as a key differentiator. This positive image significantly influences consumer trust, fosters loyalty, and shapes purchasing decisions (Balmer 2012). Research conducted by Widyastuti (2019) reveals that the implementation of green marketing strategies and the cultivation of an environmentally conscious organizational culture not only contribute to a positive green corporate image but also confer a substantial competitive advantage within the industry.

With the rising awareness among consumers regarding environmental and ethical issues, there is a rising demand for businesses to exhibit social responsibility. Beyond mere considerations of product quality and pricing, consumers are now actively seeking to support companies that resonate with their values (Cone Communications 2015). A notable transformation in consumer behavior is evident, where individuals are not only interested in the offerings but are also inclined to associate themselves with companies demonstrating green compliance. Yazdanifard and Erdoo (2011) indicated that consumers are increasingly willing to invest in a greener lifestyle and are ready to pay a premium for companies that prioritize environmental responsibility. This shift in consumer preferences underscores the necessity for businesses to incorporate ethical considerations, particularly through green initiatives, as integral components of their corporate strategies.

Fostering a green corporate image involves a synergy of environmental sustainability practices and transparent communication regarding these initiatives. The adoption of eco-friendly practices, such as minimizing carbon emissions, instituting recycling programs, and utilizing sustainable materials, serves as a tangible demonstration of a firm's dedication to environmental responsibility (Delmas and Burbano 2011). Consumers are increasingly drawn to align themselves with companies that actively contribute to environmental preservation. When companies effectively communicate these efforts through various channels such as advertising, promotions, publications, and social responsibility initiatives, they are likely to attract a substantial base of loyal customers. Isabel et al. (2016) emphasizes the significance of giving particular attention to a company's brand identity, highlighting its impact not only on employee commitment but also on overall brand performance and satisfaction.

Consumers are becoming more discerning, actively choosing products and services from companies that resonate with their values. Nielsen (2015) reports that 66% of consumers worldwide are willing to pay a premium for sustainable products. A green corporate image, therefore, not only attracts environmentally conscious consumers but also builds trust and loyalty among a broader audience (Cone Communications 2015). Transparency about sustainable practices fosters a sense of authenticity, crucial for building and maintaining trust (Gössling et al. 2018).

In a competitive market, a green corporate image can confer a significant competitive advantage. Businesses that proactively adopt sustainable practices are likely to stand out among their peers. It was also found in several studies that companies with strong environmental, social, and governance (ESG) performance often outperform their counterparts in terms of stock market and operational performance (Singh et al. 2016; Mondal and Singh 2020; Mondal et al. 2022; Kotsantonis and Serafeim 2021). This underscores the financial benefits of integrating sustainability into corporate strategy.

A favourable corporate image extends its influence to employee perceptions and engagement, a trend particularly pronounced among the younger workforce generations (Deloitte 2020). Khalid et al. (2023) found that green corporate image not only uplifts employee morale but emerges as a decisive factor in attracting top talent within a fiercely competitive job market. Although factors such as compensation and work-life balance undoubtedly wield influence, a company's commitment to environmental sustainability and effective management of business emissions surprisingly exerts robust effects on human resources (HR). Consequently, companies embracing sustainable practices not only stand to save on recruiting expenses but also mitigate turnover costs, fostering higher productivity levels among existing employees.

In the face of crises or controversies, companies with a positive corporate image, especially those associated with green initiatives, tend to fare better. Du et al. (2010) found that companies with a strong commitment to corporate social responsibility (CSR) experience better resilience during economic downturns. A green corporate image acts as a shield, helping companies weather challenges and recover more swiftly.

Meeting stakeholder expectations is essential for securing a social license to operate. Stakeholders, including consumers, investors, and communities, increasingly expect companies to contribute positively to society and the environment. Fulfilling these expectations through green initiatives not only aligns with ethical considerations but also secures the company's social license to operate (Bansal and DesJardine 2014).

4 Stakeholder Engagement and Collaboration

In the contemporary business landscape, Corporate Social Responsibility (CSR) has evolved beyond a mere philanthropic endeavour to a strategic imperative. At the heart of effective CSR lies stakeholder engagement and collaborative approaches, recognizing the interconnected relationships between businesses, communities, and the environment.

Stakeholders, encompassing customers, employees, investors, communities, and regulatory bodies, are integral to the CSR framework. Freeman's stakeholder theory (1984) asserts that organizations should consider the interests of all stakeholders, emphasizing the reciprocal relationship between business success and societal well-being. Their influence extends beyond economic considerations, shaping the ethical dimensions of corporate behavior (Freeman 1984).

Stakeholders contribute to CSR in various capacities. Employees, for instance, are crucial advocates for internal CSR practices, while customers increasingly base purchasing decisions on a company's ethical standing (Bhattacharya et al. 2008). Investors are demonstrating a growing interest in Environmental, Social, and Governance (ESG) factors, influencing corporate decisions (Hong and Kacperczyk 2009). Recognizing these dynamics, businesses engage stakeholders as active partners in their CSR initiatives, creating a collaborative ecosystem for sustainable practices.

Green growth, a pathway to sustainable development, involves fostering economic growth while minimizing environmental impact. It requires collaboration among diverse stakeholders, including businesses, governments, non-governmental organizations (NGOs), and local communities (World Bank 2012). The World Bank defines green growth as an approach that integrates economic, social, and environmental considerations to promote resilient and sustainable development. Austin and Seitanidi (2012) argued that collaborative approaches enable organizations to pool resources, share expertise, and amplify their impact, creating a synergy that transcends individual capabilities. India, ranked as the world's third-largest producer of renewable energy with 42% of its power sourced from clean sources, is leading the charge in the global shift towards net zero emissions. Amidst its remarkable economic ascent from the 13th to the fifth largest economy globally, India is setting ambitious benchmarks for sustainable development.

Government-industry collaborations play a pivotal role in fostering a regulatory environment conducive to promoting green growth. Energy Efficiency Services Limited (EESL), a joint venture of Public Sector Undertakings, in partnership with the Global Green Growth Institute (GGGI), an international inter-governmental organization, recently launched the Asia Low Carbon Buildings Transition (ALCBT) Project in India. This initiative operates under the guidance of the Ministry of Housing and Urban Affairs (MoHUA), aimed at addressing the significant rise in peak electricity demand driven primarily by the increasing demand for cooling solutions in the country. Such collaborations provide businesses with clear guidelines and incentives for adopting environmentally friendly practices.

Green supply chain management is a comprehensive initiative that extends across a company's entire operations, involving all stakeholders and transcending the confines of the factory walls. This approach, highlighted by Rao (2000), goes beyond internal processes, aiming to extend sustainability benefits to the surrounding communities. In India and across much of Asia, organizations have adopted the green supply chain approach as a prominent means of showcasing their dedication to sustainability (Bacallan 2000). Recognizing the interconnectedness of a company and its business partners within a single system for product or service delivery, any lapse in sustainability initiatives by these partners is perceived as the company's failure. To address this, leading companies have sought to seamlessly integrate sustainability initiatives throughout the entire supply chain, emphasizing the importance of a holistic approach to ensure environmental responsibility.

Investors are increasingly leveraging collaborative approaches to drive green growth through ethical investments. The Principles for Responsible Investment (PRI) is a global initiative that encourages investors to integrate Environmental, Social, and Governance (ESG) considerations into their investment decisions (PRI 2022). By joining PRI, investors actively collaborate to promote responsible and sustainable investment practices, exerting collective influence on corporate behaviour and contributing to the broader green growth agenda. All these activities are also expected to build positive brand image for the company which in turn make the investors perceive about the company's stock as less risky because all these CSR activities make the name of the company familiar. It is because of familiarity bias investors are expected to consider these stocks as less risky (Bhattacharjee and Singh 2019; Bhattacharjee et al. 2021; Singh et al. 2021).

Community engagement is vital for sustainable development, fostering inclusive practices and addressing local needs. An exemplary instance is the "Swachh Bharat Abhiyan" (Clean India Mission) in India, where the government engages communities to promote sanitation and hygiene (Government of India 2014). Through active involvement in planning and implementation, communities become stakeholders in achieving sustainable improvements in public health and environmental cleanliness. This initiative showcases the power of community engagement in aligning development goals with grassroots aspirations, creating a sense of ownership and collective responsibility for long-term impact.

4.1 Successful CSR Initiatives Through Stakeholder Engagement: Case Studies

4.1.1 Tata Group's Community Development Initiatives

The Tata Group is renowned for its comprehensive CSR initiatives that engage various stakeholders. One prominent example is the Tata Power Community Development Trust, which collaborates with local communities to address issues such as education, healthcare, and sustainable livelihoods. Stakeholder engagement is a cornerstone, involving local residents in project design and implementation, ensuring initiatives align with community needs (Bhatt 2017).

4.1.2 Infosys Foundation's Healthcare Projects

The Infosys Foundation, led by Infosys Technologies, has made significant contributions to healthcare through various projects. The foundation engages stakeholders by partnering with local healthcare providers and involving communities in health awareness programs. Their efforts have focused on preventive healthcare, benefiting both urban and rural populations. (Infosys Foundation n.d.)

4.1.3 Reliance Foundation's Rural Transformation Programs

Reliance Foundation, led by Reliance Industries, emphasizes stakeholder engagement in its CSR initiatives. The Rural Transformation Program engages with local communities to enhance agricultural practices, provide access to education, and promote healthcare. This initiative empowers stakeholders by integrating their insights into the planning and execution of projects. (Reliance Foundation n.d.).

4.1.4 Mahindra Group's Sustainability Initiatives

Mahindra Group is committed to sustainability and has engaged stakeholders in various CSR projects. The 'Project Hariyali' focuses on afforestation and environmental conservation, involving communities in tree planting drives and sustainable agricultural practices. Stakeholder engagement is key to the success of these projects, fostering a sense of ownership among local communities. (Mahindra Group n.d.).

4.1.5 Hindustan Unilever Limited's Water Conservation Programs

Hindustan Unilever Limited (HUL) has been actively engaged in CSR initiatives centered around water conservation. The 'Water for Public Good' initiative collaborates with local communities to implement water management projects, ensuring sustainable water access. Stakeholder engagement involves partnering with local NGOs and involving community members in water conservation education programs (HUL n.d.) (Figs. 4 and 5).

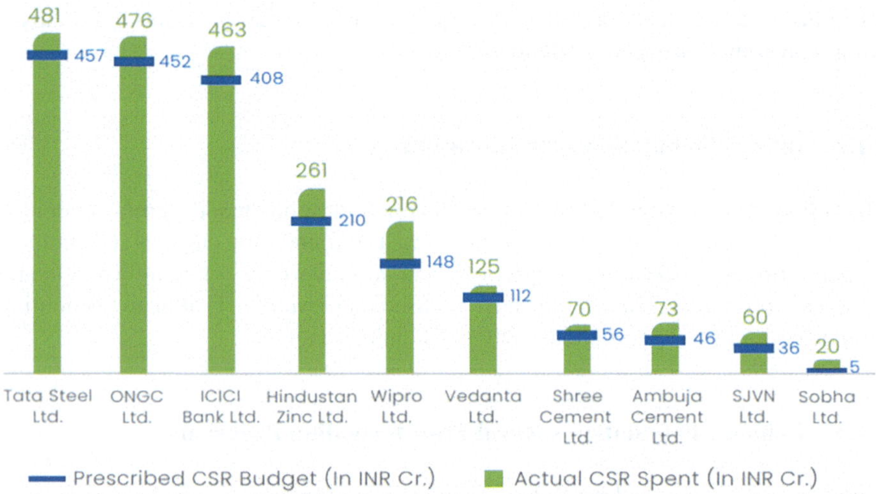

Fig. 4 Companies in India Exceeding basic CSR compliance standards. *Source* CSRBOX Research Desk

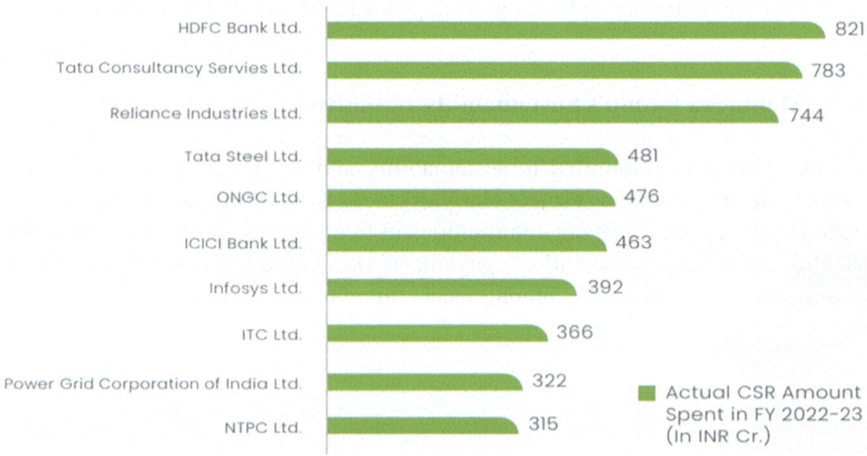

Fig. 5 Top 10 companies in India with highest CSR Spending. *Source* CSRBOX Research Desk

5 Compliance with Regulatory Standards

The business landscape in India is characterized by a dynamic regulatory framework, continually adapting to societal, economic, and environmental changes.

India's regulatory environment has witnessed significant shifts, reflecting a commitment to foster a transparent and competitive business environment. Reforms such as the Goods and Services Tax (GST) implementation, amendments in labor laws, and environmental regulations showcase the government's effort to align policies with global best practices and address emerging challenges. These changes necessitate those businesses stay agile and responsive to evolving regulatory requirements.

CSR plays a pivotal role in ensuring proactive compliance beyond mere legal obligations. It is a strategic approach where businesses voluntarily integrate ethical and responsible practices into their operations. While regulatory compliance focuses on meeting legal requirements, CSR goes further, emphasizing a commitment to societal welfare and environmental stewardship. Integrating CSR into business strategies not only aligns companies with societal expectations but also reduces the risk of non-compliance by fostering a culture of responsible business conduct.

Regulatory compliance is closely tied to a company's reputation. Adhering to established standards signals a commitment to ethical conduct and responsible business practices, enhancing trust among customers, investors, and the broader community (Eccles et al. 2014).

Compliance acts as a shield against legal risks and potential repercussions. By aligning with regulatory standards, businesses reduce the likelihood of legal disputes, penalties, and damage to their brand reputation. This risk mitigation contributes to long-term stability and viability (Khan et al. 2016; Bhattacharjee et al. 2021).

Rigorous compliance requirements often lead to improved operational efficiency. To meet regulatory standards, companies streamline their processes, adopt best practices, and implement robust governance structures. This not only ensures compliance but also enhances overall operational effectiveness (Eccles et al. 2014).

6 Challenges of Regulatory Compliance

The dynamic nature of regulations, coupled with their complexity, presents a significant challenge for businesses. Frequent changes and updates require continuous vigilance and adaptation, making it resource-intensive for companies to stay abreast of the evolving regulatory landscape (Eccles et al. 2014).

Compliance efforts often come with substantial costs, including investments in technology, human resources, and training. Small and medium-sized enterprises (SMEs) in particular, may face difficulties in allocating resources for compliance, which can affect their competitive edge in the market (Khan et al. 2016).

India's regulatory landscape is diverse, with regulations at the national, state, and local levels. Navigating this diversity can be daunting for businesses operating in multiple jurisdictions, leading to compliance challenges and potential inconsistencies in interpretation and implementation.

The relationship between CSR and regulatory compliance is symbiotic. CSR initiatives often go beyond legal requirements, addressing broader societal and environmental concerns. When businesses proactively engage in CSR, it not only enhances their reputation but also fosters a culture of responsibility, making compliance with legal standards a natural outcome (Hong and Kacperczyk 2009).

7 Investor Attraction and Long-Term Viability

In the dynamic landscape of contemporary finance, the role of investors extends beyond mere financial returns. Increasingly, investors are aligning their portfolios with Environmental, Social, and Governance (ESG) factors, emphasizing a commitment to sustainable and socially responsible business practices.

ESG factors have emerged as critical considerations in investment decisions, reflecting a shift towards sustainable and responsible investing. Investors are increasingly acknowledging that ESG factors can influence long-term financial performance and help manage risks effectively. According to the Principles for Responsible Investment (PRI), Indian investors are integrating ESG factors into decision-making processes to enhance transparency and align investments with socially responsible practices (PRI 2022).

A notable example is the rise of ESG funds in India, where asset managers incorporate ESG criteria to attract socially conscious investors. Companies adhering to strong ESG principles are perceived as more resilient, better positioned for regulatory changes, and capable of managing reputational risks (BSE 2020). Eccles et al. (2014), have demonstrated a positive correlation between strong ESG performance and financial outperformance, affirming the growing importance of these factors in the investment landscape.

Corporate Social Responsibility (CSR) serves as a conduit for attracting socially responsible investors who are increasingly seeking opportunities that align with their values. In India, companies actively engaging in Corporate Social Responsibility (CSR) initiatives can attract socially responsible investors. According to studies, CSR activities positively influence investment decisions, demonstrating a commitment to sustainable practices and ethical conduct (Gupta and Sharma 2016). Companies with robust CSR programs aligning with social and environmental goals are perceived as more attractive to socially conscious investors, fostering trust and long-term partnerships (SEBI 2019).

In the context of India, where economic growth is intricately linked with environmental and social considerations, green growth practices are integral to mitigating long-term risks. The country faces challenges such as resource depletion, pollution, and climate change impacts. In India, integrating sustainable practices is not just a moral obligation but also a strategic imperative for businesses. Green growth practices encompass initiatives like renewable energy adoption, waste reduction, and sustainable supply chain management.

Research indicates that companies demonstrating strong Environmental, Social, and Governance (ESG) practices typically achieve superior financial performance in the long run. Khan et al. (2016) found that firms with high ESG ratings are associated with lower equity capital costs and improved credit ratings, ultimately contributing to enhanced financial stability and long-term viability.

CSR serves as a powerful tool for attracting investors who prioritize sustainability and ethical conduct. Studies, such as the one conducted by Hong and Kacperczyk (2009), suggest that companies with strong CSR performance are more likely to attract socially responsible investors, fostering a positive feedback loop where responsible practices attract capital, enabling further CSR initiatives. However, there is a need to create strong awareness about the initiatives taken by the companies in this respect then only the benefits of these initiatives are expected to be harnessed by the investors (Bhattacharjee and Singh 2017; Bordoloi et al. 2020).

In India, adopting green growth practices is essential to mitigate the enduring risks linked to environmental degradation and scarcity of resources in the long term. India's National Action Plan on Climate Change (NAPCC) emphasizes the need for sustainable development strategies. Companies that align with these green growth principles not only contribute to national goals but also position themselves as leaders in a market that increasingly values environmental responsibility.

8 Conclusion

In the dynamic and fiercely competitive business landscape of India, the adoption of sustainability through Corporate Social Responsibility (CSR) transcends being a mere choice; it has become a strategic imperative for businesses aiming for long-term success. The multifaceted impact of CSR, encompassing holistic development and environmental stewardship, plays a pivotal role in shaping trust and resilience within organizations. The holistic development facilitated by CSR initiatives extends beyond philanthropy, emphasizing investments in community development, education, and healthcare. This comprehensive approach not only contributes to societal well-being but also aligns with global agendas such as the United Nations Sustainable Development Goals (UN SDGs), establishing CSR as a linchpin for fostering green growth (UNDP 2022). By intertwining economic objectives with societal and environmental well-being, businesses elevate sustainability from a mere corporate responsibility to a strategic cornerstone.

Environmental stewardship, a crucial facet of CSR, positions businesses as responsible custodians of the planet. Initiatives focusing on eco-friendly practices, renewable energy adoption, and conservation efforts not only align with international commitments like the Paris Agreement (UNFCCC 2022) but also contribute to building resilience against environmental challenges. The strategic imperative of embracing sustainability through CSR is underscored by the fact that businesses are no longer merely reactive to regulatory pressures but are actively becoming architects of positive change. The Companies Act, 2013, mandates CSR spending for eligible companies, signalling a regulatory expectation for corporate contributions to societal and environmental welfare (MCA 2013). However, businesses are going beyond compliance, recognizing that CSR is not just a regulatory requirement but a conduit for creating a positive impact and forging a competitive advantage.

In an era marked by heightened corporate consciousness, CSR has transcended its traditional role as a responsibility to become a gateway to a competitive advantage. Beyond meeting regulatory standards, businesses that embrace CSR are strategically positioning themselves as contributors to green growth—a trajectory that is not only beneficial for business but fundamentally essential for fostering a sustainable and thriving future.

9 Scope of Future Research

Future research on Corporate Social Responsibility (CSR) in India should deal with pioneering approaches for sustainable business practices. Emphasis on evaluating impact through metrics, incorporating state-of-the-art technologies, and promoting cross-sector partnerships is essential. Exploring the dynamic regulatory scenario, ensuring sustainability throughout supply chains, and understanding the economic dimensions of CSR initiatives will offer key insights. Examining shifts in consumer behavior and benchmarking against global standards will assist businesses in aligning with green growth amid competition. The research seeks to propel impactful CSR models, steering businesses toward environmental responsibility and nurturing a robust, environmentally conscious business landscape in India. There are many factors affecting the CSR activities of a company and its impact on green growth. All those factors are not equally important. In order to find out the relative weight of the factors and inter-relationship between them Social Network Analysis can be done (Singh et al. 2022). There is also a need to measure the accountability of the CSR spending and future research can be directed towards measuring accountability of such activities (Agarwal and Singh 2023).

References

Agarwal S, Singh R (2023) Customers' perception towards accountability of diagnostic centres: evidence from India. J Multidiscip Healthcare, 2947–2961

Austin JE, Seitanidi MM (2012) Collaborative value creation: a review of partnering between nonprofits and businesses. Part I: value creation spectrum and collaboration stages. Nonprofit Voluntary Sector Quart 41(5):726–758

Bacallan JJ (2000) Greening supply chain. Bus Environ 5:11–15

Balmer JMT (2012) Corporate brand management imperatives: custodianship, capital, and communication. Calif Manage Rev 54(3):6–33

Bansal P, DesJardine MR (2014) Business sustainability: It is about time. Strateg Organ 12(1):70–78

Bhatt P (2017) Tata power: empowering lives through community development

Bhattacharjee J, Singh R (2017) Awareness about equity investment among retail investors: a kaleidoscopic view. Qualit Res Financ Markets 9(4):310–324

Bhattacharjee J, Singh R, Kajol K (2021) Risk perception in respect of equity shares: a literature review and future research agenda. DLSU Bus Econ Rev 30(2):101–119

Bhattacharya CB, Sen S, Korschun D (2008) Using corporate social responsibility to win the war for talent. MIT Sloan Manag Rev 49(2):37–44

Bordoloi D, Singh R, Bhattacharjee J, Bezborah P (2020) Assessing the awareness of Islamic law on equity investment in state of Assam. India. J Islamic Financ 9(1):001–012

Brown M (2005). https://www.c2es.org/document/towards-a-climate-friendly-built-environment/

Brundtland GH (1987) Our common future: report of the world commission on environment and development. Oxford University Pres.

BSE (2020). https://www.bseindia.com/static/about/sustainability.html

Carroll AB (1979) A three-dimensional conceptual model of corporate performance. Acad Manag Rev 4(4):497–505

Carroll AB (1999) Corporate social responsibility: evolution of a definitional construct. Bus Soc 38(3):268–295

Cone Communication (2015) Capcon. https://www.capcon.it/en/2015-cone-communicationsebiquity-global-csr-study/

Delmas MA, Burbano VC (2011) The drivers of greenwashing. Calif Manage Rev 54(1):64–87

Deloitte (2020) Deloitte global millennial survey 2020

Du S, Bhattacharya CB, Sen S (2010) Maximizing business returns to corporate social responsibility (CSR): the role of CSR communication. Int J Manag Rev 12(1):8–19

Eccles RG, Ioannou I, Serafeim G (2014) The impact of corporate sustainability on organizational processes and performance (December 23, 2014). Manag Sci 6 (11):2835–2857, February 2014, Available at SSRN: https://ssrn.com/abstract=1964011 or https://doi.org/10.2139/ssrn.1964011

Elkington J (1998) Partnerships from cannibals with forks: the triple bottom line of 21st-century business. Environ Qual Manage 8(1):37–51

Freeman RE (1984) Strategic management: a stakeholder approach. Pitman, Boston

Freeman RE (2010) Strategic management: a stakeholder approach. Cambridge University Press

Garriga E, Melé D (2004) Corporate social responsibility theories: mapping the territory. J Bus Ethics 53(1–2):51–71

GDP India (2023). Forbes India. https://www.forbesindia.com/article/explainers/gdp-india/85337/1

Govt of India (2014). https://www.pmindia.gov.in/en/major_initiatives/swachh-bharat-abhiyan

GRI (2022). https://www.globalreporting.org/

Gupta N, Sharma V (2016) Exploring employee engagement–a way to better business performance. Glob Bus Rev 17. https://doi.org/10.1177/0972150916631082

Hindustan Unilever Limited (n.d.) Water conservation programs. Retrieved from https://www.huf.co.in/

Hong H, Kacperczyk M (2009) The price of sin: the effects of social norms on markets. J Financ Econ 93(1):15–36

IEA (2020) Renewables 2020. International energy agency. https://www.iea.org/reports/renewables-2020

India CSR Outlook Report (2023). https://indiacsr.in/wp-content/uploads/2023/10/India-CSR-Outlook-Report-2023.pdf

Infosys Foundation (n.d.) Healthcare projects. Retrieved from https://www.infosys.org/infosys-foundation/initiatives.html

IPCC (2018) Global warming of 1.5 °C. Intergovernmental panel on climate change. https://www.ipcc.ch/sr15/

Isabel B, Catalán S, Martínez E (2016) The importance of corporate brand identity in business management: an application to the UK banking sector. Bus Res Q 19:12–13

Jesus V et al (2016) INMA-Valencia cohort study. Reproducibility and validity of a food frequency questionnaire designed to assess diet in children aged 4–5 Years. PLoS One 11(11):e0167338. https://doi.org/10.1371/journal.pone.0167338. PMID: 27898731; PMCID: PMC5127574.

Khalid F, Sun J, Guo J, Srivastava M (2023) Green corporate image: assessing the effects of climate change management practices on corporate reputation. Corp Soc Responsib Environ Manag. https://doi.org/10.1002/csr.2663

Khan M, Khan M, Serafeim G, Yoon A (2016) Corporate sustainability: first evidence on materiality Account Rev 91(6):1697–1724., Available at SSRN: https://ssrn.com/abstract=2575912 or https://doi.org/10.2139/ssrn.2575912

Kotsantonis S, Serafeim G (2021) Four things no one will tell you about ESG data. Harv Bus Rev

MAC (2013). https://www.mca.gov.in/content/mca/global/en/acts-rules/companies-act/companies-act-2013.html

Mahindra Group (n.d.) Sustainability initiatives. Retrieved from https://www.mahindra.com/sustainability

McWilliams A, Siegel D (2001) Corporate social responsibility: a theory of the firm perspective. Acad Manag Rev 26(1):117–127

Mondal S, Singh R (2020) Is socially responsible indices weak form of efficient market? Evidences from developing economies. Asian Soc Sci 16(2):55

Mondal S, Singh R, Yadav V (2022) Testing weak form of efficient market hypothesis on socially responsible indices: comparative study between developed and emerging markets. Int J Account Financ Rev 13(1):20–34

Nielsen (2015) The sustainability imperative. https://engageforgood.com/2015-nielsen-global-sustainability-report/

Porter ME, Heppelmann JE (2014) How smart, connected products are transforming competition. Harv Bus Rev

Porter ME, Kramer MR (2011) Creating shared value. Harv Bus Rev 89(1/2):62–77

Principles for Responsible Investment (PRI) (2022) About PRI. https://www.unpri.org/annual-report-2022

Rao PK (2000) Sustainable development: economics and policy. Blackwell Publishers Inc., Oxford

Rathore M (2024) Statista. https://www.statista.com/statistics/802122/india-net-national-income-percapita/

Reliance Foundation (n.d.) Rural transformation programs. Retrieved from https://reliancefoundation.org/rural-transformation

SEBI (2019). https://www.sebi.gov.in/reports-and-statistics/publications/feb-2021/annual-report-2019-20_49071.html

Singh R, Bhattacharjee J (2019) Measuring equity share related risk perception of investors in economically backward regions. Risks 7(1):12

Singh R, Bhattacharjee J (2022) Factors affecting awareness towards investment in equity shares: a social network analysis approach. Acad Market Studies J 26(5)

Singh R, Bhattacharjee J, Kajol K (2024) Factors affecting risk perception in respect of equity shares: a social network analysis approach. Vision 28(3):386–399

Singh R, Leepsa NM, Kushwaha NN (2016) Testing the weak form of efficient market hypothesis in carbon efficient stock indices along with their benchmark indices in select countries. Iranian J Manag Studies 9(3):627–650

Stern N (2006) Stern review on the economics of climate change. HM Treasury. https://www.lse.ac.uk/GranthamInstitute/publication/the-economics-of-climate-change-the-stern-review/

The Ellen MacArthur Foundation (2017) https://www.ellenmacarthurfoundation.org/a-new-textiles-economy

UNDP (2022). https://www.undp.org/publications/undp-annual-report-2022

UNFCCC (2022). https://unfccc.int/annualreport

UNGC (n.d.) United nations global compact principles. United Nations Global Compact. https://unglobalcompact.org/
WBCSD (2018) CEO guide to the circular economy. World Business Council for Sustainable Development. https://www.wbcsd.org/Programs/Redefining-Value/Reporting-matters/Resources/Reporting-matters-2018
Widyastuti S (2019) Developing a green corporate image: an achievement for competitive advantage through organizational culture and green marketing strategy. Revista San Gregorio
World Bank (2012) Inclusive green growth: the pathway to sustainable development
World Economic Forum (2021) Global risks report 2021. https://www.weforum.org/publications/the-global-risks-report-2021/
WWF (2021) Living planet report 2020—bending the curve of biodiversity loss. World Wildlife Fund
Yazdanifard R, Erdoo MI (2011) The impact of green marketing on customer satisfaction and environmental safety

Role of Government and International Institutions in Fostering Social Responsibility and Environmental Harmony

Seda H. Bostancı and Seda Yıldırım

Abstract The world leaders have recognized the numerous challenges in achieving the 2030 Sustainable Development Goals (2030 SDGs) in the long term. The COVID-19 pandemic, the Russia-Ukraine war, and climate change make it difficult for countries to achieve the 2030 SDGs. Countries must cooperate globally to overcome economic and environmental crises. At this point, the United Nations (UN) and its related organizations are supporting countries in their efforts to combat economic and environmental crises in the long term. This study aims to provide a brief framework for the UN and its related organizations based on their roles and responsibilities for the 2030 SDGs.

Keywords Sustainability · 2030 Sustainable development goals (2030 SDGs) · United nations · Social responsibility

1 Introduction

The disregard for nature and its ecosystems has resulted in loss of biodiversity, desertification, climate change, and disruption of natural cycles (UN 2024). Natural disasters caused by climate change are basic indicators of environmental quality loss. Climate change has natural and human-induced causes. The era in which humans play a central role in shaping the Earth's biosphere and altering the world is known

S. H. Bostancı (✉)
The Faculty of Economics and Administrative Sciences, The Department of Political Science and Public Administration, Tekirdag Namık Kemal University, Kampus Street, 59030 Tekirdag, Turkey
e-mail: shbostanci@nku.edu.tr

S. Yıldırım
The Faculty of Economics and Administrative Sciences, The Department of Business Administration, Tekirdag Namık Kemal University, Kampus Street, 59030 Tekirdag, Turkey

P. Singh et al. (eds.), *Integrating Environmental Sustainability and Corporate Social Responsibility: A Move Beyond Profit*, World Sustainability Series,
https://doi.org/10.1007/978-3-031-81469-3_10

as the anthropocene (Cuckston 2021). Climate change research, government policies, international organizations, and non-governmental movements gained prominence in the 1980s. Developed countries now seem to be more aligned in addressing climate change. Environmental activists and non-governmental organizations criticize the United Nations' sustainability discourse and advocate for action to achieve environmental harmony using different and effective methods (Bostancı 2022). The "Montreal Protocol," signed in 1987 following the "Vienna Convention for the Protection of the Ozone Layer" in 1985, is the leading international agreement in preventing environmental pollution and protecting the ozone layer (Weatherhead and Andersen 2006; Tümer 2023). International environmental agreements encompass various areas such as clean water resources, energy, habitat, oceans, natural habitats, pollution control, protection of living species, and climate change mitigation (Tümer 2023). Our common future, known as the Brundtland Report of the United Nations in 1987, emphasized the critical importance of countries taking action together on the environment in order to protect environmental values and ensure the sustainability of natural resources for future generations (Yıldırım et al. 2016). Subsequently, important breakthroughs were made, such as important agreements such as the 1997 Kyoto Protocol, the 2015 Paris Climate Agreement, and the 2030 sustainable development goals determined by the UN in 2015, in which the effects of governments turned into commitments on the environment. In addition, the issues of sustainable future, environment and climate-friendly life, which can be defined as the environmental dimension of social responsibility, are also in the field of interest of non-governmental organizations.

After the 1972 United Nations Conference on the Human Environment in Stockholm, several conferences focusing on different environmental issues have been held as a continuation of this conference. One of the significant events in the environmental field is the UN Environment Conference and Sustainable Development Conference (1972), where the discussion on a sustainable future takes place (UN n.d.a). The conference was held in Rio in 1992, followed by the Johannesburg Declaration on Sustainable Development in 2002, and The United Nations Conference on Sustainable Development—Rio + 20 in 2012. In 2022, the Stockholm +50 conference titled "A healthy planet for the prosperity of all—our responsibility, our opportunity" took place. Habitat conferences, which began in 1976, are also important for making decisions on the establishment of sustainable human settlements by Un-Habitat. The Habitat II conference took place in Istanbul in 1996, and the Habitat III conference was held in Quito in 2016. Ongoing environmental conferences include the World Climate Conference, Conference on the Changing Atmosphere, UN Sustainable Development Summit, and Climate Change Conference COP, which have been held annually since 1995.

Intergovernmental Panel on Climate Change (IPCC) is a series of international conferences on combating climate change that started in 1988. IPCC has created an environment where scientific knowledge on climate change is produced and discussed (Tremmel and Robinson 2014; Vardy et al. 2017). The first World Climate Conference (WCC) in 1979 provided a major international forum devoted exclusively to climate change" (Agrawala 1998). In order to meet the 1.5 °C target in the context of global warming at COP 26 held in Glasgow in 2021, many countries pledged to achieve net zero carbon emissions and end deforestation by 2050 (Shivanna 2022).

The Unity Pact, which took on the responsibility of representing the interests of Mother Earth in the process of drafting a nature-compatible constitution in Bolivia in 2010, also undertook the task of creating a draft law based on the minutes of the World People's Conference on Climate Change and the Rights of Mother Earth (Calzadilla and Kotzé 2018). "The emergence of the global environment as a fundamental issue on the world political scene coincides with the period when non-governmental organizations were on the rise as a new important force in environmental policies" (Öztunç 2006). International NGOs on sustainability and environmental issues are very diverse. Prominent among these are; They can be listed as Greenpeace, World Wildlife Fund, 350.org, Sierra Club, Friends of the Earth, Health in Harmony. Since the scope of the study was limited to governments and UN affiliated institutions, environmental non-governmental organizations were excluded from the scope of the study.

Conferences organized by the UN and non-governmental organizations had a significant impact on the development of ideas on how humans can live a life in harmony with the environment. In this study, general information will be given on the main objectives of the conferences, which have enabled governments to make various commitments on sustainability and combating climate change, which are seen as the basic criteria of environmental adaptation. Again, international organizations are included in the scope of the study as actors that undertake duties in ensuring environmental harmony. This study focuses on UN's functions and its related organizations' functions for 2030 SDGs in the global scope. Accordingly, this study will examine the UN's function and related specialized organizations' functions for developed and developing countries in the long term based on 2030 SDGs.

2 The United Nations (UN) and Institutes: To the Sustainable Future

International organizations play a vital role in addressing global issues. Over the past two centuries, the structure of these organizations has undergone significant changes. They comprise member countries, both developed and developing, depending on their objectives (Öztunç 2006). It can be said that international organizations have taken the first important step, especially regarding the environment. At this point, the United Nations appears as the international organization that initiated the first serious actions towards environmental and sustainability problems. For example, the first global trend towards the environment began in 1949 with the United Nations Scientific Conference on the Conservation and Utilization of Resources (Öztunç 2006); The relationship between environment and humans made its official entry into the literature with the United Nations Conference on the Human Environment in 1972 (Yıldırım and Yıldırım 2020).

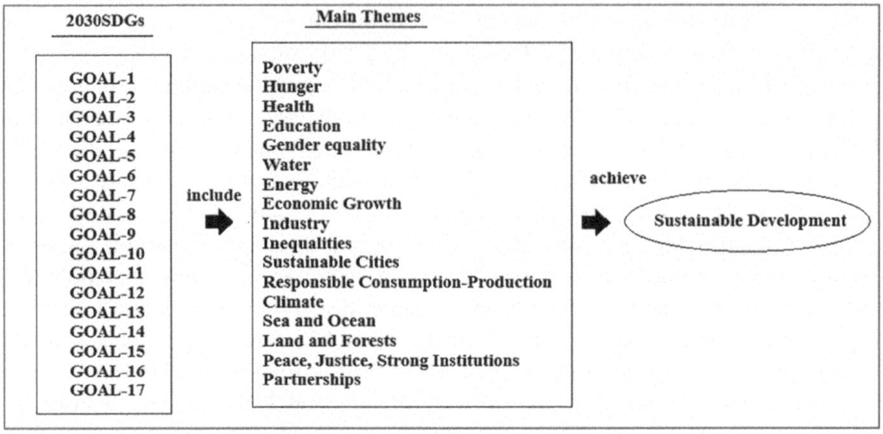

Fig. 1 2030SDGs and main themes. *Source* Created by the authors

The United Nations has brought attention to sustainable development worldwide. Traditional economic models have caused significant harm to the environment, and social welfare and equality have been overlooked (Munasinghe 2004). But, due to the polluted natural environment and the vanishing biodiversity, the depletion of natural resources, food, clean water, energy, and other vital elements that can meet the needs of future generations have become a pressing issue (Yıldırım et al. 2016). It has become clear that sustainable development is inevitable, as demonstrated by the Limits to Growth-1972 (Meadows et al. 1972), Our Common Future-1987 (The World Commission on Environment and Development (WCED), (1987), and the 2030 Sustainable Agenda-2015 (UN (2015). Thus, the United Nations has taken on the responsibility of guiding countries towards a sustainable future. The 2030 Sustainable Development Goals (2030SDGs) have been presented to countries, which include 17 basic goals to achieve by 2030 (Yildirim et al. 2022). Figure 1 summarizes 17 goals with its themes as below.

2030SGDs shows the way of achieving sustainable development in the short term. Global issues related with economic, social and environmental themes are all included in the 2030 SDGs. Countries began to work on 2030 SDGs to achieve sustainable development since 2015 (Yıldırım et al. 2023) but there will be some partnerships and collaborations between countries and institutions to achieve sustainable development in the long term. At this point, UN and its related organizations work on supporting countries to achieve sustainable development and they help and guide both of developed and developing countries to fight against climate change.

The effects of international organizations on global environmental problems can be addressed in 4 ways (White 1995; Öztunç 2006):

1. Determining the agenda for global action,
2. Realization of negotiations on global environmental problems
3. Rules for environmental themes can be developed

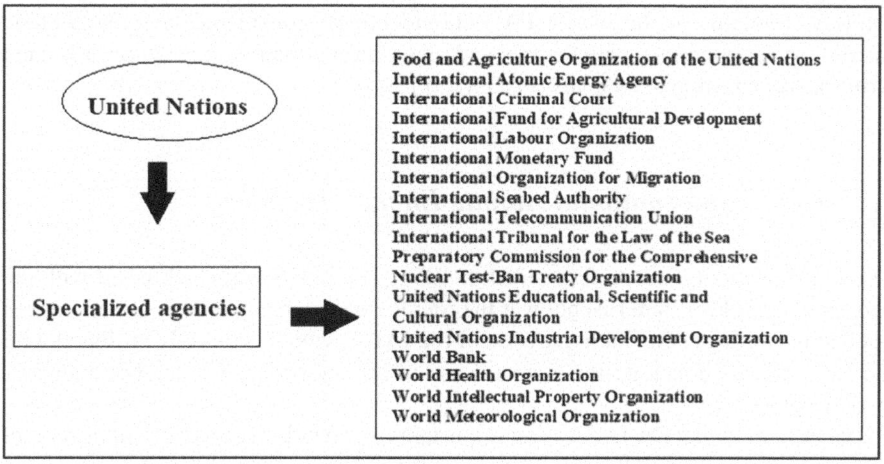

Fig. 2 UN and related specialized agencies. *Source* Created by the authors Specialized agencies related with UN are examined as below:

4. Policies can be developed for countries regarding environmental issues

Figure 2 shows the list of international organizations that collaborate with the United Nations. These organizations were established with specific objectives and play a crucial role in shaping global strategies and policies. They also provide guidance and assistance to countries in matters related to environmental sustainability.

2.1 FAO (the Food and Agriculture Organization)

FAO is directly depended with the UN. FAO was established in 1945 and it aims to ensure food security for everybody in the world. FAO is responsible for having food survey globally to determine the general picture of food security or food insecurity. Also, FAO regularly organizes congresses and meetings on food-related issues. For example, FAO launched its first survey (World Food Survey) in 1946 and there as a special meeting on food crisis at the same year (FAO n.d.a.).

However, FAO is so active organization which aims to solve global issue in 50 s and 60 s (FAO n.d.a.), it can be said that the essential importance of FAO has been recognized since 1972. United Nations Environment Program (UNEP) and FAO held UN Conference on the Human Environment in 1972. UN Conference on the Human Environment is the first official call for member countries to think about environment and future. As a result of the conference, the United Nations Environment Program (UNEP) was created (UN n.d).

FAO is at the heart of the process of realizing the UN's development goals. FAO strives to find solutions to food and agriculture-related problems, which are the primary problems in the world. FAO should not be considered as a stand-alone institution. It cooperates with the UN and other international organizations affiliated with the UN to achieve its goals (FAO n.d.b.).

2.2 IAEA (International Atomic Energy Agencies)

The IAEA is an organization founded in 1957. The IAEA was established with the approval of US President Eisenhower in order to control the fears and concerns about nuclear technology. The main purpose of the IAEA, which is directly affiliated with the UN, is to demonstrate the peaceful use of atomic energy to member countries (IAEA n.d.a).

It can be said that the IAEA is an important organization, especially in important areas in terms of 2030SDGs. Due to its institutional nature, the IAEA has a role and responsibility primarily in the field of energy. Energy is a vital element for sustainable development. Nuclear energy can be considered as a source that provides "clean and accessible energy". It has an important share in meeting the energy needs of the world population. At this point, the IAEA supports both existing and new nuclear energy programs globally and provides guidance to countries (IAEA, (n.d.b). Regarding the energy issue, it ensures that sustainable energy policies are followed within the scope of the 2030 SDGs. Thus, Goal-7 within the 2030 SDGs It can be said that it is guiding the countries for and Goal-17.

The IAEA's duties also include health-related issues. The IAEA plays a supporting role in ensuring that member states can provide adequate health services. Nuclear techniques in the field of medicine mainly represent the peaceful use of nuclear technology (IAEA n.d.c). With its peaceful practices and guidance on health, the IAEA has also undertaken tasks related to Goal-2 within the 2030 SDGs. IAEA is an important international organization for the use of nuclear technology, cooperation between countries and the development of technology.

The fight against climate change has left its mark on the agenda recently. On the path to sustainable development, it is very important to stop the negative effects of climate change. The IAEA helps countries adapt to climate change and supports countries in the application of nuclear technology to combat climate change (IAEA n.d.d). Environmental changes can change the entire planet in the long term. In order to solve future scenarios and environmental problems of countries, IAEA uses nuclear technology. The IAEA takes a scientific approach to the subject to examine the pollution occurring in terrestrial and aquatic environments and the effects on human health. IAEA helps countries dispose of nuclear waste and remediate contaminated environmental sites (IAEA n.d.e).

The IAEA also follows nuclear technology and scientific approaches to assist with food security and water security issues. Access to safe and clean water is the right of all people, but the continuity of clean and accessible water is a major problem in the face of the rapidly increasing population in the world. IAEA examines nuclear methods freshwater resources and encourages countries to use them more efficiently

(IAEA, (n.d.f). IAEA and FAO work for food security through nuclear technology. In member countries, there are many research institutes that investigating nuclear technology and applications on food and nutrition (IAEA n.d.g). In addition, the IAEA assists industry with nuclear science and technology research and applications (IAEA n.d.h).

The main duties and responsibilities of the IAEA include nuclear science and technology research and support related to this field (IAEA n.d.i). On the other hand, the IAEA appears to be an organization that helps many targets within the 2030 SDGs.

2.2.1 ICC (The International Criminal Court)

The ICC investigates and prosecutes people globally associated with the most serious crimes such as "genocide, war crimes, crimes against humanity". The ICC has a responsibility to seek international criminal justice. Essentially, the first international criminal court in the world is the ICC (ICC n.d.a). There is the Rome Statute which gives four basic authorities for ICC as below (ICC (n.d.b):

- ICC, in cases such as "genocide, attack or attempt to destroy a national, ethnic, racial or religious group"
- ICC, rape, imprisonment, enslavement of women and children, sexual slavery, torture, etc. It seeks to provide global justice for serious crimes against humanity.
- Cases of grave violation of the Geneva Conventions, war, killing of prisoners, Also deliberate attacks on hospitals, historical monuments or buildings for religious, educational, artistic, scientific or charitable purposes
- The crime of assault also falls within the jurisdiction of the ICC. Situations such as the application of armed force between states.

2.2.2 The International Fund for Agricultural Development (IFAD)

IFAD is directly related with UN, and it was established because of World Food Conference (1974). In 70 s, the food insecurity and food scarcity are global issue and UN launched a global conference on food and nutrition. It was seen that hunger issue is related with poverty in the worldwide. At this point, IFAD was established to support agricultural development projects among developing countries (IFAD n.d,a).

The 2030 Agenda advocates "leaving no one behind" and aims to advance societies that jointly maintain social-economic-environmental balance. However, global crises such as the COVID-19 crisis and the Russia-Ukraine war have created serious difficulties for countries to achieve their 2030SDGs targets. Small producers in rural areas have difficulty continuing their agricultural activities. IFAD provides aid funds by supporting small producers in rural areas. It can be said that IFAD contributes concretely to the targets within the 2030SDGs. IFAD, (n.d.b).

- IFAD's most significant contributions to the 2030 SDGs can be summarized as follows: IFAD, (n.d.b).
- IFAD supports small investors by focusing on projects to transform rural areas. Thus, it helped for Goal-1 and Goal-2 in the 2030 SDGs.
- IFAD contributes to Goal-4 with business development projects for people living in poor and rural areas.
- IFAD supports technologies that reduce carbon emissions within the scope of Goal-7.
- IFAD provides support to rural economies with the aim of higher income-economic growth and contributes to Goal-8.
- IFAD builds pathways for the market-producer channel to improve the distribution channel of rural producers and contributes to Goal-9.
- IFAD invests and supports sustainable agriculture opportunities. Thus, it contributes to Goal-13.

2.2.3 ILO (International Labour Organization)

ILO is an institution that emerged in 1919 as a symbol of universal peace. The ILO's own constitution was created in the USA in early 1919 with representatives of 9 countries (Belgium, Cuba, Czechoslovakia, France, Italy, Japan, Poland, the United Kingdom and the United States). The fact that the labor market is similar globally and the same problems are experienced globally played an important role in the establishment of the ILO. In other words, the ILO emerged to solve employment and labor problems globally. In 1946, the ILO became a specialized agency of the UN and was the first international institution to cooperate with the UN (ILO (n.d.a).

In 1920, the ILO established some basic themes to achieve its goals. These themes are as follows (ILO n.d.a) (Fig. 3).

The ILO deals with relations between government, employers and workers/ employee organizations by focusing on social and economic development. ILO

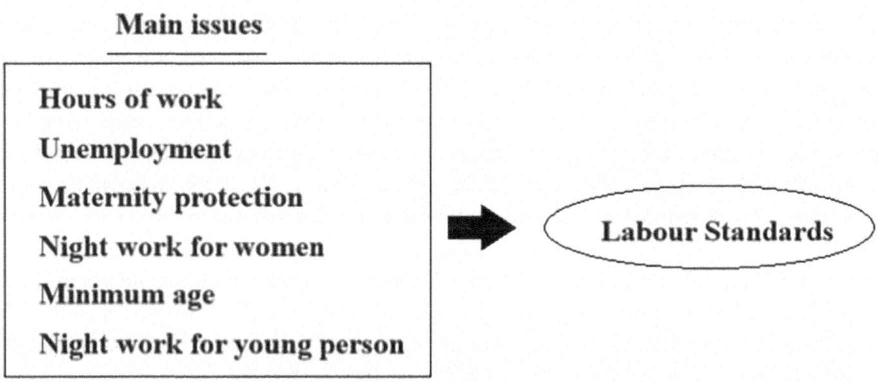

Fig. 3 ILO's labour standards. *Source* Created by the authors

guides governments and employers in developing policies that will ensure equality between men and women in employment and determining labor standards (ILO n.d.b). While realizing its objectives, the ILO also follows the 2030 Agenda and 2030 SDGs. The ILO undertakes various supporting roles towards the 17 targets within the 2030 SDGs. First of all, the goal to which ILO directly contributes is Goal-8. Because with Goal-8, it is aimed for everyone globally to obtain equal work and wages appropriate to their abilities (ILO n.d.c). It can be said that the ILO directly supports the UN and the 2030 SGDs with its objectives and practices and makes great contributions. Below we summarize the relationship between the ILO's functions and the 2030 SDGs (ILO n.d.d):

Goal-1 and ILO: The ILO works to ensure decent work for every person and supports the goal of halving poverty by 2030. ILO supports the social programs of the countries it cooperates with. Work-oriented recovery is expected with social supports, and with the contributions of the ILO, people in the countries can obtain suitable job opportunities and social security.

Goal-2 and ILO: ILO contributes to agricultural incentive practices in terms of fighting hunger for Goal-2. It provides incentives to areas such as agricultural economy, farmers and rural growth.

Goal-3 and ILO: ILO carries out studies to prevent HIV among workers in countries and to provide various social security and health services. Occupational safety and occupational health is an important issue. ILO has introduced various regulations for the use of industrial chemicals in order to protect the health of both employees and consumers.

Goal-4 and ILO: ILO attaches importance to lifelong learning. In particular, the education of groups such as children, youth and women is a priority. ILO has made various regulations to ensure a more effective transition from school to work.

Goal-5 and ILO: The ILO strives to end gender discrimination in employment by promoting equality of opportunity and pay parity.

Goal-6 and ILO: Both production depends on water within natural resources, and people can be productive as long as they have access to clean water. That's why clean and accessible water is a priority issue.

Goal-7 and ILO: While creating jobs for people is a priority with ILO's practices, it is an accepted fact that there will be no employment without energy. Therefore, investment in clean energy sources and sustainable energy is important. ILO provides guidance for young investors and entrepreneurs in terms of green jobs.

Goal-8 and ILO: Employment and job creation are an ILO priority. It is the first and directly related target for the ILO.

Goal-9 and ILO: One of the necessary elements for employment is a correct infrastructure and local government. The ILO supports sustainable industry and emphasizes the importance of infrastructure.

Goal-10 and ILO: The ILO advocates equality in employment and job creation. It focuses on gender equality in employee policies in countries and produces solutions considering the benefit of the country in immigration policies.

Goal-11 and ILO: ILO provides support to people in developing countries in areas such as affordable housing, employee safety and health in public transport, and sustainable home and workplace construction.

Goal-12 and ILO: ILO, as part of the UN, is an institution focused on sustainable consumption-production. ILO promotes safe production techniques in workplaces to prevent environmental quality deterioration.

Goal-13 and ILO: It is seen that ILO encourages employment projects that prioritize low carbon and resource efficiency with its "Green Jobs" project. Green jobs are important in the fight against climate change.

Goal-14 and ILO: The ILO cooperates with FAO on fisheries and food security issues. The ILO has made various agreements for the formalization of fisheries and provides support to small island developing states.

Goal-15 and ILO: ILO carries out work health and work safety studies in the forestry sector.

Goal-16 and ILO: The ILO is an organization that sets labor standards internationally and strives to protect labor/internal rights around the world. Fair employment is important for safe and peaceful societies.

Goal-17 and ILO: ILO, an important institution that provides intercontinental cooperation, carries out nearly 700 active programs and projects in more than 130 countries, according to the recent data.

2.2.4 IMF (International Monetary Fund)

It is a global organization that helps member countries for sustainable development and growth. Working with a focus on financial stability, the IMF supports countries in the context of employment and economic prosperity (IMF n.d.a). Table 1 shows how IMF contribute to countries in recent years. In other words, Table 1 shows key themes for IMF's policy development in the global economy (IMF n.d.b).

2.2.5 The International Organization for Migration (IOM)

IOM emerged in 1951 because of World War II. Due to the war, policies were developed on European settlement planning and immigrant transportation for millions of people. Essentially, it started to serve as an international organization in 1989. Historically, IOM's early activities were mostly focused on logistics, but today IOM's activities have evolved into international migration policies (IOM n.d.a).

2.2.6 The International Seabed Authority (ISA)

It is an organization of the United Nations. It is directly affiliated with the UN. According to International Maritime Law, it makes suggestions and contributions to the maritime law of member countries. Regulations regarding the activities and

Table 1 IMF and its Funding Areas **Source:** Created by the authors

Key areas	Functions
Artificial intelligence	IMF supports the development and research of AI technology
Climate change	IMF supports low-carbon and resilient growth policies in combating climate change, while also considering economic growth
Fintech	Fintech is important for the development of the economy, but it also involves some risks. IMF continues to conduct research on the subject
Comprehensive surveillance review	The IMF makes some financial and economic recommendations in its surveillance results for countries
Fragile and conflict-affected states (FCS)	The IMF offers long-term economic policy so that vulnerable countries can benefit from the macroeconomic experiment
Fiscal policies	The IMF is the leading institution in fiscal policy and strategy and management globally
Income inequality	Inequality In order to support economic growth, the IMF is against all kinds of inequality
Gender	While equality between men and women is an important development goal in itself, women's participation in the economy is also a part of the growth and stability equation
Governance and anti-corruption	Investments and developments in the field of governance will be beneficial for the fight against corruption and sustainable development
COVID-19	The IMF continues its support to countries vulnerable to the devastating effects of COVID-19
Integrated policy framework	IMF works to ensure stability in international capital flows
Low-Income countries	The IMF continues its aid to low-income countries and increases this aid during global crisis periods
Sustainable development goals	IMF's work and practices are compatible with sustainable development goals
State debt	The IMF provides financial stability assistance to vulnerable and low-income countries
Special drawing rights	Special Drawing Rights are an additional IMF asset granted to member countries
Trade	The IMF provides support for the continuity of international trade

payment activities of the countries party to UNCLOS in the maritime zone are determined according to ISA. Member states of UNCLOS are also accepted as members of ISA (ISA n.d.a).

2.2.7 International Telecommunication Union (ITU)

It is the authorized and specialized agency of the United Nations on information and communication. Both UN member states, companies, universities and international organizations are included in ITU. It is considered to be the oldest institution of the UN, since the use of telegraph began in 1865. ITU's aim is to facilitate international connectivity. Relevant procedures, practices and policies in this context are available globally from ITU (ITU n.d.).

2.2.8 The International Tribunal for the Law of the Sea (ITLOS)

ITLOS was established according to the UN Maritime law convention of 1982. It essentially functions as a judicial body. It has jurisdiction over all disputes regarding contracts. Here, maritime legal agreements such as maritime jurisdictions, maritime transportation, and protection of marine creatures fall within the jurisdiction of ITLOS (ITLOS n.d.).

2.2.9 International Civil Service Commission

The International Civil Service Commission is an independent and expert body established by the UN. As an institution that observes the balance between needs and conditions, it deals with policies regarding public services. By creating a "Common System", UN creates a common personnel standard for salaries, allowances and other rights of all personnel working for UN. The justifications for the common system are as follows: (ICSC n.d.):

- Efficiency/economy of scale
- Equality/justice
- Cost efficiency/control

2.2.10 UNESCO (United Nations Educational, Scientific and Cultural Organization)

Founded in 1945, UNESCO is an organization that helps education, science and culture and sustainable development. UNESCO creates the conditions for respect for common values between civilizations and cultures globally. If we look at UNESCO in terms of sustainable development goals, the following elements stand out: (UN 2013):

- Quality education and lifelong learning all over the world
- Realization of sustainable development in the light of scientific knowledge and policies
- Producing solutions to social and ethical problems
- Protecting and disseminating cultural diversity
- Promoting the information society

2.2.11 UNIDO (United Nations Industrial Development Organization)

UNIDO is an organization affiliated with the UN that works to promote industrial growth and development. Within the 2030 SDGs, UNIDO continues its activities directly related to Goal-9. With Goal-9, the goal of sustainable industry is tried to be achieved. UNIDO's corporate vision is to reduce poverty and eliminate hunger through sustainability and stability in industry. UNIDO's focal elements are as follows (UNIDO n.d):

- Helping enterprises,
- Using renewable energy to reduce industrial-based CO_2 gas
- Protecting the scarce resources of businesses in developing countries

2.2.12 World Bank

The World Bank is a global partnership and organization working to reduce poverty and achieve a sustainable future in developing countries. World Bank also provides financial support for developing countries. The World Bank, which also provides information support, organizes many conferences. The main focuses of the World Bank are as follows (World Bank n.d.a):

- Results: Obtaining measurable results for developing countries,
- Reform: World Bank is always making improvements in the way of doing business,
- Open Development: Publicly accessible databases, research results and information are provided.

Historically, the work and activities of the World Bank have changed according to global trends. For example, In the late 90 s the World Bank focused mostly on conflict prevention and reconstruction in post-conflict countries. In the mid-2000s, World Bank emerged as an information bank. With the emergence of the Sustainable Development Goals in 2015, the World Bank started to carry out its work within the scope of 2030SDGs (World Bank n.d.b).

2.2.13 WIPO (the World Intellectual Property Organization)

WIPO is a global forum and a self-financing agency of the UN. WIPO operates primarily for intellectual property services (WIPO n.d.).

2.2.14 WMO (the World Meteorological Organization)

WMO is the authorized organization of the UN, which studies the atmospheric changes, weather and climate change of our planet. International cooperation is essential in the fight against climate change. WMO works within the framework of weather and climate issues and makes policy recommendations to countries (WMO n.d.).

2.2.15 WHO (World Health Organization)

UN member states receive support from WHO to achieve global health goals. WHO produces health policies for society, both for those working in the health sector and for public health (WHO n.d.).

In addition, other international organizations that cooperate with the United Nations are shown separately in Figs. 4, 5 and 6 (UN n.d.c).

3 Conclusions

Climate change is becoming a more pressing issue due to its recent impact on the environment. The frequency of natural disasters is increasing, and the Earth's temperature is rising rapidly. Policy makers and world leaders have agreed to implement sustainable policies to ensure the safety of the Earth. However, there are numerous challenges in achieving long-term sustainable policies and implementations. The United Nations plays a key role in addressing sustainability for the future. Sustainable development aims to balance economic, social, and environmental concerns for both the present and the future. The United Nations has multiple organizations working to achieve the 2030 Sustainable Development Goals (SDGs) and to guide both developed and developing countries in implementing sustainable policies. The specialized agencies of the UN are particularly important and are directly linked to the UN. Upon reviewing these specialized agencies, it becomes clear that they are all related to sustainable development principles, and most of them strongly support the 2030 SDGs through their functions.

The UN and its specialized agencies are all important and vital for achieving 2030 SDGs and reaching a sustainable future in the long term. Some of them guide developed and developing countries for legal issues and some of them guide developed and

Fig. 4 UN and intergovernmental organizations with having permanent offices Headquarters. *Source* Created by the authors

developing countries for financial issues. Most of them launch useful conferences and give regular reports and statistics for its working area.

The concept of a shared future that includes everyone is an important starting point for achieving environmental balance. The survival of modern cities relies on extensive construction and energy, technology, and workforce. These components can also be viewed as progress and growth. However, every resource consumed diminishes the prospects for a sustainable future, from energy derived from fossil fuels to the use of plastic and paper. While most governments worldwide promote growth as the driving force of their countries, reducing or maintaining consistent growth opens up new avenues for ideas in the realm of sustainability. Ensuring environmental balance is a form of social responsibility.

United Nations (Intergovernmental Organizations
(not maintaining Permanent Offices at Headquarters)

African Development Bank
Agency for the Prohibition of Nuclear Weapons in Latin America and the Caribbean
Andean Community
Andean Development Corporation
ASEAN's Macroeconomic Research Office
Asian Development Bank
Asian Forest Cooperation Organization (AFoCO)
Asian Infrastructure Investment Bank
Association of Caribbean States
Association of Southeast Asian Nations
Black Sea Economic Cooperation Organization
Central American Bank for Economic Integration
Central Asia Regional Economic Cooperation Institute (CAREC)
Central European Initiative
Collective Security Treaty Organization
Common Fund for Commodities
Commonwealth of Independent States
Community of Portuguese-speaking Countries
Community of Sahelo-Saharan States
Conference of Ministers of Justice of the Ibero-American Countries
Conference on Interaction and Confidence-building Measures in Asia
Council of Europe
Customs Cooperation Council (World Customs Organization)
Developing Eight Countries Organization for Economic Cooperation
East African Community
Economic Community of Central African States
Economic Community of West African States
Economic Cooperation Organization
Energy Charter Conference
Eurasian Development Bank
Eurasian Economic Community
Eurasian Group on Combating Money Laundering and Financing of Terrorism
European Organization for Nuclear Research (CERN)
European Public Law Organization
Fund for the Development of the Indigenous Peoples of Latin America and the Caribbean
Global Dryland Alliance
Global Fund to Fight AIDS, Tuberculosis and Malaria
Global Green Growth Institute
Group of Seven Plus
Hague Conference on Private International Law
Indian Ocean Commission
Indian Ocean Rim Association
International Anti-Corruption Academy

International Centre for Migration Policy Development
International Chamber of Commerce
International Civil Defence Organization (ICDO)
International Conference on the Great Lakes Region of Africa
International Council for the Exploration of the Sea
International Fund for Saving the Aral Sea
International Humanitarian Fact-Finding Commission
International Hydrographic Organization
International Institute for the Unification of Private Law (UNIDROIT)
International Network for Bamboo and Rattan Organization
International Solar Alliance
International Think Tank for Landlocked Developing Countries
Islamic Development Bank Group
Italian-Latin American Institute
Latin American and Caribbean Economic System
Latin American Integration Association
Latin American Parliament
New Development Bank
OPEC Fund for International Development
Organization of African, Caribbean and Pacific States
Organization of Eastern Caribbean States
Organization for Security and Cooperation in Europe
Organization for Democracy and Economic Development – GUAM
Pacific Islands Development Forum
Pacific Islands Forum
Pan African Intergovernmental Agency for Water and Sanitation for Africa
Regional Centre on Small Arms and Light Weapons in the Great Lakes Region, the Horn of Africa and Bordering States
Shanghai Cooperation Organization
Small Island Developing States Dock (SIDS DOCK)
South Asian Association for Regional Cooperation
South Centre
Southern African Development Community
Union for the Mediterranean
Union of South American Nations
West African Economic and Monetary Union

Fig. 5 UN and intergovernmental organizations with not having permanent offices Headquarters. *Source* Created by the authors

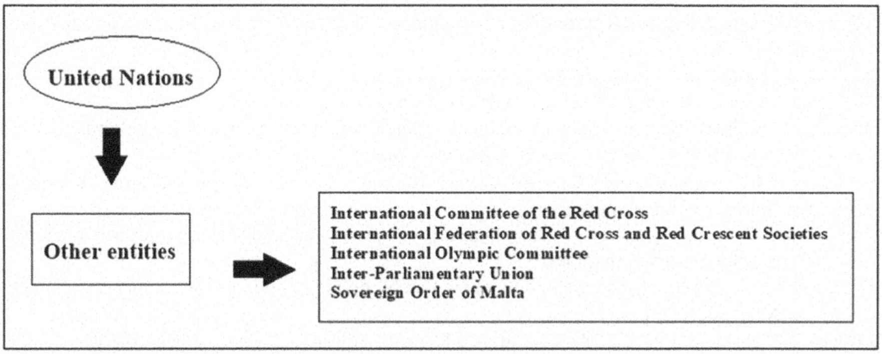

Fig. 6 UN and related other organizations. *Source* Created by the authors

References

Agrawala S (1998) Context and early origins of the intergovernmental panel on climate change. Clim Change 39(4):605–620
Bostancı S (2022) Climate and water policies of local governments. JOEEP J Emerg Econ Policy 7(1):395–410
Calzadilla PV, Kotzé LJ (2018) Living in harmony with nature? A critical appraisal of the rights of Mother Earth in Bolivia. Trans Environ Law 7(3):397–424
Cuckston T (2021) Accounting and conservation to live in harmony with nature, we must organise nature. Soc Environ Account J 41(1–2):1–7
FAO (n.d.a.) About FAO, https://www.fao.org/about/about-fao/en/
FAO (n.d.b.) FAO and UN system partnerships, https://www.fao.org/partnerships/fao-un-system/en/
IAEA (n.d.a) History, https://www.iaea.org/about/overview/history
IAEA (n.d.b) Energy, https://www.iaea.org/topics/energy
IAEA (n.d.c) Health, https://www.iaea.org/topics/health
IAEA (n.d.d) Climate change, https://www.iaea.org/topics/climate-change
IAEA (n.d.e) Environment, https://www.iaea.org/topics/environment
IAEA (n.d.f) Water, https://www.iaea.org/topics/water
IAEA (n.d.g) Food and agriculture, https://www.iaea.org/topics/food-and-agriculture
IAEA (n.d.h) Industry, https://www.iaea.org/topics/industry
IAEA (n.d.i) Nuclear science, https://www.iaea.org/topics/nuclear-science
ICC (n.d.a) About the court, https://www.icc-cpi.int/about/the-court
ICC (n.d.b) How the court works. https://www.icc-cpi.int/about/how-the-court-works
ICSC (n.d.a) History, https://icsc.un.org/Home/History
IFAD (n.d.a) History, https://www.ifad.org/en/history
IFAD (n.d.b) 17 goals to transform our world, https://www.ifad.org/en/ifad-and-the-sdgs
ILO (n.d.a) History of the ILO, https://www.ilo.org/global/about-the-ilo/history/lang--en/index.htm
ILO (n.d.b). https://www.ilo.org/global/about-the-ilo/how-the-ilo-works/lang--en/index.htm
ILO (n.d.c) Decent work and the 2030 Agenda for sustainable development, https://www.ilo.org/global/topics/sdg-2030/lang--en/index.htm
ILO (n.d.d) ILO and the global goals, https://www.ilo.org/global/topics/sdg-2030/ilo-global-goals/lang--en/index.htm
IMF (n.d.a) About the IMF, https://www.imf.org/en/About
IMF (n.d.b) Topics, https://www.imf.org/topics
IOM (n.d.a) IOM history, https://www.iom.int/iom-history
ISA (n.d.a) About ISA, https://www.isa.org.jm/about-isa/

ITU (n.d.) About international telecommunication union (ITU), https://www.itu.int/en/about/Pages/default.aspx
ITLOS (n.d.) Latest news, https://www.itlos.org/en/main/latest-news/
ICSC (n.d.a) ICSC in Brief, https://icsc.un.org/Home/Brief
Munasinghe M (2004) Sustainomics: a trans-disciplnary framework for making development more sustainable, https://isecoeco.org/pdf/sustainomics.pdf
Öztunç Ö (2006) Uluslararası Çevre Politikalarında Birleşmiş Milletlerin Rolü, Ankara Üniversitesi Sosyal Bilimler Enstitüsü, Ankara
PennStateUniversity Libraries (2021) Sustainable development, https://guides.libraries.psu.edu/sustainable-development/international
Shivanna KR (2022) Climate change and its impact on biodiversity and human welfare. Proc Indian Natl Sci Acad 88:160–171
Tremmel JC, Robinson K (2014) Climate ethics: Environmental justice and climate change. Bloomsbury Publishing
Tümer EK (2023) Uluslararası çevre anlaşmalarının belirleyicileri: Politika difüzyonu ve mekânsal etkiler. Ege Üniversitesi, Sosyal Bilimler Enstitüsü Doktora Tezi, İzmir
UN (2015) Transforming our world: the 2030 agenda for sustainable development, resolution adopted by the general assembly on 25 September 2015, https://documents.un.org/doc/undoc/gen/n15/291/89/pdf/n1529189.pdf?token=YnVSEkCE5tgtrzYpdi&fe=true
UN (n.d.a) United nations conference on the human environment, 5–16 June 1972, Stockholm, Conferences | Environment and sustainable development, https://www.un.org/en/conferences/environment/stockholm1972
UN (n.d.b) Specialized agencies, https://www.un.org/en/about-us/specialized-agencies
UN (n.d.c) Intergovernmental and other organizations, https://www.un.org/en/about-us/intergovernmental-and-other-organizations
UN (2013) UNESCO: united nations educational, scientific and cultural organization, https://www.un.org/youthenvoy/2013/08/unesco-united-nations-educational-scientific-and-cultural-organization/
UN (2024) Harmony with nature, http://www.harmonywithnatureun.org/
UNIDO (n.d) Our mandate, our vision, our work, https://www.unido.org/about-us/who-we-are
The World Commission on Environment and Development (WCED) (1987) Report of the world commission on environment and development: our common future, https://sustainabledevelopment.un.org/content/documents/5987our-common-future.pdf
Weatherhead EC, Andersen SB (2006) The search for signs of recovery of the ozone layer. Nature 441(7089):39–45
White ND (1995) The law of international organisations. Manchester University Press, Manchester
WIPO (n.d.) Inside WIPO, What is WIPO?, https://www.wipo.int/about-wipo/en/
WHO (n.d.) Who we are, https://www.who.int/about/who-we-are
WMO (n.d.) Overview, https://wmo.int/about-wmo/overview
World Bank (n.d.a) What We Do, https://www.worldbank.org/en/about/what-we-do.print
World Bank (n.d.b) Explore history, https://www.worldbank.org/en/archive/history
Vardy M, Oppenheimer M, Dubash NK, O'Reilly J, Jamieson D (2017) The intergovernmental panel on climate change: challenges and opportunities. Annu Rev Environ Resour 42:55–75
Yıldırım S, Yıldırım DÇ (2020) Achieving sustainable development through a green economy approach. In: Patti S, Trizzino G (Eds), Advanced integrated approaches to environmental economics and policy: emerging research and opportunities, pp 1–22. IGI Global. https://doi.org/10.4018/978-1-5225-9562-5.ch001
Yildirim S, Demirtas I, Yildirim DC (2022) A review of alternative economic approaches to achieve sustainable development: the rising digitalization and degrowth Post COVID-19. In: Castanho R (Ed.), Handbook of research on sustainable development goals, Climate Change, and Digitalization (pp 288–307). IGI Global. https://doi.org/10.4018/978-1-7998-8482-8.ch018
Yıldırım S, Yıldırım DÇ, Gedikli A (2016) Sustainable consumption trends in the world in the context of green economy and sustainability. In: Erdoğdu M, Arun T, Ahmad I (Eds), Handbook

of research on green economic development initiatives and strategies, pp 65–84. IGI Global. https://doi.org/10.4018/978-1-5225-0440-5.ch004

Yıldırım S, Bostancı SH, Yıldırım DÇ (2023) Parameters for the study of climate refugees. In: Singh P, Ao B, Yadav A (eds) Global climate change and environmental refugees. Springer, Cham. https://doi.org/10.1007/978-3-031-24833-7_11

"AI-Driven Green Growth and Carbon–Neutral Solutions"—The CSR Spectrum Redefined

Tushar Soubhari, P. P. Jayaprakasan, C. V. Shobha, Anil Kumar Mannarath, and Sudhansu Sekhar Nanda

Abstract **Purpose**-In the quest for sustainability and environmental responsibility, businesses are increasingly using cutting-edge technologies to reshape their Corporate Social Responsibility (CSR) strategies. This study redefines corporate sustainability initiatives for the modern era by examining the transformative role of Artificial Intelligence (AI) in promoting carbon–neutral solutions and driving green growth within the context of CSR. It starts by looking at how corporate social responsibility (CSR) is developing and how important it is for businesses to address environmental issues, especially the pressing need to reduce carbon emissions and become carbon neutral. Further, it suggests how AI-driven solutions could transform corporate social responsibility initiatives; highlighting how these solutions could improve resource efficiency, boost operational effectiveness, and ease the shift to sustainable businesses. The research shows how AI technologies are changing the scope of corporate social responsibility (CSR) toward green growth and carbon neutrality through a thorough analysis of case studies and examples from various industries and geographical areas. AI-driven innovations are enabling businesses to achieve dual goals of reducing carbon emissions and promoting economic growth and social impact. These innovations range from energy-efficient operations and integration of renewable energy sources to smart transportation systems and sustainable supply

T. Soubhari (✉)
N.S.S. College, Manjeri, Manjeri, Kerala, India
e-mail: tusharsoubhari1988@gmail.com

P. P. Jayaprakasan
TM Government College, Tirur, University of Calicut, Malappuram, Kerala, India
e-mail: jpahaan13@gmail.com

C. V. Shobha
St. Joseph's College (Autonomous), Devagiri, Calicut, Kerala, India
e-mail: shobhacv@devagiricollege.org

A. K. Mannarath
Government College, Malappuram, Kerala, India

S. S. Nanda
MIT World Peace University, Pune, Maharashtra, India

P. Singh et al. (eds.), *Integrating Environmental Sustainability and Corporate Social Responsibility: A Move Beyond Profit*, World Sustainability Series,
https://doi.org/10.1007/978-3-031-81469-3_11

chains. The study also examines the moral issues and difficulties surrounding the use of AI in CSR projects, highlighting the significance of stakeholder involvement, accountability, and openness in guaranteeing ethical AI use for environmental sustainability. Finally, it is evident from this study that businesses can improve their competitive edge and make a major contribution to the global endeavor to create a future that is more resilient, equitable, and environmentally sustainable by utilizing AI to support sustainability and mitigate climate change. **Originality/Value**-The paper is the first of its kind not ever published elsewhere and is original in nature.

Keywords Sustainability · CSR · AI · Carbon–neutral · Resilient · Equitable · Climate change

1 Introduction

Business adoption of sustainable practices and environmental stewardship have become more and more important in recent years. Companies everywhere are facing pressure to rethink their Corporate Social Responsibility (CSR) strategies considering the impending threat of climate change and the necessity of moving towards a low-carbon economy. The convergence of sustainability and artificial intelligence (AI) in this context offers a game-changing chance to promote green growth and carbon–neutral solutions. By providing creative solutions to urgent environmental problems, the convergence of AI-driven technologies and environmental sustainability is changing the landscape of corporate social responsibility (Kim 2022; Wirba 2023; Yu et al. 2022). AI-driven initiatives hold immense potential to redefine the scope of corporate social responsibility (CSR) and drive businesses toward a more sustainable future. These initiatives can range from optimizing energy consumption and reducing carbon emissions to promoting sustainable supply chains and fostering circular economies. This research investigates, within the context of corporate social responsibility, the paradigm shift resulting from AI-driven green growth initiatives (Zhao 2022; Javaid et al. 2022; Shayan et al. 2022; Dwivedi et al. 2022). In addition to increasing operational effectiveness and profitability, businesses can help the global effort to reduce global warming and attain carbon neutrality by utilizing AI technologies. This work aims to examine the revolutionary effects of AI-driven solutions on sustainability practices and the ramifications for companies, legislators, and society at large through the prism of corporate social responsibility (CSR). A turning point in the development of corporate social responsibility has been reached with the advent of AI-driven green growth and carbon–neutral solutions, which present new opportunities and difficulties for companies to overcome. The integration of sustainability into core business strategies is becoming increasingly important for companies. As a result, the CSR spectrum needs to be redefined through AI-driven initiatives to promote environmental resilience, stimulate economic growth, and guarantee a sustainable future for future generations (Cowls et al. 2021; Dwivedi et al. 2022; Javaid et al. 2022).

Even though AI-driven green growth and carbon–neutral solutions are becoming more popular in the context of corporate social responsibility (CSR), there are still some research gaps that need to be filled:

- Although AI's potential to spur sustainability initiatives is acknowledged, little is known about the full range of applications of AI and how well they work to support carbon neutrality and green growth. The wide range of AI technologies and their unique contributions to CSR goals require more investigation.
- There are frequently insufficient in-depth case studies in the literature that show how AI-driven CSR initiatives are implemented in the real world for environmental sustainability. The analysis of successful AI projects across industries, geographies, and company sizes is still lacking empirical research to pinpoint best practices and lessons learned.
- Initiatives for green growth powered by AI have not had a full ethical and societal analysis done. Most of the time, issues like algorithmic bias, data privacy, and the social effects of AI adoption in sustainability initiatives go unanswered. These ethical issues and how they affect CSR practices should be investigated in future studies.
- A knowledge vacuum exists regarding the potential integration of AI-driven green growth projects with conventional CSR frameworks and tactics. It is necessary to research to examine the opportunities and difficulties that arise from integrating AI technologies with traditional corporate social responsibility practices.
- Analysing the effect of AI-driven CSR initiatives on environmental sustainability is made more difficult by the absence of standardized measurement and evaluation metrics. To assess the efficacy and results of AI-driven green growth initiatives, future research should concentrate on creating reliable frameworks and metrics.
- Investigations into the creation of legal and regulatory structures that oversee the moral application of AI in environmental projects are necessary. This covers factors that are relevant to AI-driven CSR initiatives, such as data governance, accountability, transparency, and regulatory compliance.

Hence, our understanding of AI-driven green growth and carbon–neutral solutions within the CSR spectrum will advance only if these research gaps are filled. Researchers can help businesses address environmental challenges and advance corporate social responsibility (CSR) goals by bridging these gaps and developing more sustainable strategies.

1.1 Need and Significance of the Study

There has never been a more urgent need to address climate change and move toward a low-carbon economy. Businesses urgently need to implement sustainable practices and help achieve carbon neutrality considering the world's rising temperatures, extreme weather, and ecological degradation. This research will investigate the critical role that AI-driven green growth initiatives can play in tackling these issues.

Secondly, when it comes to promoting environmental sustainability, corporations are crucial. Businesses must adopt corporate social responsibility (CSR) strategies that give environmental stewardship top priority because they are significant sources of carbon emissions and resource consumption. For businesses to meet their sustainability targets, they must comprehend how AI technologies can reshape the CSR landscape and enable carbon–neutral solutions. Thirdly, by maximizing resource use, cutting waste, and boosting energy efficiency, artificial intelligence (AI) technologies have the potential to completely transform sustainability initiatives (Dwivedi et al. 2022; Santos 2022; Chen 2021). Businesses can achieve significant carbon emission reductions and encourage green growth across multiple sectors by utilizing AI-driven solutions. Examining the effects of AI within the framework of corporate social responsibility can yield a significant understanding of the revolutionary possibilities of these innovations. Furthermore, there are several chances for creativity and cooperation where sustainability and AI converge. Businesses can propel technological advancements, cultivate industry partnerships, and open new markets for sustainable goods and services by funding AI-driven initiatives. The main areas where AI-driven solutions can spur innovation and change the CSR environment will be determined by this study. Besides this, the global relevance of studying AI-driven green growth and carbon–neutral solutions is evident in the growing recognition of sustainability among businesses worldwide. This study will contribute to the global discourse on sustainability by analyzing case studies and best practices from various geographical regions. The insights gained from this research will apply to industries and borders. Lastly, the policy choices and regulatory frameworks that support sustainability and carbon neutrality can benefit from the study's conclusions. Policymakers can create standards, rules, and incentives that incentivize companies to invest in green technologies and adopt sustainable practices by emphasizing the potential of AI-driven solutions.

1.2 Research Problem Identified from the Study

Determining how AI-driven projects can change the CSR landscape is the main research challenge, especially when it comes to encouraging green growth and carbon–neutral solutions. Important inquiries consist of:

(a) What strategies are businesses using to lower carbon emissions and promote green growth with AI technology?
(b) What are the opportunities and difficulties in putting AI-driven CSR initiatives into practice?
(c) In what ways can artificial intelligence-powered solutions help achieve carbon neutrality and further sustainable development objectives?

1.3 *Research Objectives Formulated*

The principal aim of the research is to examine how artificial intelligence (AI) can support green growth initiatives in the framework of corporate social responsibility (CSR). This entails examining the various ways artificial intelligence (AI) is being used to improve resource usage, lower carbon emissions, and promote sustainable practices in a variety of sectors and industries. Evaluating the effects of AI-driven solutions for reaching carbon neutrality targets is another goal. The study will investigate how AI technologies, through enhancing energy efficiency, facilitating the integration of renewable energy sources, and enabling carbon offsetting strategies, can expedite the shift towards a low-carbon economy. Next, finding best practices and case studies of businesses that have successfully adopted AI-driven CSR initiatives for green growth and carbon–neutral solutions is the goal of the study. The study aims to extract insights, lessons learned, and success factors from real-world examples that can guide future corporate social responsibility strategies and initiatives. Furthermore, the study will evaluate the potential and obstacles related to the integration of AI technologies in advancing sustainability initiatives. This entails looking into matters like algorithmic bias, data privacy, regulatory compliance, and the socioeconomic effects of AI-driven CSR projects. The study's final goal is to offer suggestions on how companies, legislators, and other stakeholders can redefine the scope of corporate social responsibility for environmental sustainability by utilizing AI-driven solutions. This could include frameworks for strategically integrating AI into CSR strategies, best practices guidelines, and policy recommendations.

By tackling these goals, the research aims to expand knowledge and comprehension of the revolutionary possibilities of AI-driven green growth and carbon–neutral CSR solutions. In the end, the research hopes to support the creation of resilient and sustainable business practices that support international initiatives to combat climate change and accomplish sustainable development goals (SDG).

1.4 *Methodology Decoded*

The entire methodology of how this study was conducted has been illustrated in Fig. 1 as given below:

2 Silver Bullets and the Carbon Neutral Solutions

Carbon–neutral and AI-driven green growth solutions solve environmental issues and advance equitable and sustainable development globally, all while bringing economic advantages through increased productivity, cost savings, and innovation.

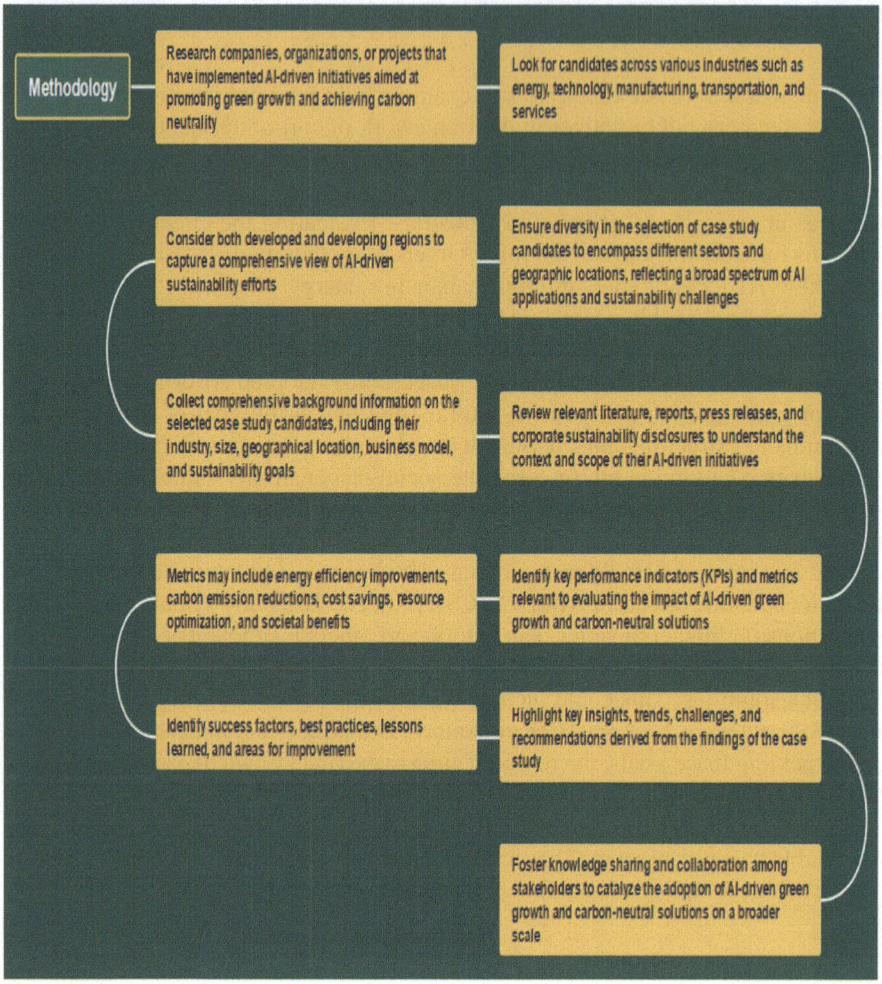

Fig. 1 Shows the methodology decoded for the study purpose. *Source* Developed by Authors for the study purpose

2.1 *The Indian Sustainable Equation*

- AI aids in maximizing the grid's integration of renewable energy sources like solar and wind power, which is important given India's lofty ambitions for renewable energy. The switch to a greener energy mix is facilitated by AI algorithms, which estimate energy consumption, control variability, and improve grid stability (Panda and Roy 2023).

- Precision farming methods powered by AI increase crop yields, maximize water use, and minimize chemical inputs in the agriculture industry. Robotic crop monitors and soil sensors, for example, are AI-enabled products that enable Indian farmers to make data-driven decisions that improve sustainability and productivity (Talaviya et al. 2020; Karunathilake et al. 2023; Dhanaraju et al. 2022).
- Sustainable urban planning is necessary given India's growing urbanization. AI makes it easier to create smart cities by controlling traffic, enhancing waste management, and optimizing building energy use. AI is being used by cities like Bengaluru and Pune to solve urban problems and advance environmental sustainability (Syed et al. 2021).
- Deforestation, water shortages, and air pollution are some of the environmental issues that India faces. Artificial intelligence (AI)-driven sensors, satellite imagery analysis, and predictive models facilitate early warning systems and real-time monitoring, therefore aiding in the mitigation of environmental degradation and preservation of ecosystems (Chen et al. 2023; Shivaprakash et al. 2022; Jain et al. 2023).
- To lessen its reliance on fossil fuels and fight air pollution, India wants to switch to electric cars (EVs). AI helps EV adoption by expediting the transition to sustainable transportation by forecasting demand patterns, optimizing the deployment of charging infrastructure, and enhancing battery performance (Hussain et al. 2023; Goel et al. 2021; Alanazi 2023).

There have been various case examples highlighting the above-said points validated, as shown below (From Figs. 2, 3, 4 and 5).

2.2 Sustainability Challenges and AI Corrections

By increasing resilience, promoting more inclusive and equitable development routes, and optimizing resource usage, artificial intelligence (AI) technologies provide creative answers to sustainability problems (Khakurel et al. 2018; Nahar 2024; Zhao 2022; Gikunda 2024; Jain et al. 2023); as discussed in Table 1.

3 Key Performance Indicators and Metrics Relevant to AI Impact on Green Growth and Carbon–Neutral Solutions

The following KPIs and metrics shown in Table 2 provide a comprehensive framework for evaluating the impact of AI-driven green growth and carbon–neutral solutions across various dimensions, including energy efficiency, carbon emissions reduction, renewable energy integration, resource optimization, operational efficiency, supply chain sustainability, stakeholder engagement, and social impact (Danish and Senjyu 2023; Chen et al. 2023; Liu et al. 2022).

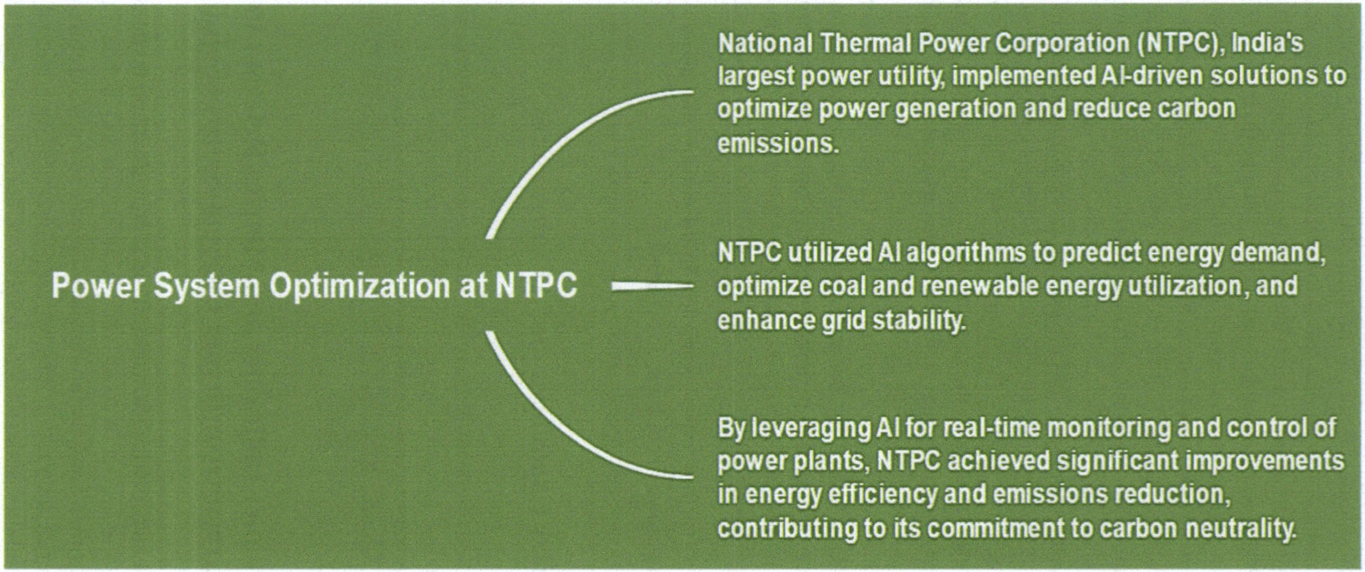

Fig. 2 Shows the case example of power system optimisation. *Source* Developed by Authors for the study purpose

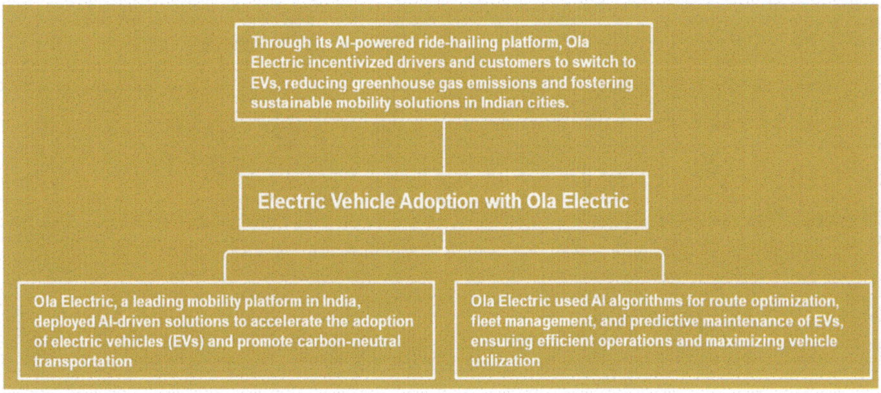

Fig. 3 Shows the case example of electric vehicle adoption. *Source* Developed by Authors for the study purpose

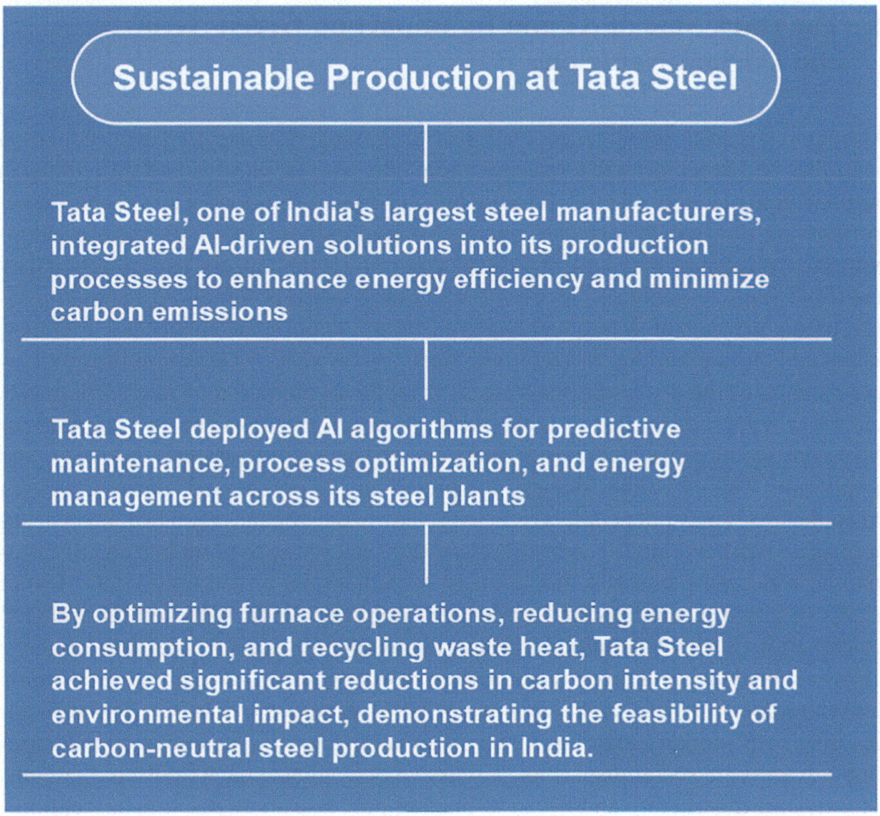

Fig. 4 Shows the case example of sustainable production. *Source* Developed by Authors for the study purpose

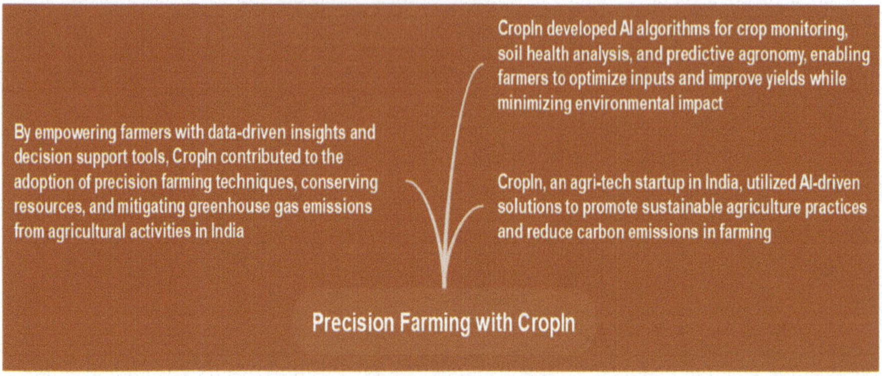

Fig. 5 Shows the case example of precision farming. *Source* Developed by Authors for the study purpose

4 The Dark Side of AI and Its Impact on Green- Growth and Carbon-Neutrality

Although artificial intelligence (AI) has enormous promise to advance carbon neutrality and green growth strategies, there may also be hazards and difficulties in putting it into practice. AI models frequently need a large amount of processing power to train and run, which increases energy use and carbon emissions. AI algorithm-powered data centers may have significant environmental impacts, particularly if they use fossil fuels to generate electricity. The quick speed at which AI technology is developing may result in frequent hardware component replacements and upgrades, which adds to the buildup of electronic trash, or "e-waste." Depletion of resources and damage to the ecosystem may result from the exploitation of rare earth metals and other resources needed to make AI hardware. Large data sets are required for AI-driven solutions, which raises questions regarding data security, privacy, and monitoring. The gathering, storing, and processing of private and sensitive data for AI applications may put people's right to privacy in danger and result in data breaches or misuse. Artificial intelligence (AI) algorithms are prone to prejudice and discrimination, replicating and sustaining social prejudices seen in training data. Unfair access to green technology and advantages might result from biased AI systems' exacerbation of social inequities and environmental injustices. Artificial intelligence (AI)-)-enabled task and process automation has the potential to disrupt the economy and eliminate jobs, especially in manual labor-intensive industries. The difficulties displaced workers may encounter in switching to new positions or industries might exacerbate social disparities. Excessive dependence on AI-powered systems for crucial infrastructure and decision-making processes may result in susceptibilities and isolated sources of failure (Dwivedi et al. 2022; Danish and Senjyu 2023). Energy grids, transportation networks, and environmental monitoring systems may all be significantly and cascaded affected by malfunctions, hacks, or algorithmic

Table 1 Shows the sustainability challenges and corrections offered by AI technologies

S.no	Sustainability challenges	AI corrections
1	***Energy Waste and Inefficiency***: Waste and increased carbon emissions result from inefficient energy use in houses, businesses, and transportation	Energy management systems powered by artificial intelligence (AI) examine sensor and meter data to find areas for energy savings and optimize energy use in real-time. By modifying heating, cooling, lighting, and equipment operations in response to occupancy trends, meteorological predictions, and energy tariffs, these systems cut down on energy waste and carbon emissions
2	***Resource Depletion and Overconsumption***: Overuse of natural resources, including minerals, land, and water, results in depletion, deterioration of the ecosystem, and extinction of species	Resource allocation may be optimized and waste can be reduced with the help of AI-powered resource management systems, which also track resource utilization and forecast demand patterns. Precision agricultural methods powered by AI decrease chemical inputs, maximize water use, and enhance soil health, while supply chain analytics powered by AI spot possibilities to increase resource efficiency and implement circular economy projects
3	***Climate Change Impacts and Extreme Events***: Climate change causes extreme weather events like hurricanes, floods, and wildfires to occur more frequently and with greater severity, resulting in financial losses and human suffering	AI-powered climate models forecast and lessen the effects of climate change by analyzing enormous volumes of data from satellites, weather stations, and ocean buoys. In addition to supporting adaptation tactics to increase resilience in vulnerable populations, these models also give early warnings for extreme weather occurrences and influence disaster planning and response operations
4	***Waste Generation and Pollution***: Ecosystems, human health, and environmental quality are at risk due to rising waste creation and pollution levels	To reduce landfill disposal and enhance resource recovery, AI-driven waste management systems improve garbage collecting, sorting, and recycling procedures. Smart air, water, and soil pollution levels are tracked by AI-powered sensors, drones, and satellites, which allows for early source identification and focused repair activities
5	***Loss of Biodiversity and Habitat Destruction***: Deforestation, habitat degradation, and overfishing cause ecosystem services and biodiversity to be lost, endangering fragile ecosystems and global biodiversity hotspots	To map and monitor ecosystems, identify vulnerable species, and prioritize conservation areas, AI-powered conservation technologies use satellite images, acoustic data, and biodiversity records. These instruments support sustainable management practices, habitat restoration initiatives, and land-use planning choices that safeguard biodiversity and vital habitats
6	***Inequitable Access to Resources and Opportunities***: Social inequality and environmental injustices are both made worse by unequal access to opportunities, resources, and services	AI-driven social impact evaluations evaluate the fair distribution of resources and benefits and identify people at risk by analyzing environmental, socioeconomic, and demographic data. To solve environmental inequities and advance social fairness, AI-powered decision support systems enable inclusive policymaking, community participation, and participatory planning procedures

Table 2 Depicts the KPIs and Sustainable Solutions derived from promoting carbon neutrality in India and Abroad

S.no	Key performance metrics	Sustainable solutions (India)	Sustainable solutions (Abroad)
1	**Energy Efficiency**: • Energy intensity is the amount of energy used for each output or activity unit • Energy Savings: Measuring the amount of energy used less due to optimization powered by artificial intelligence • Energy productivity is the measure of how well resources are used by dividing economic output (such as GDP) by energy consumption	Energy consumption in buildings, transportation, and industry is optimized using AI-driven energy management systems. Large energy savings may be achieved by using automation and predictive analytics to spot inefficiencies and modify energy usage patterns in real-time	Power distribution networks are more dependable and efficient when using AI-powered smart grid technology. Artificial intelligence (AI) algorithms integrate renewable energy sources, balance supply and demand, and optimize grid operations to increase energy efficiency and lower system losses

(continued)

Table 2 (continued)

S.no	Key performance metrics	Sustainable solutions (India)	Sustainable solutions (Abroad)
2	**Carbon Emissions Reduction**: • The amount of carbon emissions per unit of output or activity is called the carbon intensity • The whole quantity of greenhouse gas emissions (CO2e) connected to a good, service, company, or person is known as their "carbon footprint." • Emission Reduction Targets: Amount of carbon emissions to be reduced as a percentage from a baseline or target year	Integration of solar and wind power into the energy mix is made easier by AI applications in forecasting and grid optimization for renewable energy. AI helps minimize carbon emissions and lessen reliance on fossil fuels by making the most of sustainable energy sources	Carbon footprinting tools and AI-powered climate models help firms measure and reduce their carbon footprints. AI-driven energy management solutions minimize emissions, maximize energy use, and help businesses achieve their sustainability objectives

(continued)

Table 2 (continued)

S.no	Key performance metrics	Sustainable solutions (India)	Sustainable solutions (Abroad)
3	**Renewable Energy Integration**: • The percentage of renewable energy sources (such as solar and wind) in the energy mix is known as renewable energy penetration or integration • The quantity of energy generated by renewable sources, expressed in megawatt-hours (MWh) or kilowatt-hours (kWh), is known as renewable energy generation • The capacity factor is a measurement of a renewable energy system's actual energy output about its greatest potential output	Forecasting models powered by artificial intelligence maximize the grid's integration of renewable energy sources like wind and solar energy. Artificial Intelligence assists utilities in managing grid stability and optimizing the use of sustainable energy sources by forecasting patterns of energy generation and variations in demand	Renewable energy sources are more effectively integrated into electricity distribution networks thanks to AI-powered smart grid technology. Artificial intelligence (AI) algorithms enhance energy storage systems, balance supply and demand in real time, and increase the dependability and effectiveness of renewable energy integration

(continued)

Table 2 (continued)

S.no	Key performance metrics	Sustainable solutions (India)	Sustainable solutions (Abroad)
4	**Resource Optimization**: • Resource Utilization Efficiency: Metric measuring how well resources, such as land, raw materials, or water, are used • Waste Reduction: The percentage of waste creation that is decreased via AI-powered optimization • The percentage of resources recovered and recycled from waste streams is known as the resource recovery rate	Renewable energy sources like solar and wind power are optimally integrated into the grid using forecasting models driven by artificial intelligence. AI helps utilities maintain grid stability and optimize the use of sustainable energy sources by forecasting trends in energy output and demand variations	The integration of renewable energy sources into electricity distribution networks is improved by smart grid technology driven by artificial intelligence. Artificial intelligence (AI) algorithms enhance the efficiency and dependability of renewable energy integration, optimize energy storage systems, and balance supply and demand in real time

(continued)

Table 2 (continued)

S.no	Key performance metrics	Sustainable solutions (India)	Sustainable solutions (Abroad)
5	**Operational Efficiency**: • Production Yield: An indicator of production efficiency that is calculated as output per unit of input • Process optimization is the percentage of production processes that are improved via AI-driven optimization • Reduction of Downtime: The percentage of time spent reducing unscheduled stops or downtime in production or operational procedures	Predictive maintenance technologies powered by artificial intelligence maximize the efficiency of industrial machinery and infrastructure. Artificial Intelligence (AI) helps the industrial, transportation, and infrastructure sectors detect maintenance requirements ahead of time, minimize downtime, and increase operational efficiency by evaluating sensor data and equipment health parameters	Process optimization solutions driven by AI improve operational efficiency in sectors including manufacturing, logistics, and energy generation. Artificial intelligence algorithms minimize expenses and enhance efficiency by optimizing supply chain operations, logistical routes, and manufacturing schedules

(continued)

Table 2 (continued)

S.no	Key performance metrics	Sustainable solutions (India)	Sustainable solutions (Abroad)
6	**Supply Chain Sustainability**: • Total greenhouse gas emissions related to the supply chain, including emissions from Scopes 1, 2, and 3 • Supply Chain Transparency: Level of insight into the activities of the supply chain, including the adherence of suppliers to sustainable standards • The measure of the robustness and capacity of a supply chain to adjust to shocks, including natural catastrophes or geopolitical events, is known as supply chain resilience	Logistics of transportation, supplier relationships, and inventory management are all optimized by AI-driven supply chain analytics. Artificial Intelligence (AI) assists in finding possibilities for waste reduction, emissions mitigation, and sustainable sourcing methods in supply chains by evaluating data from sensors, satellites, and transaction records	Businesses can track and trace items across the supply chain to ensure compliance with social and environmental standards thanks to AI-powered supply chain sustainability systems. Supply chain risks and weaknesses are found by AI algorithms, which makes proactive risk management and resilience-building strategies possible

(continued)

Table 2 (continued)

S.no	Key performance metrics	Sustainable solutions (India)	Sustainable solutions (Abroad)
7	**Customer and Stakeholder Engagement**: • Customer satisfaction: quantifying how happy customers are with environmentally friendly goods or services • Stakeholder involvement: The degree of cooperation and involvement with consumers, suppliers, workers, and communities • Reputation and Brand Value: How the public views the company's sustainability initiatives and its standing as a leading brand	AI-powered consumer interaction solutions increase customer service, customize marketing efforts, and improve user experiences for businesses of all sizes. Chatbots and virtual assistants driven by AI allow companies to engage with clients in real-time, respond to questions, and provide tailored suggestions	Stakeholder engagement solutions driven by AI make it easier to collaborate and communicate with a variety of stakeholders, including suppliers, consumers, workers, and communities. To promote openness and confidence in corporate operations, artificial intelligence (AI) algorithms examine social media interactions, sentiment analysis, and feedback to find patterns, preferences, and places for development

(continued)

Table 2 (continued)

S.no	Key performance metrics	Sustainable solutions (India)	Sustainable solutions (Abroad)
8	**Social Impact and Equity**: • Community Benefits: Metric measuring the beneficial social effects on neighborhood communities, such as the creation of jobs, economic growth, and higher living standards • Equality and Inclusion: Evaluation of the distributional effects of AI-powered solutions, guaranteeing fair access to advantages and prospects for a range of demographics • Environmental Justice: An assessment of how much AI-driven efforts prioritize the needs of underprivileged groups and alleviate environmental injustices	AI-powered healthcare technologies facilitate early illness identification and increase access to high-quality medical care. Education systems driven by AI improve student learning results and offer individualized learning experiences	Applications of AI in humanitarian aid and disaster response speed up the process of assessing and responding to natural disasters and humanitarian emergencies. Social impact projects powered by AI tackle issues including healthcare access, poverty, and inequality

mistakes in AI systems. Complex systems with AI-driven decision-making may provide moral risks and ethical conundrums. Algorithms that maximize energy efficiency, for instance, could put financial savings ahead of human welfare or environmental sustainability, resulting in trade-offs and unexpected outcomes. The fast advancement of artificial intelligence technology presents obstacles to regulatory frameworks and governance procedures. It might be difficult for policymakers to keep up with technology advancements, which could result in oversight and regulation gaps for AI applications in carbon neutrality and environmental sustainability.

Proactive steps to reduce risks, improve accountability, and encourage responsible AI research and deployment are necessary to address the negative aspects of AI in green growth and carbon-neutrality solutions. As part of this, ethical AI standards must be adopted, algorithmic decision-making must be transparent and accountable, and AI governance frameworks must incorporate justice and environmental sustainability. Furthermore, encouraging multidisciplinary cooperation and stakeholder involvement may guarantee that AI technologies be used responsibly to accomplish sustainable development objectives (Rohde et al. 2024; Vinuesa et al. 2020; Jain et al. 2023).

5 Recommendations to Different Stakeholders

In the context of corporate social responsibility (CSR), many proposals seek to direct various stakeholders in utilizing AI-driven solutions for green growth and carbon neutrality, encouraging cooperation, innovation, and group action towards a more sustainable future (as shown below).

5.1 Recommendation to Policymakers

- Policymakers should promote innovation by providing funds, incentives, and enabling laws to assist the advancement of AI technology for sustainable development.
- Regulatory Frameworks: Create rules and guidelines that guarantee the ethical creation and use of AI-driven solutions while addressing issues with sustainability of the environment, privacy, and ethics.
- Collaborative Platforms: Promote cooperation in AI-driven sustainability projects by bringing together government agencies, academic institutions, businesses, and civil society. This will encourage innovation and information exchange.

5.2 Recommendation to Businesses

- Integrate Sustainability into Business Strategy: Align AI-driven efforts with environmental goals and carbon-neutrality targets by integrating sustainability indicators and goals into company plans.
- Increase accountability and transparency in AI-driven sustainability projects by giving stakeholders access to information on data usage, decision-making procedures, and environmental effects.
- Partnerships and Collaboration: By pooling resources and expertise, work with other companies, academic institutions, and non-governmental organizations to co-create and execute AI-driven solutions for carbon neutrality and green growth.

5.3 Recommendation to Non-Governmental Organizations (NGOs) and Civil Society

- Advocacy and Awareness: Encourage public discussion on the role of AI in sustainable development, raise stakeholder awareness, and champion ethical AI principles, environmental sustainability, and social responsibility in AI-driven projects.
- Community Engagement: To ensure inclusion and address the needs and goals of the local community, include grassroots groups and communities in the design, execution, and oversight of AI-driven sustainability initiatives.
- Monitoring and Supervision: Keep an eye on how AI-powered sustainability programs are being implemented, evaluate their effects, and hold legislators and companies responsible for their pledges to carbon neutrality and green growth.

5.4 Recommendation to Academia and Research Institutions

- Interdisciplinary Research: Investigate novel strategies and ways for resolving environmental issues by conducting interdisciplinary research at the nexus of artificial intelligence, environmental science, and sustainability.
- Capacity Building: To give professionals, researchers, and students the abilities and information required to create and implement AI-driven solutions for sustainable development, and offer training, instruction, and capacity-building programs.
- Exchange of Knowledge: To advance knowledge of AI-driven green development and carbon–neutral solutions, and disseminate research findings, best practices, and case studies via publications, conferences, and cooperative platforms.

5.5 Recommendation to Investors and Financial Institutions

- Responsible Investment: Take into account the sustainability performance and environmental effects of businesses creating AI-driven solutions when incorporating environmental, social, and governance (ESG) factors into investment decision-making processes.
- Impact investing aims to link investment portfolios with sustainability objectives and impact goals by providing support and capital to startups, ventures, and initiatives that are building AI-driven solutions for carbon neutrality and green growth.
- Engagement and Communication: Interact with investees and portfolio firms to promote responsibility, openness, and ongoing development in AI-driven sustainability projects, promoting communication and cooperation on social and environmental challenges.

6 Implications of the Study

To solve environmental issues and advance sustainable development, the study emphasizes the significance of incorporating AI-driven solutions within the CSR spectrum. Considering the ethical, social, and managerial ramifications, corporations, policymakers, and the general public must collaborate to fully use AI's promise for carbon neutrality and green growth.

6.1 Policy Implications

- Regulatory Frameworks: To promote green growth and carbon neutrality, policymakers must create and put into effect rules and guidelines that control the creation, application, and utilization of AI-driven solutions. These rules ought to take into account environmental effects, data privacy issues, and ethical issues.
- Incentive Mechanisms: By offering tax exemptions, grants, subsidies, and other financial incentives, governments may encourage companies to participate in AI-driven sustainability projects. These systems may promote creativity and the uptake of environmentally friendly technology.
- International Cooperation: To address the world's environmental concerns, policymakers should encourage international cooperation and collaboration. In AI-driven sustainability initiatives, the exchange of best practices, information, and resources may be facilitated through multilateral agreements and collaborations.

6.2 *Social Implications*

- Equity and Access: It's important to make sure that AI-driven green development projects assist underprivileged groups as well as the general public. Fair access to green technology, training courses, and job opportunities should be encouraged by policy.
- Ethical Considerations: Handling ethical issues about prejudice, discrimination, and transparency is necessary before society can adopt AI-driven solutions. To promote responsibility and confidence in AI-driven sustainability projects, stakeholder participation is crucial.
- Education and Awareness: Investing in education and awareness campaigns can empower individuals and communities to understand the potential benefits and risks of AI-driven green growth solutions. Efforts to engage the public can promote informed decision-making and active involvement in the development of sustainable practices and policies.

6.3 *Managerial Implications*

- Aligning AI-driven sustainability efforts with company strategy, values, and long-term goals is a critical task for business executives. Innovation, competitiveness, and shareholder value may all be generated by incorporating sustainability into fundamental company operations and decision-making processes.
- Risk management: Businesses implementing AI-driven sustainability projects must evaluate and control the risks involved, including those related to technology, legal compliance, and reputation. Strong frameworks for managing risks and backup strategies can lessen any negative effects.
- Collaborative Partnerships: The success of AI-driven sustainability efforts depends on stakeholders working together across sectors and value chains. Creating strategic alliances with vendors, clients, government organizations, and non-profits may help us make use of complementary networks, resources, and experience.

7 Conclusion

The study has redefined the CSR spectrum by highlighting the transformative potential of artificial intelligence (AI) in addressing environmental challenges and promoting sustainable development. Through a comprehensive analysis of AI-driven initiatives, the study has underscored the importance of integrating AI into corporate social responsibility (CSR) strategies to advance green growth and achieve carbon neutrality. Policymakers, companies, and society at large can address environmental concerns, promote sustainable development, and create a more resilient and fair future for future generations by using the potential of AI-driven innovation. To achieve green growth and carbon–neutral solutions, the study indicated includes a clear and comprehensive agenda for investigating the synergies between artificial intelligence (AI) and corporate social responsibility (CSR). The research intends to demonstrate the diverse ways in which artificial intelligence (AI) may support environmental sustainability by concentrating on many areas, including resource optimization, carbon emission reduction, and sustainable behaviors across multiple industries. Furthermore, given the urgency of tackling climate change, research exploring AI's involvement in achieving carbon neutrality objectives is very pertinent and important. The project intends to speed the shift to a low-carbon economy by analyzing how AI technologies might improve energy efficiency, encourage renewable energy integration, and facilitate carbon offsetting schemes. For companies wishing to undertake AI-driven CSR activities, the study's depth is increased by the incorporation of real-world case studies and best practices, which offer insightful information and practical advice. To fully comprehend the socioeconomic effects of AI-driven sustainability initiatives, it is important to address any roadblocks and issues that may arise throughout the integration process, such as algorithmic bias and data privacy concerns. The research's ultimate goal is to increase understanding while also offering firms, stakeholders, and policymakers useful frameworks and proposals for redefining the parameters of corporate social responsibility (CSR) in the context of environmental sustainability. The research considerably advances the worldwide endeavor to tackle climate change and advance a more sustainable future by encouraging resilient and sustainable business practices that are in line with international efforts and the Sustainable Development Goals (SDGs). Businesses may successfully incorporate AI-driven solutions by integrating sustainability with company strategy, enhancing transparency, and cultivating partnerships. These suggestions can encourage creativity, improve responsibility, and help achieve carbon neutrality objectives. NGOs and civil society may be empowered to advance moral AI principles and guarantee inclusive and sustainable AI-driven projects through advocacy, community engagement, and monitoring initiatives. The public's understanding of and involvement in sustainability programs may be increased by these activities. AI-driven sustainability solutions may be better understood and implemented with the help of interdisciplinary research, information exchange, and capacity building. These initiatives can provide experts and scholars with the information and abilities required to successfully handle environmental concerns. Investment portfolios may be aligned with sustainability targets and AI-driven projects through the use of engagement techniques, impact investing, and responsible investment practices.

Additional Resources

1. Reports and Publications:

World Economic Forum (WEF)—"Harnessing Artificial Intelligence for the Earth" (2018).

United Nations Environment Programme (UNEP)—"Digital Transformation for Sustainable Development" (2020).

OECD—"AI and the Environment: Policy Frameworks and Case Studies" (2021).

2. Academic Papers:

Flach, T., & Altmann, J. (2020). "AI for the Earth: Leveraging AI to Monitor, Mitigate, and Adapt to Climate Change". AI and Ethics, 1(4), 357–369.

Zeng, S., et al. (2021). "Artificial Intelligence for Sustainable Development: A Review and Bibliometric Analysis". Sustainability, 13(10), 5642.

Khan, M. A., & Qamar, A. (2022). "Integration of Artificial Intelligence in Corporate Social Responsibility: A Review and Future Research Agenda". Journal of Business Ethics, 175, 221–238.

3. Industry Reports and Case Studies:

IBM Institute for Business Value—"AI and Sustainability: How Artificial Intelligence can Drive Sustainable Business Practices" (2020).

Deloitte—"AI for Good: Using Artificial Intelligence to Achieve Sustainable Development Goals" (2021).

Microsoft—"AI for Earth: Case Studies and Impact Stories" (available on their website).

4. Government and International Organization Publications:

European Commission—"AI for Climate Action: Harnessing Artificial Intelligence to Address Climate Change" (2020).

International Energy Agency (IEA)—"Net Zero by 2050: A Roadmap for the Global Energy Sector" (2021).

United Nations Development Programme (UNDP)—"AI and Sustainable Development: Opportunities and Challenges" (2022).

References

The Advantages of Data-Driven Decision-Making | HBS Online (2019) Business insights blog. https://online.hbs.edu/blog/post/data-driven-decision-making

AI alliance launches as an international community of leading technology developers, researchers, and adopters collaborating together to advance open, safe, responsible AI. (n.d.). IBM Newsroom. https://newsroom.ibm.com/AI-Alliance-Launches-as-an-International-Community-of-Leading-Technology-Developers,-Researchers,-and-Adopters-Collaborating-Together-to-Advance-Open,-Safe,-Responsible-AI

AI in renewable energy: harnessing data-driven insights for sustainable power generation (2023) Times of India Blog. https://timesofindia.indiatimes.com/readersblog/aitalks/ai-in-renewable-energy-harnessing-data-driven-insights-for-sustainable-power-generation-54219/

AI-Driven Sustainability Solutions in a Changing World (2024). Omdena. https://www.omdena.com/blog/ai-driven-sustainability-solutions-in-a-changing-world

Alanazi F (2023) Electric vehicles: benefits, challenges, and potential solutions for widespread adaptation. Applied Sciences. https://doi.org/10.3390/app13106016

Chen L, Chen Z, Zhang Y, Liu Y, Osman AI, Farghali M, Hua J, Al-Fatesh AS, Ihara I, Rooney D, Yap P (2023) Artificial intelligence-based solutions for climate change: a review. Environ Chem Lett.https://doi.org/10.1007/s10311-023-01617-y

Chen JM (2021) Carbon neutrality: toward a sustainable future. Innovation 2(3):100127. https://doi.org/10.1016/j.xinn.2021.100127

Corporate Social Responsibility: The Good, the Bad and the Ugly (2009) Soc Bus Rev 4(2):163–166.https://doi.org/10.1108/17465680910977226

Cowls J, Tsamados A, Taddeo M, Floridi L (2021) The AI gambit: leveraging artificial intelligence to combat climate change—opportunities, challenges, and recommendations. AI & Soc 38(1):283–307. https://doi.org/10.1007/s00146-021-01294-x

Danish MSS, Senjyu T (2023) Shaping the future of sustainable energy through AI-enabled circular economy policies. Circ Econ 2(2):100040. https://doi.org/10.1016/j.cec.2023.100040

Dhanaraju M, Chenniappan P, Ramalingam K, Pazhanivelan S, Kaliaperumal R (2022) Smart farming: internet of things (IoT)-based sustainable agriculture. Agriculture 12(10):1745. https://doi.org/10.3390/agriculture12101745

Dwivedi YK, Hughes L, Kar AK, Baabdullah AM, Grover P, Abbas R, Andreini D, Abumoghli I, Barlette Y, Bunker D, Kruse LC, Constantia I, Davison RM, De R, Dubey R, Fenby-Taylor H, Gupta B, He W, Kodama M, … Wade M (2022) Climate change and COP26: are digital technologies and information management part of the problem or the solution? An editorial reflection and call to action. Int J Inf Manag 63:102456. https://doi.org/10.1016/j.ijinfomgt.2021.102456

Gikunda K (2024) Harnessing artificial intelligence for sustainable agricultural development in Africa: opportunities, challenges, and impact. arXiv (Cornell University). https://doi.org/10.48550/arxiv.2401.06171

Goel S, Sharma R, Rathore AK (2021) A review of barriers and challenges of electric vehicles in India and vehicle to grid optimization. Transp Eng. https://doi.org/10.1016/j.treng.2021.100057

Hussain N, Khan SA, Nguyen DK, Stocchetti A, Corbet S (2023) Board-level governance and corporate social responsibility: a meta-analytic review. J Econ Surv. https://doi.org/10.1111/joes.12603

Jain H, Dhupper R, Shrivastava A, Kumar D, Kumari M (2023) AI-enabled strategies for climate change adaptation: protecting communities, infrastructure, and businesses from the impacts of climate change. Comput Urban Sci 3(1). https://doi.org/10.1007/s43762-023-00100-2

Javaid M, Haleem A, Singh RP, Suman R, Gonzalez ES (2022) Understanding the adoption of Industry 4.0 technologies in improving environmental sustainability. Sustain Oper Comput 3:203–217. https://doi.org/10.1016/j.susoc.2022.01.008

Karunathilake EMBM, Le AT, Heo S, Chung YS, Mansoor S (2023) The path to smart farming: innovations and opportunities in precision agriculture. Agriculture 13(8):1593. https://doi.org/10.3390/agriculture13081593

Khakurel J, Penzenstadler B, Porras J, Knutas A, Zhang W (2018) The rise of artificial intelligence under the lens of sustainability. Technologies 6(4):100. https://doi.org/10.3390/technologies6040100

Kim RC (2022) Rethinking corporate social responsibility under contemporary capitalism: five ways to reinvent CSR. Bus Ethics Environ Respon 31(2):346–362. https://doi.org/10.1111/beer.12414

Lee CC, Hussain J (2022) Carbon neutral sustainability and green development during energy consumption. Innov Green Develop 1(1):100002. https://doi.org/10.1016/j.igd.2022.100002

Liu Z, Sun Y, Xing C, Liu J, He Y, Zhou Y, Zhang G (2022) Artificial intelligence powered large-scale renewable integrations in multi-energy systems for carbon neutrality transition: Challenges and future perspectives. Energy and AI. https://doi.org/10.1016/j.egyai.2022.100195

Nahar S (2024) Modeling the effects of artificial intelligence (AI)-based innovation on sustainable development goals (SDGs): applying a system dynamics perspective in a cross-country setting. Technological forecasting & social change/technological forecasting and social change. https://doi.org/10.1016/j.techfore.2023.123203

Panda A, Ray KK (2023) Equity market performance: the role of environmental protection and corporate social responsibility efforts. Bus Strategy Dev 6(4):739–747. https://doi.org/10.1002/bsd2.274

Rohde F, Wagner JR, Meyer A, Reinhard P, Voß M, Petschow U, Mollen A (2024) Broadening the perspective for sustainable artificial intelligence: sustainability criteria and indicators for Artificial Intelligence systems. Curr Opinion Environ Sustain.https://doi.org/10.1016/j.cosust.2023.101411

Santos AR (2022) The Importance of artificial intelligence in start-up, automation, and scalation of business for entrepreneurs. Int J Appl Eng Technol 4:1–5. https://www.scirp.org/reference/referencespapers?referenceid=3624559

Santos FD, Ferreira PL, Pedersen JST (2022) The climate change challenge: a review of the barriers and solutions to deliver a Paris solution. Climate. https://doi.org/10.3390/cli10050075

Shayan NF, Mohabbati-Kalejahi N, Alavi S, Zahid MA (2022) Sustainable development goals (SDGs) as a framework for corporate social responsibility (CSR). Sustainability 14(3):1222. https://doi.org/10.3390/su14031222

Shivaprakash KN, Swami N, Mysorekar S, Arora R, Gangadharan A, Vohra K, Jadeyegowda M, Kiesecker JM (2022) Potential for artificial intelligence (AI) and machine learning (ML) applications in biodiversity conservation, managing forests, and related services in India. Sustainability 14(12):7154. https://doi.org/10.3390/su14127154

Syed A, Sierra-Sosa D, Kumar A, Elmaghraby A (2021) IoT in smart cities: a survey of technologies, practices and challenges. Smart Cities. https://doi.org/10.3390/smartcities4020024

Talaviya T, Shah D, Patel N, Yagnik H, Shah M (2020) Implementation of artificial intelligence in agriculture for optimisation of irrigation and application of pesticides and herbicides. Artif Intell Agric 4:58–73. https://doi.org/10.1016/j.aiia.2020.04.002

Vinuesa R, Azizpour H, Leite I, Balaam M, Dignum V, Domisch S, Felländer A, Langhans SD, Tegmark M, Nerini FF (2020) The role of artificial intelligence in achieving the Sustainable Development Goals. Nature Commun. https://doi.org/10.1038/s41467-019-14108-y

Wirba AV (2023) Corporate social responsibility (CSR): the role of government in promoting CSR. J Knowl Econ. https://doi.org/10.1007/s13132-023-01185-0

Yu F, Jiang D, Wang T (2022) The impact of green innovation on manufacturing small and medium enterprises corporate social responsibility fulfillment: the moderating role of regional environmental regulation. Corp Soc Responsib Environ Manag 29(3):712–727. https://doi.org/10.1002/csr.2231

Zhao J, Gómez Fariñas B (2022) Artificial intelligence and sustainable decisions. Europ Bus Organ Law Rev 24(1):1–39. https://doi.org/10.1007/s40804-022-00262-2

Zhao C, Ju S, Xue Y, Ren T, Ji Y, Chen X (2022) China's energy transitions for carbon neutrality: challenges and opportunities. Carbon Neutrality 1(1). https://doi.org/10.1007/s43979-022-00010-y

Zhao C (2022) Carbon neutrality: aiming for a net-zero carbon future. Carbon Neutrality 1(1). https://doi.org/10.1007/s43979-022-00013-9

Zhao W (2023) Understanding CSR campaigns through the lens of culture values and moral emotion. J Creat Commun 18(3):309–327. https://doi.org/10.1177/09732586221143183

Improving Social Responsibility Through Gender Diverse Boards: Challenges and Strategies

Sunita

Abstract Corporate social responsibility (CSR) and sustainability are crucial for the long-term success of the business. Companies must go beyond focusing solely on profits and address the needs of all stakeholders and ensure environment sustainability. Boards of directors, particularly women directors play a vital role in enhancing firms' social responsibility, as they tend to prioritise the concerns of stakeholders and environmental well-being. Despite this, women continue to be in low proportion on boards globally. The present chapter explores how gender-diverse boards contribute to improved CSR and sustainability. It also examines the inhibitors for achieving gender parity on boards are proposes measures to increase women's representation on boards, to enable the companies to transcend profit-driven motives and support environmental sustainability.

Keywords Gender diverse boards · Corporate social responsibility · Sustainable development · Inhibitors · Facilitators

1 Introduction

Traditionally the motive of businesses was to create profits and provide return to the shareholders on their investments. It was the fundamental goal of the companies to maximise wealth for its owners. However, over the years the shift has been taken place because of the regulatory developments across the countries. Social awareness too compelled the firms to become socially responsible. It became clear that to ensure their long-term survival, firms have to take care of the interests of all stakeholders in addition to running its business transparently and ethically.

Stakeholder theory (Evan and Freeman 1988) of corporate governance also emphasise that firms' long term success depends on satisfying all of its stakeholders. In addition to stakeholder theory, Legitimacy theory by Suchman (1995) advocates

Sunita (✉)
Department of Commerce, University of Delhi, PGDAV College, New Delhi, India
e-mail: sunita@pgdav.du.ac.in

P. Singh et al. (eds.), *Integrating Environmental Sustainability and Corporate Social Responsibility: A Move Beyond Profit*, World Sustainability Series,
https://doi.org/10.1007/978-3-031-81469-3_12

that a social contract exists between business and society and that a company's actions shall be desirable, proper, or appropriate from the point of view of norms, values, and beliefs prevailing in society.

It is well known that businesses derive several benefits from the society and the environment therefore the corporations need to provide the returns as well to society and environment (Gautam and Singh 2010). Firms, as societal members therefore, bear a responsibility to the communities and nations in which they operate. Society relies on businesses to achieve economic objectives such as creating employment opportunities, increasing per capita income, and ensuring self-sufficiency in basic needs. Additionally, corporations are expected to meet social needs by taking care of their employees, protecting the environment, and supporting local communities. With rising environmental concerns and the imperative for inclusive societal development, governments are increasingly expecting businesses to fulfill their share of responsibility. Corporate Social Responsibility (CSR) as defined by the World Business Council for Sustainable Development refers as "the continuing commitment by business to behave ethically and contribute to economic development while improving the quality of life of the workforce and their families as well as of the local community and society at large"(Jhunjhunwala 2024).

Corporate Social Responsibility (CSR) involves a company's commitment to managing its social, environmental, and economic effects responsibly. CSR initiatives encompass a broad range of activities, such as reducing carbon footprints, ensuring fair labor practices, engaging in philanthropy, and adhering to ethical business practices. Social responsibility is a well-established concept, and in today's competitive business landscape, integrating it into corporate strategies is crucial for promoting sustainability. Companies are expected to consider the social, environmental, and economic impacts of their operations and decision-making processes. This approach not only fosters a positive corporate image but also contributes to the long-term success of the company and the well-being of society. The more CSR concepts are embraced and integrated into business processes, the easier it becomes to benefit from innovative thinking and address inevitable challenges. As businesses more deeply integrate these practices within their value chains, they gain greater opportunities to influence the approaches of their partners and stakeholders. CSR also entails transparency and taking into account the expectations and desires of stakeholders (Gautam and Singh 2010).

Corporate boards play a crucial role in meeting the corporate social responsibility of a business as most of the important decisions required to manage the firms are being taken by the board of directors (Muttakin et al. 2018).

2 Gender Diversity on Boards

Diversity is a vital and strategic element for a business. When it comes to diversity on corporate boards, two types of diversities have been talked about namely– demographic and cognitive. Difference in education, knowledge, values and perception is known as cognitive diversity. When the difference is there on the basis of age, gender, ethnicity, and race, it refers to the demographic diversity (Haldar et al. 2015).

Board composition, particularly gender diversity on boards, has garnered significant attention from various stakeholders, policy makers and researchers over the past few decades (Catalyst 2001; Hyland and Marcellino 2002; Kirsch 2018). The presence of women on boards is viewed as crucial for promoting the gender equality, full utilization of economic resources and supporting sustainable development (Yogendrarajah 2013). The United Nations Sustainable Development Goals, specifically fifth goal (SDG-5) also emphasise the increasing significance of gender equality across various dimensions, such as business, academia, politics and the social arena. It aims to eradicate poverty, protect the planet, and ensure peace and prosperity for all people by 2030. Gender equality is not only a women's issue but should involve the active participation of both men and women. It is a prerequisite for sustainable development that prioritises the well-being of all people. The United Nations has stressed that ending all discrimination against women and girls is vital for a sustainable future, and it is confirmed that empowering women improves economic growth and development (United Nations 2015).

Governance failures across the world have also drawn attention towards effectiveness of board composition (Aguilera et al. 2021; Kersley et al. 2019) as these failures affect the lives of numerous stakeholders. While addressing the board's effectiveness and performance, researchers have emphasized the relevance of gender diversity on boards (Daily et al. 2003). Research shows that women serve as catalysts for change and their involvement in decision-making processes significantly influence the board strategy (Nielsen and Huse 2010).

Among the several types of diversity, gender composition of boards is one of the important issues faces by businesses in present scenario. It is considered significant not only from the diversity perspective but also for firm's financial and social performance (Rao and Tilt 2016). Gender diversity at senior management and corporate board levels differs for various countries and sectors. According to study based on global data for Fortune 500 companies, the proportion of women is low in the corporate boards all over the world and in India it is even lower as compared to other countries (Kamath and Lakhangaonkar 2021).

Several countries have taken either compulsory or voluntary measure to improve the proportion of women on corporate boards. Countries, such as Norway, Finland, Belgium, France, Italy, and Germany etc. have imposed quotas to increase the women's representation on corporate boards (Aguilera et al. 2021; Kersley et al. 2019). India has also introduced gender quotas for corporate boards in 2013 under Companies Act 2013 to improve gender diversity on boards (The Companies Act 2013). Figure 1, shows the presence of women on boards in select countries (World Economic Forum 2024).

Despite a rise in the proportion of females on boards, they continue to be underrepresented on boards across the world with less than 20% women directors on Indian boards. It depicts that gender parity on boards is still a far off goal for countries like India and Japan.

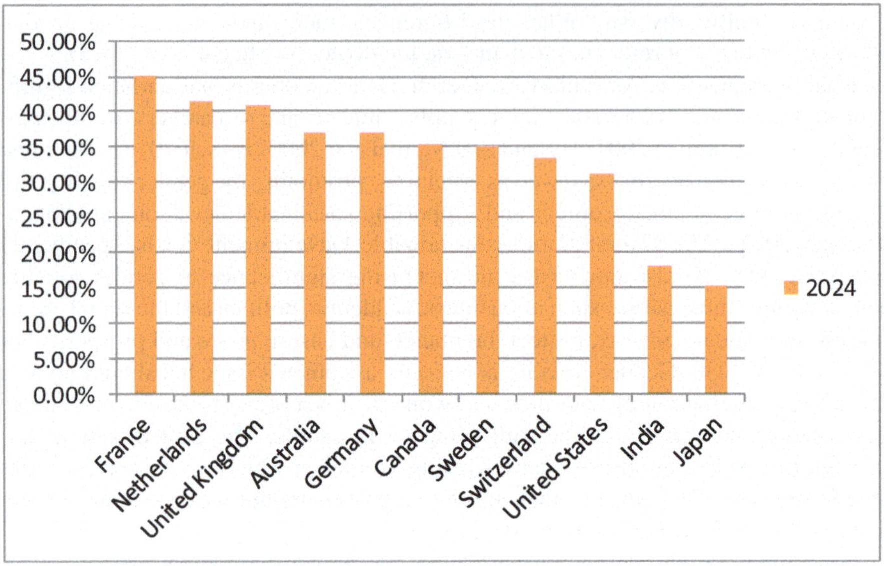

Fig. 1 Women on boards in select countries. *Source* Developed by Author. Data Source: Global Gender Gap Report 2024, World Economic Forum

Initiatives of policy makers and regulators clearly show that women's representation on boards may have a significant effect on the governance of firms in several ways (Adams and Ferreira 2009) and this underrepresentation of females on boards need to be addressed. Next section discusses the importance of women directors as identified form the literature.

3 Women Directors Benefit Firms and Economy

Research indicates that having a gender-diverse board enhances decision making by incorporating diverse opinions and viewpoints (Daily et al. 2003). Proponents of diversity argue that it brings a variety of perspectives and problem-solving approaches, improves communication, and fosters more detailed critical analysis of issues, which leads to better decisions. Increasing the proportion of females on corporate boards also enriches intellectual resources by integrating a broader pool of human capital. It gives firms a competitive edge over others and improves their performance. It will also results in optimal utilisation of human resources and supports sustainable development of economy (Adams and Flynn 2005; Bilimoria 2000; Campbell and Mínguez-Vera 2008; Dezsö and Ross 2012; Wang and Clift 2009). Numerous studies suggest that gender diverse boards bring benefits to firms and economy and this rationale is aligned with corporate governance theories as well.

From the perspective of agency theory (Jensen and Meckling 1976), independent directors are vital for corporate governance and serve the shareholders' interests effectively. It has been found that women directors often do not belong to the "old boys' club' therefore tend to have a similar impact as independent directors have on governance. Studies have established a positive relationship between gender diversity on boards and board effectiveness (Adams and Ferreira 2009; Carter et al. 2003; Choudhury 2015; Handa and Singh 2013; Sarkar and Selarka 2016). Contrary to the tokenism theory (Kanter 1977), which suggests that women face hurdles in influencing group decisions, women directors influence the boards' decisions and with higher proportion of women, boards become more active (Elstad and Ladegard 2012; Handa and Singh 2013) and generate the ability to engage in cognitive conflict. Women directors are more likely to ask tough questions and succeed in controversial situations (Adams and Ferreira 2009; Choudhury 2015; Elstad and Ladegard 2012; Tsiklashvili et al. 2019).

In addition to improving the governance of a firm the women directors, Just like good stewards, are shown to maximize the shareholders' wealth (Stewardship Theory of Corporate Governance) (Donaldson and Davis 1991). Numerous studies (Carter et al. 2003; Kramaric and Miletic 2017; Noland et al. 2016; Panitapu 2015; Sarkar and Selarka 2016; Schmid and Urban 2018) find a positive relationship between the percentage of women directors and firm performance and value. This positive relationship is often attributed to the diverse leadership styles of women compared to their male counterparts, which strengthens the performance of the firm (Desvaux et al. 2017; Sarkar and Selarka 2016; Thomas et al. 2020). However, The stakeholders' theory of corporate governance (Evan and Freeman 1988) argues against the shareholders' primacy and asserts that all stakeholders are important to the business and should be taken care off in managing its affairs (Fig. 2).

It is found that gender diversity on boards enhances understanding of customers and the marketplace, improves creativity and innovation, makes problem-solving and decision-making effective that leads to the improved global relationships (Bagati and Carter 2010; Burke 1997; Carter et al. 2003). Therefore, gender diverse boards not only improve firm performance and value, but also better address the interests of all stakeholders (Adams and Ferreira 2009; Burke 1993, 1997; Kramaric and Miletic 2017).

Board's sensitivity towards women's issues improves with the higher presence of female directors on boards (Burke 1997) and there is a positive impact on the recruitments, advancement, development and retention of women in the company (Kitterød and Teigen 2018; Lazzaretti et al. 2013; Thomas et al. 2020). Moreover, appointing women directors is beneficial for the firm as they are often more prudent and effective in managing conflicts and stressful situations (Tsiklashvili et al. 2019). Investing in female talent not only promotes gender parity equality but also enhances a nation's competitiveness. Developing female talent can help sustain growth, improve competitive advantage, and address leadership, diversity, and innovation issues (Bagati and Carter 2010).

Resource Dependency theory of corporate governance (Pfeffer and Salancik 1978) also endorses appointing more women directors on boards, businesses are entities that

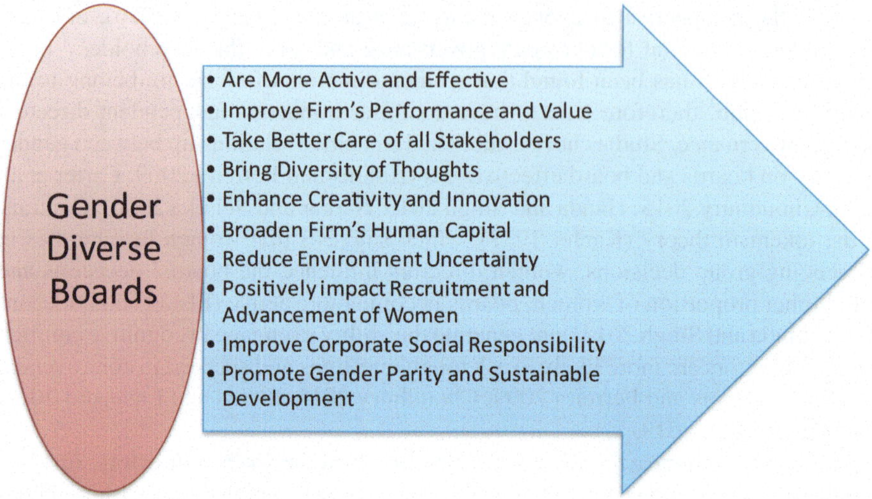

Fig. 2 Significance of gender diverse boards. *Source* Developed by Author

need to exchange and obtain resources for survival, creating dependencies between businesses and other external units. The board of directors help in managing the environmental uncertainties through their linkages with the environment or network governance (Hillman et al. 2000). Directors also bring essential resources such as skills, information, suppliers, and legitimacy, which reduce uncertainty (Gales and Kesner 1994; Nielsen and Huse 2010). Diverse leadership skills are also crucial for firms to thrive in uncertain and competitive environments that supports the appointment of more female directors on boards (Hillman et al. 2000). However, the contribution of women directors to boards and their performance depends on various factors, including board leadership and culture (Nielsen and Huse 2010).

As Legitimacy theory posits that a social contract exists between businesses and society (Suchman 1995) and therefore a company's actions shall be desirable, and appropriate according to the societal norms, values, and beliefs. Research shows that the presence of women directors enhances board governance and corporate social responsibility. They also positively impact other women's success in the company (Kitterød and Teigen 2018; Lazzaretti et al. 2013). Unlike male leaders, female leaders often adopt a holistic approach, which emphasises caring, nurturing, power-sharing (Amudha et al. 2016). They more frequently apply leadership behaviors of people development, role-modeling, inspiration, rewarding success, and participative decision-making than their male counterparts (Desvaux et al. 2017; Thomas et al. 2020). Inclusive boards demonstrate legitimacy through their commitment to ethical conduct and social responsibility.

Firms face pressure from stakeholders and society to provide equal and respectful treatment to all employees and increase the representation of women in senior positions. More women in top leadership positions not only benefits firms but also promote gender equality in society and contributes to the sustainable development of the economy (Imberg et al. 2023).

In light of benefits of female directors on boards discussed above, it can be stressed upon that firms can foster corporate social responsibility through improving the gender equality on boards.

4 Few Females on Boards..Why?

Women face almost similar obstacles across the world. These obstacles in corporate and social environments affect their access to top leadership positions and boards. The barriers either originate within organisation or from social roles outside it, especially those related to family responsibilities (Schipani et al. 2008). At lower level, women join the firm in large number however their proportion decreases as they rise up the career ladder due to an invisible hindrance termed as "Glass Ceiling". The term "Glass ceiling" is made up of traditions and stereotypical attitudes which makes it tough for women to reach the top positions (Hymowitz and Schellhardt 1986; Oakley 2000; Zimmer 2015). These glass ceiling barriers had been classified differently by different authors and organisations (Diehl and Dzubinski 2016; Gabaldon et al. 2016; Ismail and Ibrahim 2008; Lathabhavan and Balasubramanian 2017; Oakley 2000; Syed et al. 2018; Taparia and Lenka 2022a). This section discusses the factors that inhibit females to reach the higher echelons.

4.1 Personal Inhibitors

Studies across countries show that women face several barriers at the personal level that create a roadblock in their pathway to top leadership positions and boards. As posited by Person-centered theories men and women have different skills, traits, and behavior based on their socialisation process and behavioral attributes (Kiaye and Singh 2013). Men are more linked with income provision, dominance, aggression, autonomy, achievement, exhibition, and endurance, while women are linked to homemaking, nurturing, deference, and humility (Konrad et al. 2000). As per the research worldwide, women are perceived as lacking the qualities, ambition, and self-confidence required for leadership roles (Amudha et al. 2016; Maguire and Kleiner 1993; Nilufer Azeez and Priyadarshini 2018; Taparia and Lenka 2022a). Women are found as challenge averse in Indian as well as western context (Hurn 2013; Jain and Mukherji 2010). This lack of self-confidence among women hampers their the career progress of women to boards (Amudha et al. 2016). In addition to this, women due to their introvert nature, refrain from promoting themselves through networking or highlighting their skills, and willingness to take on the higher positions (Amudha et al. 2016; Burke 1993; Hurn 2013; Kiaye and Singh 2013; Stamper and McGowan 2022).

It is necessary for individuals who aspire to become directors to acquire good stocks of human capital (Kesner 1988; Kiaye and Singh 2013; Vohra 2020). However, as per Human capital theory, women make only the minimum investments in pre-employment skills and training (Akande 1993; Kiaye and Singh 2013; Reavley 1993) and lack the required education and qualifications to become directors (Tsiklashvili et al. 2019). Consequently, the presence of women in management and boards is affected (Akande 1993; Kiaye and Singh 2013; Reavley 1993).

A study based on 17 countries, including the USA and India (Boone et al. 2013), found that self-imposed or personal barriers such as family priorities, work-life balance, and life-cycle stage are the main reasons women and men do not pursue top positions. Another study based on India (Rath et al. 2019) found marriage a significant barrier to women's career advancement. However, Kiaye and Singh (2013) in context of Durban opposed the existence of individual barriers and confirmed that the majority of women feel confident. They are ambitious, competitive, and suited for leadership roles. Another Indian study based on 100 large listed firms in the country (Haldar et al. 2020) denied that women lack the right human capital for directorships.

4.2 Societal Inhibitors

Social roles and stereotypical attitudes significantly hinder the women's rise to leadership positions as posited by social role theory (Kiaye and Singh 2013; Nilufer Azeez and Priyadarshini 2018). Due to the social norms, women often find it challenging to manage the responsibilities at home and at work (Conceição et al. 2020; Landry et al. 2020; Saxena et al. 2009; Stroh et al. 1996). Women are forced to make suboptimal career choices. They often quit their jobs due to motherhood, unpaid care work, and other family responsibilities (Chawla and Sharma 2016; Kiaye and Singh 2013; Rath et al. 2019; Tripathy 2018; Valk and Srinivasan 2011). Consequently, the men's careers tend to follow linear and uninterrupted paths, while women's careers are often interrupted and non-linear (Conceição et al. 2020; Landry et al. 2020; Ravindran and Baral 2014; Saxena et al. 2009; Stroh et al. 1996). This disparity affects employers' decisions to invest in female employees, and leads to a low presence of female directors on boards (Saxena et al. 2009).

These gendered perceptions and societal roles, are seen as barriers across the world, including India. However in India, women face additional barrier of conflicting values as they are expected to be a nurturer and this role always conflict with their professional aspirations to reach higher positions (Chawla and Sharma 2016). These societal attitudes are often internalised by women themselves and significantly inhibit their career progression to boards (Rath et al. 2019; Terjesen and Singh 2008; Woetzel et al. 2015).

Societal barriers often lead to organisational barriers, reinforce each other and create a challenging environment for women's career progress (Boone et al. 2013; Taparia and Lenka 2022a).

4.3 *Organisational Inhibitors*

As advocated by Organisation theories, the success of women depends a lot on the organisational environment and culture more than their traits and skills (Kiaye and Singh 2013). A supportive and inclusive culture facilitates the advancement of women to top (Jain and Mukherji 2010), while a hostile workplace environment adversely affect the career of women. Studies in the Western world and India show that women face discrimination in salaries, incentives, recruitment, and promotion that impede their career progress (Amudha et al. 2016; Cooper Jackson 2001; Maguire and Kleiner 1993; Tavakolian 1993; Tripathy 2018). A study based on India (Tripathy 2018) finds that women often experience harassment at the workplace and are more likely to be asked personal questions in job interviews. Women also encounter a lack of sensitivity and support from senior management (Amudha et al. 2016; Rath et al. 2019; Tripathy 2018). They face a lack of recognition, respect and support from male colleagues (Amudha et al. 2016; Burke and Mckeen 1995; Nilufer Azeez and Priyadarshini 2018; Rath et al. 2019; Tripathy 2018). Women are often deprived of critical information at the right time (Maguire and Kleiner 1993; Rath et al. 2019). Due to these challenges women face a "leaky talent pipeline" and only few of them reach higher positions as compared to men. These challenges proved even more threatening for women with children due to the lack of workplace flexibility and adequate and affordable childcare known as Motherhood Penalty (Carli and Eagly 2016; Hurn 2013; Valk and Srinivasan 2011).

Even if women succeed in rising up in their career, they face gender-based discrimination in performance evaluations too. Unlike men, the success of women is often attributed to luck or chance, and failure to a lack of competence or skills. This biased interpretation of success or failure impedes women's career development (Macarie and Moldovan 2012). Moreover companies tend to associate performance with time commitment, disadvantaging women (Amudha et al. 2016; Nilufer Azeez and Priyadarshini 2018; Radovic-Markovic et al. 2013).

Apart from these inhibitors, women face the "sticky floor syndrome" with lack of growth opportunities (Amudha et al. 2016; Carli and Eagly 2016), and unequal opportunities for training and development as compared to their male counterparts (Amudha et al. 2016; Oakley 2000; Tsiklashvili et al. 2019). Females face a lack of international and visibility exposure that affect their career growth (Amudha et al. 2016; Kiaye and Singh 2013).

Moreover women face a lack of female mentors or role models in the organisation (Burke and Mckeen 1995; Hurn 2013; Marrujo and Kleiner 1992; Woetzel et al. 2018). In India, a noticeable shortage of female role models and lack of opportunities to connect with influential mentors impact women's chances of attaining the top executive positions (Boone et al. 2013; Chawla and Sharma 2016; Valk and Srinivasan 2011).

As Social exclusion theories posit, women are often excluded from the informal networks. One such network is "Old Boys' Network", that is an invisible discrimination women face during their career progression (Amudha et al. 2016; Chawla and

Sharma 2016; Hurn 2013; Marrujo and Kleiner 1992). Due to this network men get opportunities to rise in their career whereas the women's career growth is resisted. Not only this, Gender-specific Roles and Societal Stereotypes which are carried over to the firms affect the decisions and contribute to glass ceiling (Terjesen and Singh 2008). Women due to their communal traits and leadership styles are often perceived as misfits for leadership positions. This leads to discrimination against women at the time of leaders' selection, recruitment, and performance evaluation (Akande 1993; Gupta et al. 2018; Jani and Deshpande 2019; Kiaye and Singh 2013; Reavley 1993). Being a patriarchal society in India, stereotypical attitudes and gendered perceptions pose more challenges for women as compared to Western countries (Chawla and Sharma 2016; Rath et al. 2019). Research within Indian context indicates that women are less likely to be considered for top leadership positions because such roles do not align with traditional female gender stereotypes (Mohan 2014; Panitapu 2015; Tsiklashvili et al. 2019).

All of these barriers create a glass ceiling effect that prevents women from advancing to boards despite having the necessary education and skills for higher positions (Burke 1993; Cooper Jackson 2001; Maguire and Kleiner 1993; Tavakolian 1993). Scholars have suggested some measures also to shatter the glass ceiling and improve women's representation in top positions and boards.

5 Measures to Improve Women's Presence on Boards

To increase the women's representation on boards it is required to address the barriers at personal, societal and organisational levels.

At personal level, it is suggested that women are capable and willing to excel so they need to shift their own mindset to overcome self-imposed barriers first (Nilufer Azeez and Priyadarshini 2018). Societal beliefs and stereotypes must also be changed to improve the gender parity at all levels. However, this change will take time, so it is crucial for women to identify their aims in life and plan accordingly (Marrujo and Kleiner 1992). Women need to work harder than men to reach the same level (Kiaye and Singh 2013), and focus on their productivity instead of discrimination (Hacker and Kleiner 1993; Schilling 2012; Sur and Kleiner 1995).

Women are also required to develop self-awareness, recognise their strength and abilities, and communicate the same to organisations (Amudha et al. 2016; Kurup et al. 2011; Marrujo and Kleiner 1992; Tripathy 2018). They should adopt an appropriate communication style (Amudha et al. 2016; Tripathy 2018) and be courageous enough to address any discrimination they face (Tripathy 2018). Higher education and professional experience can also aid in reaching top positions (Kiaye and Singh 2013; Vohra 2020).

Literature across countries suggests that women should manage the dual burden of home and work by prioritising the responsibilities (Kurup et al. 2011; Marrujo and Kleiner 1992), and employing success strategies like time management, proper planning, delegation of work (Chawla and Sharma 2016; Schilling 2012).

In society, societal support in removing stereotypes and raising awareness is also essential (Kitterød and Teigen 2018; Lazzaretti et al. 2013; Nilufer Azeez and Priyadarshini 2018) to facilitate more women to reach corporate boards (Chawla and Sharma 2016). Moral and domestic support from parents and spouse (Chawla and Sharma 2016; Kurup et al. 2011; Marrujo and Kleiner 1992; Nilufer Azeez and Priyadarshini 2018; Ravindran and Baral 2014) and equitable sharing of unpaid work by family members also proved crucial to achieve gender parity and women empowerment (Woetzel et al. 2015) (Fig. 3).

Not only individuals and society rather companies should also take some measures to have gender diverse boards and derive their benefits. Firms should promote gender parity through fair policies and practices (Kiaye and Singh 2013; Sur and Kleiner 1995). Studies based on India suggest that companies should implement policies like fair practices, equal pay, safe work environments to improve the gender diversity (Chawla and Sharma 2016; Kiaye and Singh 2013; Tripathy 2018; Woetzel et al. 2015).

Research in the context of Western World as well as India shows that a positive and inclusive organisational culture that embraces empathy and meritocracy is crucial for women to reach leadership positions and boards (Akande 1993; Bagati and Carter 2010; Burke 1997; Chawla and Sharma 2016; Vohra 2020). Firms must broaden their leadership models to effectively address future challenges (Catalino and Marnane 2019).

Commitment and Support of Senior Management in meeting targets of gender equality at all levels play significant role in the rise of women to corporate boards (Amudha et al. 2016; Nilufer Azeez and Priyadarshini 2018; Sneader and Yee 2019;

Fig. 3 Gender diversity on boards: inhibitors and improvement measures. *Source* Developed by Author

Tsiklashvili et al. 2019; Woetzel et al. 2018). Firms should support their female employees through flexible working policies, adequate childcare, and career comeback initiatives (Akande 1993; Amudha et al. 2016; Chung and Horst 2018; Hurn 2013; Rawat 2014; Saxena et al. 2009; Tsiklashvili et al. 2019; Woetzel et al. 2018). These policies ease the work-life balance of women and enable them to reach on boards (Taparia and Lenka 2022b; Valk and Srinivasan 2011).

Fair Performance evaluations and targeted training and development programs can also help women advance in their careers in India as well as Western countries (Akande 1993; Boone et al. 2013; Kiaye and Singh 2013; Saxena et al. 2009; Tavakolian 1993; Tripathy 2018; Woetzel et al. 2015).

Further, Sur and Kleiner (1995), in the context of the USA, suggested networking as an facilitator for women's career growth however, the effect of networks on women's progress is mixed in the Indian context (Kurup et al. 2011; Vohra 2020). Businesses are required to implement gender sensitisation and awareness programmes to change traditional attitudes of employees (Akande 1993; Marrujo and Kleiner 1992; Rawat 2014; Taparia and Lenka 2022a; Tavakolian 1993; Yoon and Kleiner 1993). Female role models and company-wide mentoring programmes help women develop the leadership qualities and break into the old boys' network that leads to shattering of the glass ceilings to reach top positions (Akande 1993; Burke 1997; Chawla and Sharma 2016; Huang et al. 2019; Hurn 2013; Kiaye and Singh 2013; Vohra 2020).

Government and regulatory bodies play a crucial role in improving the gender diversity on boards. They should actively implement policies and legislation to enhance women's proportion at all stages of the talent pipeline, particularly on boards (Kiaye and Singh 2013). Following the steps of some Western countries, others may set some national targets to increase women's presence on boards and encourage the corporations to elevate more women to the top in alignment with these targets (Tsiklashvili et al. 2019; Woetzel et al. 2018). Literature indicates that reducing the gender gap in business and on corporate boards requires concentrated and active efforts from all stakeholders.

6 Conclusion

Based on the above discussion, it can be concluded that having gender diversity on boards offers numerous benefits to firms such as diverse opinions, better decision making, improved performance and value. A diverse board also focuses on governance and long term goals of businesses. The inclusion of more women on boards can enhance corporate social responsibility, as they advocate for policies that benefit both the firm and all stakeholders that leads to business sustainability. Moreover, gender equality and women's empowerment are one of the key components of the United Nations Sustainable Development Goals. Therefore, it is essential to minimise obstacles to women's career growth and take measures to increase the presence of women directors on boards in order to promote social responsibility and sustainability.

References

Adams RB, Ferreira D (2009) Women in the boardroom and their impact on governance and performance. J Financ Econ 94(2):291–309. https://doi.org/10.1016/j.jfineco.2008.10.007

Adams SM, Flynn PM (2005) Local knowledge advances women's access to corporate boards. Corporate Govern Int Rev 13(6):836–846. https://doi.org/10.1111/j.1467-8683.2005.00474.x

Aguilera RV, Kuppuswamy V, Anand R (2021) What happened when india mandated gender diversity on boards. Harv Bus Rev. https://hbr.org/2021/02/what-happened-when-india-mandated-gender-diversity-on-boards

Akande A (1993) The glass ceiling: women and mentoring in management and business. Equal Oppor Int 12(4):14–21

Amudha R, Cresenta Shakila Motha L, Selvabaskar S, Alamelu R, Surulivel S (2016) Glass ceiling and glass escalator—an ultimate gender divide in urban vicinity. Indian J Sci Technol 9(27):97620. https://doi.org/10.17485/ijst/2016/v9i27/97620

Bagati D, Carter NM (2010) Leadership gender gap in India inc. Myths and Realities. https://www.catalyst.org/wp-content/uploads/2019/01/Leadership_Gap_in_India_Inc._Myths_and_Realities.pdf

Bilimoria D (2000) Building the business case for women corporate directors. In: Burke RJ, Mattis MC (Eds), Women on corporate boards of directors: international challenges and opportunities, pp 25–40. Springer Netherlands. https://doi.org/10.1007/978-90-481-3401-4_3

Boone J, Veller T, Nikolaeva K, Keith M, Kefgen K, Houran J (2013) Rethinking a glass ceiling in the hospitality industry. Cornell Hospitality Quart 54(3):230–239. https://doi.org/10.1177/1938965513492624

Burke RJ (1993) Women on corporate boards of directors. Equal Opport Int 12(6):5–13. https://doi.org/10.1108/eb010613

Burke RJ (1997) Women on corporate boards of directors: a needed resource. Women Corporate Manag 37–43. https://doi.org/10.1007/978-94-011-5610-3_5

Burke RJ, Mckeen CA (1995) Proportion of men employees in organisations and the experiences of female managers and professionals. Equal Opport Int 14(1,2):27

Campbell K, Mínguez-Vera A (2008) Gender diversity in the boardroom and firm financial performance. J Bus Ethics 83(3):435–451. https://doi.org/10.1108/MRR-02-2016-0039

Carli LL, Eagly AH (2016) Women face a labyrinth: an examination of metaphors for women leaders. Gender Manag 31(8):514–527. https://doi.org/10.1108/GM-02-2015-0007

Carter DA, Simkins BJ, Simpson WG (2003) Corporate governance, board diversity, and firm value. Financ Rev 38(1):s33–53. https://doi.org/10.1111/1540-6288.00034

Catalino N, Marnane K (2019) When women lead, workplaces should listen. Mckinsey Quarterly. https://www.mckinsey.com/~/media/McKinsey/FeaturedInsights/Leadership/When women lead workplaces should listen/When-women-lead-workplacess-should-listen-vF.pdf

Catalyst (2001) Catalyst census of women board directors. https://www.catalyst.org/wp-content/uploads/2019/02/2001_Catalyst_Census_Women_Board_of_Directors.pdf

Chawla S, Sharma RR (2016) How women traverse an upward journey in Indian industry: multiple case studies. Gender Manag 31(3):181–206. https://doi.org/10.1108/GM-06-2015-0050

Choudhury B (2015) Gender diversity on boards: beyond quotas. Europ Bus Law Rev 26(1):229–243. https://doi.org/10.54648/eulr2015012

Chung H, van der Horst M (2018) Women's employment patterns after childbirth and the perceived access to and use of flexitime and teleworking. Human Relations 71(1):47–72. https://doi.org/10.1177/0018726717713828

Conceição P, Hall J, Hsu Y-C, Jahic A, Kovacevic M, Mukhopadhyay T, Ortubia A, Rivera C, Tapia H (2020) 2020 human development perspectives TACKLING SOCIAL NORMS a game changer for gender inequalities. http://hdr.undp.org/en/content/

Cooper Jackson J (2001) Women middle managers' perception of the glass ceiling. Women Manag Rev 16(1):30–41. https://doi.org/10.1108/09649420110380265

Daily CM, Dalton DR, Cannella AA (2003) Corporate governance: decades of dialogue and data. Acad Manag Rev 28(3):371–382. https://doi.org/10.5465/AMR.2003.10196703

Desvaux G, Devillard S, Labaye E, Sancier-Sultan S, Kossoff C, Zelicourt A de (2017) Women Matter Time to accelerate Ten Years of insights into gender diversity. https://www.mckinsey.com/featured-insights/gender-equality/women-matter-ten-years-of-insights-on-gender-diversity

Dezsö CL, Ross DG (2012) Does female representation in top management improve firm performance? a panel data investigation. Strateg Manag J 33(9):1072–1089. https://doi.org/10.1002/smj

Diehl AB, Dzubinski LM (2016) Making the invisible visible: a cross-sector analysis of gender-based leadership barriers. Hum Resour Dev Q 27(2):181–206

Donaldson L, Davis JH (1991) Stewardship theory or agency theory: CEO governance and shareholder returns. Aust J Manag 16(1):49–64. https://doi.org/10.1177/031289629101600103

Elstad B, Ladegard G (2012) Women on corporate boards: key influencers or tokens? J Manage Govern 16(4):595–615. https://doi.org/10.1007/s10997-010-9165-y

Evan WM, Freeman RE (1988) A stakeholder theory of the modern corporation: Kantian capitalism. https://doi.org/LK, https://worldcat.org/title/681168145

Gabaldon P, De Anca C, Mateos De Cabo R, Gimeno R (2016) Searching for women on boards: an analysis from the supply and demand perspective. Corporate Govern Int Rev 24(3):371–385. https://doi.org/10.1111/corg.12141

Gales LM, Kesner IF (1994) An analysis of board of director size and composition in bankrupt organizations. J Bus Res 30(3):271–282. https://doi.org/10.1016/0148-2963(94)90057-4

Gautam R, Singh A (2010) Corporate social responsibility practices in India: a study of Top 500 companies. Glob Bus Manag Res 2(1):41–56

Gupta VK, Han S, Mortal SC, Silveri S, Turban DB (2018) Do women CEOs face greater threat of shareholder activism compared to male CEOs? A role congruity perspective. J Appl Psychol 103(2):228–236. https://doi.org/10.1037/apl0000269

Hacker M, Kleiner BH (1993) Success strategies for women. Equal Oppor Int 12(2):1–4. https://doi.org/10.1108/eb010594

Haldar A, Datta S, Shah S (2020) Tokenism or realism? Gender inclusion in corporate boards. Equal Div Inclusion 39(6):707–725. https://doi.org/10.1108/EDI-04-2019-0126

Haldar A, Shah R, Nageswara Rao SVD (2015) Gender diversity in large listed Indian companies. Corporate Ownership Control 12(3CONT5):573–580. https://doi.org/10.22495/cocv12i3c5p8

Handa R, Singh B (2013) Strengthening board rooms through women: needed resource for effective governance. Chartered Secret J Corporate Professionals XLIII(10):1202–1207. https://www.icsi.edu/media/webmodules/linksofweeks/OCT2013.pdf

Hillman AJ, Cannella AA, Paetzold RL (2000) The resource dependence role of corporate directors: Strategic adaptation of board composition in response to environmental change. J Manage Stud 37(2):235–256. https://doi.org/10.1111/1467-6486.00179

Huang J, Krivkovich A, Starikova I, Yee L, Zanoschi D (2019) Women-in-the-workplace-2019. Featured Insights. https://www.mckinsey.com/featured-insights/gender-equality/women-in-the-workplace-2019

Hurn BJ (2013) Are cracks now appearing in the boardroom glass ceiling? Ind Commer Train 45(4):195–201. https://doi.org/10.1108/00197851311323475

Hyland MMC, Marcellino PA (2002) Examining gender on corporate boards: a regional study. Corporate Govern Int J Bus Soc 2(4):24–31. https://doi.org/10.1108/14720700210447669

Hymowitz C, Schellhardt TD (1986) The glass ceiling: Why women can't seem to break the invisible barrier that blocks them from the top jobs. Wall Street J 24(1):1573–1592

Imberg M, Shaban M, Szabo N (2023) Global gender diversity 2023. https://altrata.com/reports/global-gender-diversity-2023

Ismail M, Ibrahim M (2008) Barriers to career progression faced by women: evidence from a Malaysian multinational oil company. Gender Manag 23(1):51–66

Jain N, Mukherji S (2010) The perception of ' glass ceiling ' in Indian organizations: an exploratory study. South Asian J Manag 17(1):23–39
Jani H, Deshpande RC (2019) International journal of business, management and allied sciences (Ijbmas) the dynamic career pathway from glass ceiling to glass cliff-a perception of male counterparts on women leadership in banking industry. Int J Buss Mang Allied Sci 6(2):2349–4638. https://doi.org/10.33329/ijbmas.6219.40
Jensen MC, Meckling WH (1976) Theory of the firm: Managerial behavior, agency costs and ownership structure. J Financ Econ 3(4):305–360. https://doi.org/10.1016/0304-405X(76)90026-X
Jhunjhunwala S (2024) Corporate governance creating value for stakeholders. Palgrave Macmillan. https://doi.org/10.1007/978-981-99-2707-4
Kamath K, Lakhangaonkar S (2021) A study on the representation of women on corporate boards in India Vis-à-vis the Global Scenario. PalArch's J Archaeol Egypt/ Egyptology 18(7):2698–2706
Kanter RM (1977) Men and women of the corporation. Basic Books
Kersley R, Klerk E, Boussie A, Longworth BS, Natzkoff JA, Ramji D (2019) The CS gender 3000 in 2019: tThe changing face of companies. In Credit Suisse (Issue October). https://www.credit-suisse.com/about-us-news/en/articles/news-and-expertise/cs-gender-3000-report-2019-201910.html
Kesner IF (1988) Directors' characteristics and committee membership: an investigation of type, occupation, tenure, and gender. Acad Manag J 31(1):66–84. https://doi.org/10.5465/256498
Kiaye RE, Singh AM (2013) The glass ceiling: a perspective of women working in Durban. Gender Manag 28(1):28–42. https://doi.org/10.1108/17542411311301556
Kirsch A (2018) The gender composition of corporate boards: a review and research agenda. Leadersh Quart 29(2):346–364. https://doi.org/10.1016/j.leaqua.2017.06.001
Kitterød RH, Teigen M (2018) Bringing managers back in: support for gender-equality measures in the business sector. Nordic J Working Life Studies 8(3). https://doi.org/10.18291/njwls.v8i3.109545
Konrad AM, Ritchie JE, Lieb P, Corrigall E (2000) Sex differences and similarities in job attribute preferences: a meta-analysis. Psychol Bull 126(4):593–641. https://doi.org/10.1037/0033-2909.126.4.593
Kramaric TP, Miletic M (2017) Critical mass in the boardroom of Croatian banks. South East Europ J Econ Bus 12(1):22–37. https://doi.org/10.1515/jeb-2017-0002
Kurup A, Chandrashekar S, Muralidharan K (2011) Woman power in corporate India In conversation with Kiran Mazumdar Shaw, Chairperson & MD. Biocon Ltd. IIMB Manag Rev 23(4):223–233. https://doi.org/10.1016/j.iimb.2011.08.006
Landry M, Vyas A, Malhotra G, Nagaraj N (2020) Adolescents' development of gender equity attitudes in India. Int J Adolesc Youth 25(1):94–103. https://doi.org/10.1080/02673843.2019.1590852
Lathabhavan R, Balasubramanian SA (2017) Glass Ceiling and women employees in Asian organizations: a tri-decadal review. Asia-Pacific J Bus Administr 9(3):232–246. https://doi.org/10.1108/APJBA-03-2017-0023
Lazzaretti K, Godoi CK, Camilo SPO, Marcon R (2013) Gender diversity in the boards of directors of Brazilian businesses. Gender Manag 28(2):94–110. https://doi.org/10.1108/17542411311303239
Macarie FC, Moldovan O (2012) Gender discrimination in management. Theoretical and empirical PERSPECTIVES. Transylvanian Rev Administr Sci 35:153–172. http://ssrn.com/abstract=2098052
Maguire TM, Kleiner BH (1993) Formal and informal organisational barriers to women and how to overcome them. Equal Oppor Int 12(5):22–25
Marrujo RE, Kleiner BH (1992) Why women fail to get to the top. Equal Opport Int 11(4):1–5. https://doi.org/10.1108/eb010576
The Companies Act, 1 (2013). https://www.mca.gov.in/content/mca/global/en/acts-rules/companies-act/companies-act-2013.html

Mohan N (2014) A review of the gender effect on pay, corporate performance and entry into top management. Int Rev Econ Financ 34:41–51. https://doi.org/10.1016/j.iref.2014.06.005

Muttakin MB, Khan A, Mihret DG (2018) The effect of board capital and CEO power on corporate social responsibility disclosures. J Bus Ethics 150(1):41–56. https://doi.org/10.1007/s10551-016-3105-y

Nielsen S, Huse M (2010) Women directors' contribution to board decision-making and strategic involvement: the role of equality perception. Eur Manag Rev 7(1):16–29. https://doi.org/10.1057/emr.2009.27

Nilufer Azeez PV, Priyadarshini RG (2018) Glass ceiling factors affecting women career advancement in IT industry in India. IOP Conference series: materials science and engineering, 390(1). https://doi.org/10.1088/1757-899X/390/1/012021

Noland M, Moran T, Kotschwar B (2016) Is gender diversity profitable? Evidence from a global survey (No. 16–3; Working Paper Series). https://doi.org/10.2139/ssrn.2729348

Oakley JG (2000) Gender-based barriers to senior management positions: understanding the scarcity of female CEOs. J Bus Ethics 27(4):321–334. https://doi.org/10.1023/A:1006226129868

Panitapu MK (2015) Women empowerment in India. SSRN Electron J. http://ssrn.com/abstract=2658666

Pfeffer J, Salancik GR (1978) The external control of organizations: a resource dependence perspective. SSRN Electron J. ssrn: https://ssrn.com/abstract=1496213

Radovic-Markovic M, Salamzadeh A, Razavi SM (2013) Women in business and leadership: critiques and discussions. SSRN Electron J. http://ssrn.com/abstract=2349488

Rao K, Tilt C (2016) Board composition and corporate social responsibility: the role of diversity, gender, strategy and decision making. J Bus Ethics 138(2):327–347. https://doi.org/10.1007/s10551-015-2613-5

Rath TS, Mohanty M, Pradhan BB (2019) An alternative career progression model for Indian women bank managers: a labyrinth approach. Women's Studies Int Forum 73:24–34. https://doi.org/10.1016/j.wsif.2019.01.005

Ravindran B, Baral R (2014) Factors affecting the work attitudes of Indian re-entry women in the it sector work-life balance job satisfaction. Vikalpa 39(2):31–42

Rawat PS (2014) Patriarchal beliefs, women's empowerment, and general well being. Vikalpa 39(2):43–55

Reavley MA (1993) The employment status of women in Canada: a general overview and description of theoretical approaches used to study this phenomenon. Equal Oppor Int 12(1):1–6

Sarkar J, Selarka E (2016) Women on board and performance of family firms: evidence from India. https://doi.org/10.2139/ssrn.2730551

Saxena R, Bhatnagar D, Kannan G, Gupta V, Mavalankar D, Dev R, Vohra N, Bhatia A, Nanavaty R, Deo ANS, Shivashankar S, Gupta V, Sircar S, Gore I, Chakraborty C, Kaul A, Mishra S (2009) Gender and workplace experience. Vikalpa 34(4):79–117. https://doi.org/10.1177/0256090920090407

Schilling E (2012) Non-linear careers: desirability and coping. Equal Div Inclusion 31(8):725–740. https://doi.org/10.1108/02610151211277590

Schipani CA, Dworkin TM, Kwolek-Folland A, Maurer VG (2008) Pathways for women to obtain positions of organizational leadership: the significance of mentoring and networking (No. 1117; Ross School of Business Working Paper). http://ssrn.com/abstract=1281466

Schmid T, Urban D (2018) The economic consequences of a "Glass-Ceiling": women on corporate boards and firm value. SSRN Electron J. https://doi.org/10.2139/ssrn.2344786

Sneader K, Yee L (2019) One is the loneliest number. Mckinsey Quart. https://www.mckinsey.com/featured-insights/gender-equality/one-is-the-loneliest-number

Stamper CL, McGowan RA (2022) Furthering the metaphor of the leadership labyrinth: different paths for different people. J Leadership Organ Studies 29(4):443–460. https://doi.org/10.1177/15480518221106063

Stroh LK, Brett JM, Reilly AH (1996) Family structure, glass ceiling, and traditional explanations for the differential rate of turnover of female and male managers. J Vocat Behav 49

Suchman MC (1995) Managing legitimacy strategic and institutional approaches. Acad Manag Rev 20(3):571–610. https://doi.org/10.5465/amr.1995.9508080331

Sur G, Kleiner BH (1995) Sex discrimination in employment : everyone's problem. Equal Opport Int 14(6/7):54–60. https://doi.org/10.1108/eb010650

Syed J, Ali F, Hennekam S (2018) Gender equality in employment in Saudi Arabia: a relational perspective. Career Dev Int 232:163–177

Taparia M, Lenka U (2022a) An integrated conceptual framework of the glass ceiling effect. J Organ Effect 9(3):372–400. https://doi.org/10.1108/JOEPP-06-2020-0098

Taparia M, Lenka U (2022b) Family friendly policies: a double-edged sword? Ind Commer Train 54(2):293–316. https://doi.org/10.1108/ICT-05-2021-0034

Tavakolian HR (1993) Break on through to the other side of the glass ceiling. Equal Oppor Int 12(6):14–20

Terjesen S, Singh V (2008) Female presence on corporate boards: a multi-country study of environmental context. J Bus Ethics 83(1):55–63. https://doi.org/10.1007/s10551-007-9656-1

Thomas R, Cooper M, Cardazone G, Urban K, Bohrer A, Long M, Yee L, Krivkovich A, Huang J, Prince S, Kumar A, Coury S (2020) Women in the workplace 2020. https://wiw-report.s3.amazonaws.com/Women_in_the_Workplace_2020.pdf

Tripathy M (2018) Breaking the code of gender barrier to leverage the position of women in Indian workplaces. Space Culture India 6(3):180–193. https://doi.org/10.20896/saci.v6i3.328

Tsiklashvili N, Turmanidze T, Beridze T (2019) Women's role in organizations and public opinion. Problems Perspect Manag 17(2):510–517. https://doi.org/10.21511/ppm.17(2).2019.39

United Nations (2015) Goal 5 gender-equality. https://www.undp.org/sustainable-development-goals#gender-equality.

Valk R, Srinivasan V (2011) Work-family balance of Indian women software professionals: a qualitative study. IIMB Manag Rev 23(1):39–50. https://doi.org/10.1016/j.iimb.2010.10.010

Vohra N (2020) Women on boards in india numbers, compositions, experiences and inclusion of women directors. https://www.nseinfobase.com/indianboards/files/IIM_Ahmedabad_FICCI_PRIME_WOB_report.pdf

Wang Y, Clift B (2009) Is there a "business case" for board diversity? Pac Account Rev 21(2):88–103. https://doi.org/10.1108/01140580911002044

Woetzel J, Madgavkar A, Gupta R, Mankiya J, Ellingrud K, Gupta S, Krishnan M (2015) The power of parity: advancing women's equality in india. https://www.mckinsey.com/~/media/McKinsey/FeaturedInsights/Employment and Growth/The power of parity Advancing womens equality in India/MGI India parity_Full report_November 2015.ashx

Woetzel J, Madgavkar A, Sneader K, Tonby O, Lin D-Y, Lydon J, Sha S, Krishnan M, Ellingrud K, Gubieski M (2018) The power of parity: advancing women's equality in asia pacific focus: India. https://www.mckinsey.com/featured-insights/gender-equality/the-power-of-parity-advancing-womens-equality-in-india-2018

World Economic Forum (2024) Global gender gap 2024

Yogendrarajah R (2013) Women empowerment through decision making. Int J Econ Bus Manag 3(1). http://ssrn.com/abstract=2431352

Yoon J, Kleiner BH (1993) What companies can do for women. Equal Oppor Int 12(1):14–17

Zimmer B (2015) The phrase 'Glass Ceiling' stretches back decades. Wall Street J (Online), 1. http://search.ebscohost.com/login.aspx?direct=true&db=bth&AN=101892840&site=ehost-live

Towards Sustainability: Exploring Sustainable Innovation in Chinese Agricultural Food Packaging

Zhibin Tao

Abstract Sustainable innovation in food packaging has become a pressing concern for environmental and business sustainability, due to the ongoing growth of the global food industry. Nevertheless, research on sustainable innovation in food packaging is scarce in China, which holds a significant market share. Hence, this study seeks to investigate the driving factors and mechanisms of sustainable innovation in agricultural food packaging in China, addressing a critical research gap. Through the examination of survey data obtained from 482 Chinese companies specializing in agricultural product packaging, the study has uncovered a number of significant findings. Initially, the pressure from regulatory, market demand, and competition not only directly boosts sustainable innovation in food packaging but also spurs further development through the mediating role of knowledge acquisition. Additionally, these pressures have a beneficial influence on a firm's acquisition of knowledge, which in turn supports the attainment of sustainable innovation in food packaging. The research presented in this paper offers both significant theoretical knowledge and practical evidence in the realm of sustainable innovation within the food packaging industry. It enhances our comprehension of how regulation, market requirements, and competitive factors impact corporate innovation practices, contributing fresh empirical backing to enhance existing theoretical frameworks. From a practical perspective, this study provides practical advice for Chinese companies in the agricultural food packaging sector, aiding in the improvement of their sustainable competitive edge. Additionally, it serves as a cornerstone for the industry's green growth, promoting the creation of a more secure, healthy, and sustainable future society.

Keywords Food packaging sustainable innovation · Regulatory pressure · Market demand pressure · Competition pressure · Knowledge acquisition

---

The original version of the chapter has been revised. A correction to this chapter can be found at https://doi.org/10.1007/978-3-031-81469-3_15

Z. Tao (✉)
Faculty of Business and Law, School of Accounting, Economics and Finance, University of Portsmouth, Portsmouth, UK
e-mail: up2150948@myport.ac.uk

P. Singh et al. (eds.), *Integrating Environmental Sustainability and Corporate Social Responsibility: A Move Beyond Profit*, World Sustainability Series,
https://doi.org/10.1007/978-3-031-81469-3_13

1 Introduction

In recent times, with the escalation of global environmental concerns and the growing awareness of consumers regarding food safety and environmental conservation, there has been a significant focus on the sustainable development of agricultural businesses (Li and Chen 2024; Kalfas et al. 2024; Oukil et al. 2024). Within this context, the role of food packaging, crucial for the distribution and preservation of agricultural products, has gained increased importance (Ahmad et al. 2024; Trotta et al. 2024). Specifically in China, a major player in both agricultural production and consumption worldwide, the exploration of innovative and sustainable food packaging in agricultural enterprises has emerged as a prominent issue (JIAO et al. 2018).

In recent studies, the significance of packaging has emerged as a crucial factor in addressing the primary challenges of sustainable food consumption and has piqued the curiosity of researchers (Jurić et al. 2024; Abdulla et al. 2024). Food packaging serves the purpose of safeguarding food from environmental contamination and various external factors (such as smells, vibrations, dirt, temperature fluctuations, physical harm, light exposure, microorganisms, and moisture). It plays a critical role in maintaining food quality and safety, as well as prolonging shelf life and reducing food wastage (Varghese et al. 2023). Despite this, packaging is frequently perceived as an additional economic and environmental burden rather than being recognized for its worth in reducing waste (Guillard et al. 2018).

When food is disposed of, the packaging is also discarded, resulting in additional environmental challenges (Sundqvist-Andberg and Åkerman 2021). Conventional food packaging is frequently composed of plastic, causing problems like environmental contamination and depletion of resources. Food plastic packaging has emerged as a significant issue because of its disastrous environmental consequences and potential risks to human health (Hassoun et al. 2023). Research has highlighted that following the intrinsic single and very brief use of food packaging, 40% ultimately find their way into landfills, amounting to 9 million tons of plastic packaging waste, destined to accumulate in the ground (Guillard et al. 2018). 32% seeps out from the collection and sorting systems and eventually makes its way into the soil and the sea (Jambeck et al. 2015). If production and utilization persist following the current linear model and no actions are implemented by 2050, the quantity of plastic in the ocean may surpass that of fish (Agenda 2016). Global apprehensions regarding marine plastic debris are notably heightened, with the United Nations 2030 Agenda explicitly articulating this apprehension by designating floating plastic trash density as an input gauge for its targeted aim 14.1 (Santi et al. 2022). Apart from marine contamination, the dispersion of plastic refuse in natural surroundings can result in land pollution and waterway obstructions (Santi et al. 2022), which subsequently may trigger floods and waterborne illness propagation, leading to a substantial number of fatalities annually in developing nations (Harvey 2019). Consequently, agricultural establishments urgently must explore innovative, environmentally friendly sustainable food packaging solutions to meet market demands while diminishing harmful effects on the ecosystem (Mateus et al. 2023; Al Mahmud et al. 2024; Lau and Wong 2024).

Innovative sustainable packaging for food is a necessity and an opportunity for agri-food companies (Versino et al. 2023; Han et al. 2018). According to Guillard et al. (2018), the use of sustainable packaging innovations can not only decrease food waste and maintain food quality but also contribute to addressing food safety concerns such as foodborne illnesses and chemical contamination. With the rising awareness of environmental conservation, the demand for eco-friendly products is on the rise (Lindh et al. 2016). As a result, agri-food companies can improve their market reputation and competitiveness by developing packaging materials, utilizing biodegradable alternatives, and introducing sustainable packaging options (Keränen et al. 2021).

Chinese agricultural enterprises have long relied on traditional methods for food packaging, causing a notable strain on the environment. It is crucial to address this pressing issue promptly. The current standard of food packaging often results in excessive waste and environmental harm, highlighting the need for a transition to sustainable substitutes. Despite this necessity, there is a lack of research in China's agricultural industry focused on developing innovative and eco-friendly food packaging solutions, particularly in the realm of agricultural product processing and packaging. However, in light of the rising demand for sustainable development worldwide and the growing emphasis on arable land and environmental conservation, it is crucial for agricultural businesses to focus on innovation in sustainable food packaging. Leading the way in this evolving sustainable development initiative, this research project aims to bridge existing research gaps by comprehensively investigating the motivators for Chinese agricultural enterprises to embrace innovative sustainable food packaging. The primary goal is to offer the industry visionary perspectives and practical solutions.

The second part of this study involves conducting discussions and developing the literature review and research hypotheses. Part three presents the research methods, while the results of the empirical analysis are detailed in Part four. Part five focuses on discussing the research conclusions, which include elucidating both theoretical and practical contributions, as well as addressing the study's limitations and suggesting avenues for future research.

This study is on the cutting edge as it explores an underexplored aspect of the Chinese agricultural food industry. By analyzing the mechanisms behind innovative sustainable food packaging, this research offers novel insights that could reshape current industry norms. Additionally, the results could lead to progressive policies and tactics to promote sustainable food packaging innovation not only in China but globally. The significance of this study lies in addressing the growing environmental issues associated with conventional food packaging, presenting a modern and progressive method to harmonize agriculture with environmental sustainability.

By conducting innovative research, this study seeks to drive a revolutionary change in the realm of food packaging innovation, providing concrete answers to the environmental challenges encountered by Chinese agricultural food businesses. Emphasizing practical results and its potential influence on the worldwide conversation about eco-friendly food packaging, this research stands at the forefront of the field and offers considerable potential for prompting beneficial transformations in the quest for sustainable growth in the food packaging sector, specifically in the realm of Chinese agriculture.

2 Literature Review and Research Hypothesis Development

2.1 Institutional Theory

The investigation in this paper uses the framework of institutional theory to analyze the factors contributing to sustainable innovation in food packaging. Numerous scholars have acknowledged the significance of institutional theory in elucidating organizational actions (e.g., Lammers and Barbour 2006; Khassawneh and Elrehail 2022). According to institutional theory, firms can improve their ability to grow and survive in a competitive landscape by fulfilling the requirements of stakeholders (Risi et al. 2023).

Firms are inevitably limited by the institutional surroundings in which they function (Paul and Powell Walter 1983). The institutional surroundings not only mold and strengthen the business beliefs of the company but also compel it to adhere to external regulations, standards, and principles (Oliver 1991). According to institutional theory, businesses are devoted to acquiring legitimacy, which involves recognizing and acknowledging their institutional surroundings, and this exerts substantial influences and stresses on their organizational conduct (Suchman 1995; Paul and Powell Walter 1983).

This research refers to the three institutional pressure forms identified by DiMaggio and Powell (1983)—coercive, normative, and mimetic. These pressures generate hypotheses concerning the factors driving sustainable innovation in food packaging.

2.2 Regulatory Pressure as Coercive Pressure

Government agencies establish regulations that impose coercive pressure on companies (Martínez-Ferrero and García-Sánchez 2017). To obtain governmental legitimacy, companies must comply with these laws and regulations (Prajogo et al. 2012; Chen et al. 2018). Regulatory pressure is the binding force of written laws, regulations, and policies issued by the government (Berrone et al. 2013). Non-compliance with strict environmental protection laws and regulations can result in significant costs for companies (Sarkis et al. 2010). Moreover, as per the findings of Liao (2018), governmental influence significantly impacts the credibility of corporations, with regulatory measures being a key determinant in this aspect. Businesses generally adhere to the principles and standards set forth by legislation, while also prioritizing their personal agendas. Simultaneously, the enforcement of restrictive measures on enterprises serves to nurture a shared perception of promoting ecological advancements (Campbell 2007).

Regulatory pressure from the government has the potential to promote sustainable innovation in food packaging within the agricultural sector. The initial emphasis on the attributes and significance of sustainable innovation in food packaging has garnered the attention of government officials, leading to the implementation of regulatory measures. Sustainable innovation plays a critical role in decreasing the production of packaging waste, reducing resource consumption, and addressing environmental pollution, ultimately benefiting the community at large. As these advancements offer widespread societal advantages, it is imperative for the government to prioritize and stimulate innovation within this domain through regulatory frameworks (Martínez-Ferrero and García-Sánchez 2017; Masocha and Fatoki 2018).

Additionally, depending only on voluntary efforts from businesses to implement innovative sustainable food packaging is inadequate. Although some companies may choose to adopt sustainable packaging voluntarily, not all will follow suit. In these instances, governmental regulations can enforce the adoption of these innovative practices by all companies, ultimately guiding the industry towards a more sustainable path. The government can promote the adoption of environmentally friendly and sustainable packaging innovations by implementing regulations, standards, and incentive programs, which will enhance enterprises' commitment to eco-friendly practices and compliance (Croley 2008; Zhang et al. 2022). Consequently, institutional theory offers a well-suited framework for examining sustainable innovation in the realm of corporate food packaging (Peters 2022). As such, governmental regulations exert significant influence on fostering sustainable innovation in food packaging among agricultural businesses, leading to improved environmental conservation and societal accountability via the stimulation of corporate creativity.

Furthermore, recent studies have indicated that environmental policy aids companies in surmounting organizational resistance, fostering a receptive mindset towards fresh concepts, stimulating imaginative solutions, and allocating resources towards eco-friendly advancements, like enhancements in sustainable technology (Eiadat et al. 2008). Research conducted by Menguc et al. (2010) revealed that heightened governmental scrutiny, enforcement of strict emissions controls via coercive means, and enforcement of legal repercussions prompt businesses to actively pursue eco-friendly innovations as a means of mitigating political and financial risks. Furthermore, the implementation of environmental regulations underscores the government's growing dedication to safeguarding the environment (Zhang and Zhu 2019). As a result, companies have the opportunity to expand their expertise across multiple sectors, tackling environmental challenges and enhancing product sustainability from inception to disposal. By doing so, they stand to not only secure preferential treatment from authorities but also enhance their competitive edge (Cui and Wang 2022). In light of this, the present study puts forth the following research proposition:

H1: Regulatory pressure has a positive impact on sustainable innovation in food packaging.

2.3 *Market Demand Pressure as Normative Pressure*

Normative pressure is derived from clientele and non-governmental organizations (Berrone et al. 2013; Delgado-Ceballos et al. 2012; Zhu 2016), and is intricately linked to conforming to societal ethical norms (Zhang et al. 2015). Differing from compulsory coercive pressure, normative pressure comprises gentle limitations on enterprises (Zhu 2016). Within this type of pressure, ethical norms and societal expectations steer businesses towards honoring applicable environmental regulations and guidelines and participating in environmentally friendly innovation (Krell et al. 2016). Simultaneously, professionalization pressure also stems from normative pressure (DiMaggio and Powell 1983); market demand can serve as a powerful catalyst for organizations' environmental efforts and constitute the fundamental normative pressure (Zhu and Sarkis 2007).

The behavior of each corporation is restricted by the societal norms, rules, and anticipations of external stakeholders, serving as a means of social endorsement (Cavusoglu et al. 2015). Although laws and institutions with their coercive and punitive powers are absent, normative influence continues to shape corporate creativity (Liao 2018). Furthermore, resistance to societal pressures may result in hindrances to business dealings with customers, exclusion from industry groups, or missed prospects for future trade (Liao 2018).

In their study, Bansal and Roth (2000) discovered that companies often implement eco-friendly practices in response to the demands of environmentally conscious consumers. According to Zhang et al. (2015), suppliers are frequently under significant pressure to meet the environmental standards set by consumers, particularly those in the global market. In order to enhance their environmental performance and meet the expectations of consumers, suppliers, and partners, businesses are increasingly embracing green and sustainable innovation initiatives. Additionally, research has shown that companies may be hesitant to adopt innovative environmental practices if they do not perceive pressure from their clients (Zhu and Geng 2013).

Consumer demand usually acts as the central normative force, playing a crucial role in driving corporate environmental innovation (Babiak and Trendafilova 2011). This study posits that the pressure from the market, which spurs environmental innovation efforts, originates from the eco-friendly purchasing behaviors of customers. With the rising consciousness of environmental conservation among the public, there is an increasing inclination towards eco-friendly and sustainable products (Kamalanon et al. 2022; Kumar et al. 2022), along with an expanding recognition of the importance of food safety. As a result, it is becoming more common in the industry for goods to fulfill eco-friendly, safe, and environmentally conscious roles by utilizing sustainable and green advancements (Wu et al. 2023; Tao et al. 2023). Customer requests for protecting the environment and ensuring food safety directly motivate companies to innovate sustainable packaging solutions. Therefore, the subsequent hypothesis is put forward:

H2: Sustainable food packaging innovation is positively influenced by customers' green consumerism.

2.4 Competition Pressure as Mimetic Pressure

Businesses often respond to uncertainty by adopting mimetic pressure (DiMaggio and Powell 1983; Martínez-Ferrero and García-Sánchez 2017). Companies are increasingly focusing on environmental practices, green initiatives, and sustainable innovations to gain a competitive edge in today's market, which is saturated with organizations offering top-notch products, excellent customer service, and various competitive benefits (Kim 2018; Xiang et al. 2022). Ismail (2015) and Sin et al. (2016) also stress the importance of businesses monitoring and adapting to their competitors' environmental strategies in the competitive green landscape.

Market competition compels businesses to continuously pursue innovation (Yuan et al. 2022), including sustainable advancements in food packaging (Keränen et al. 2021). The adoption of environmentally friendly packaging materials and technologies can boost a company's brand reputation and address consumer desires for eco-friendly products, enabling them to excel in the competitive market landscape (Williams et al. 2016; Viscecchia et al. 2021). Moreover, competition in the market drives businesses to seek ongoing innovation, spurring advancements in products and technologies (Choua et al. 2011). In today's society, sustainability has emerged as a significant societal concern. Businesses that engage in sustainable innovation in order to meet market demands not only earn greater social recognition and support, but also enhance their investments and commitment to sustainable innovation. As a result, they gain increased competitive advantages and an improved corporate reputation in the increasingly competitive market of the future (Montiel and Delgado-Ceballos 2014; Montiel 2008). Hence, the following hypothesis is put forward:

H3: Competition pressure has a positive impact on sustainable food packaging innovation.

2.5 The Mediating Role of Knowledge Acquisition

According to the resource-based view, organizations consist of a range of assets, capabilities, data, expertise, and other components under their control, which are essential for the development and implementation of strategies (Madhani 2010). Competitive advantage is derived from the resources held by companies (Wernerfelt 1984). In the context of environmental innovation strategy, various forms of knowledge play a vital role as key resources (Marzucchi and Montresor 2017). Recent literature emphasizes the importance for companies to not only gather fresh knowledge within their organization, but also acquire novel market and technical information from external sources in order to innovate new services, products, and technologies (Crossan 1996).

Acquiring knowledge is the process by which organizations recognize their surrounding environment and extract insights from it. It encompasses obtaining and converting external knowledge or various internally generated knowledge (Kuhn et al. 1995; Carley 1986; Gaines 2013). The wider array of external knowledge a firm obtains, the more diverse its knowledge base becomes (Liao 2018). This external knowledge can enhance the organization's existing knowledge, serving as the groundwork for enhancing its innovative capacity (Liao 2018). In line with the perspectives of McGraw and Harbison-Briggs (1989); Van Beveren (2002); Brulé and Blount (1989), companies possessing advanced knowledge acquisition abilities are expected to exhibit superior learning results, enabling them to promptly adapt to shifts in the external landscape and market requirements. The acquisition of knowledge serves as a remedy to the challenge of inadequate resources, expertise, and capacities encountered by businesses during the innovation phase, facilitating the seamless progression of innovative endeavors. Additionally, effective acquisition of knowledge may improve the transfer, communication, and sharing of internal knowledge within the organization, boosting innovation effectiveness and reinforcing the organization's commitment to innovation (Kuhn et al. 1995; Vătămănescu et al. 2023; Sahoo et al. 2023).

Government regulations have pushed businesses to seek knowledge in environmental protection and sustainable development. This includes staying current on environmental protection technology, innovations in sustainable packaging, and regulatory standards (Liao 2018; Xu et al. 2012). Companies must constantly update their knowledge to adhere to government policies on environmental protection. At the same time, business innovation is driven by consumer demand for sustainable and environmentally friendly packaging. To meet this demand, businesses must comprehend consumer preferences and market trends, acquire information regarding environmentally friendly packaging, and constantly update their products and technologies based on market needs (Esper et al. 2010; Li 2014).

In order to excel in the fiercely competitive market, companies must consistently gather pertinent information and expertise regarding the industry and their rivals. This information aids businesses in comprehending market patterns, innovative endeavors of competitors, and the most recent advancements in environmentally-friendly packaging technologies and materials (Bloodgood 2019; Yli-Renko et al. 2001). To summarize, acquiring knowledge can assist businesses in comprehending environmental protection regulations, market needs, and competitive landscape, as well as obtaining cutting-edge environmental protection technologies and innovative trends. This, in turn, facilitates enhanced decision-making and implementation in the realm of sustainable innovation in food packaging (Bloodgood 2019; Xie et al. 2018). From this premise, the subsequent research hypothesis is posited:

H4: Knowledge acquisition is positively influenced by regulatory pressure.
H5: Knowledge acquisition benefits from market demand pressure.
H6: Positive effects on knowledge acquisition are seen with competition pressure.
H7: Sustainable food packaging innovation is positively influenced by knowledge acquisition.

H8: The impact of regulatory pressure on sustainable food packaging innovation is mediated by knowledge acquisition.
H9: Knowledge acquisition serves as a mediator between market demand pressure and sustainable food packaging innovation.
H10: The relationship between competition pressure and sustainable food packaging innovation is mediated by knowledge acquisition.

3 Methodology

3.1 Scale Development and Design

Figure 1 demonstrates the research model utilized in the current investigation, comprising five distinct variables considered in this study. By consulting past research and popular research measures, adjustments and enhancements were made to create a measurement tool tailored specifically for the objectives of the study.

In order to guarantee the scientific validity and precision of the research scale, several steps have been implemented. Initially, thorough literature reviews were conducted to acquire a deep comprehension of the underlying forces behind sustainable innovation in food packaging. Building upon this groundwork, an in-depth analysis and assessment of current research scales will be carried out, followed by enhancements and adjustments to ensure comprehensive and precise representation of the driving factors of sustainable innovation in food packaging within the framework of China's institutional environment.

Additionally, the research includes inviting professionals and industry professionals in associated sectors, along with workers from corporations, to participate in expert interviews and group discussions to gather their perspectives and insights on the present sustainable advancements in food package design. The viewpoints of these individuals will enhance the depth of the research framework, boosting its precision and dependability. Finally, in order to guarantee a strong translation,

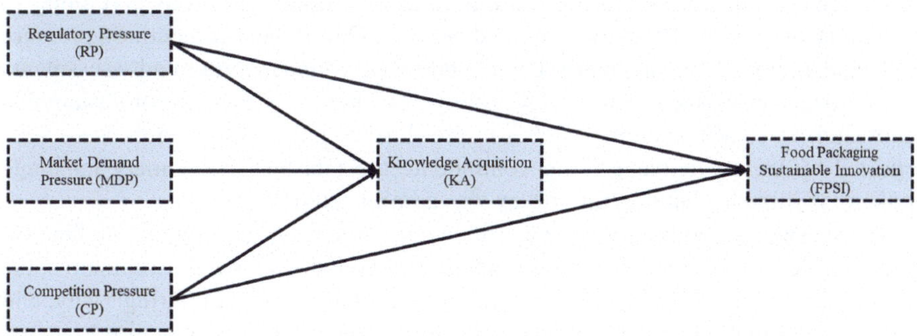

Fig. 1 Research theoretical model. *Source* author's analysis

we adhered to the forward–backward method specified by Sousa and Rojjanasrirat (2011). At the same time, a team of expert translators was utilized for both forward and backward translations to ensure the precision and efficacy of the framework.

Prior to distributing the final research scale, assessed the initial questionnaire items by surveying 100 companies as part of the pre-survey test. The reliability and validity of the scale were examined using the data gathered, resulting in the removal of measurement questions that did not comply with the standards. Subsequent to the research process and feedback from participants, modifications and refinements were made to the measurement items.

Appendix 1 displays the particular measurement items across various dimensions. Conversely, a seven-point Likert scale is utilized in this study for evaluation (ranging from 1 = strongly disagree to 7 = strongly agree). Importantly, the empirical analysis section that ensues will elaborate on the validity and rigor of this scale.

3.2 Sample and Data Collection

Through a variety of channels, participants were recruited for this study via written invitations, online survey platforms, and telephone outreach. A survey was carried out on companies in the Chinese agricultural food packaging sector, with enterprises in eight key regions of China such as Beijing, Shanghai, Guangdong, Jiangsu, Shanxi, Zhejiang, Shandong, and Hebei representing the Northern, Eastern, and Southern regions of the China. The diversity of the sample set is enriched by the varying latitudes and geological features of the cities chosen (Lin and Qiao 2023). To summarize, the distribution of questionnaires considered the diversity of the sample in terms of regional and geographical variances, alongside policy distinctions. The utilization of distinct survey locations leads to greater sample representativeness and minimizes sample selection bias (Lo and Jim 2010).

In order to disentangle the influence of common method bias, this study followed the approaches utilized in prior research by gathering data from participants on three separate occasions, with a six-week interval between each session (Podsakoff et al. 2003; Iqbal et al. 2023). During the initial survey round, questions pertaining to regulatory pressure (RP), market demand pressure (MDP), and competition pressure (CP) were included. Subsequently, the second survey round incorporated assessments related to knowledge acquisition (KA), followed by measures focusing on sustainable innovation in food packaging (FPSI) in the third round. All responses were self-reported by the participants. To accommodate the fact that the primary language spoken in China is Chinese, the survey was administered in Chinese.

The survey questionnaires were distributed in three ways: (1) conducting face-to-face interviews where respondents completed and handed in the questionnaires; (2) having respondents fill out and return the questionnaires by mail during visits; and (3) utilizing email. The participants were predominantly CEOs or senior management of companies with in-depth knowledge of their organizations. A total of 526 questionnaires were distributed, leading to the receipt of 498 questionnaires after three rounds of surveying, resulting in a recovery rate of 94.677%. Of these, 16 questionnaires were deemed invalid, yielding a valid questionnaire recovery rate of 96.787%. As a result, 482 Chinese agricultural food product packaging companies partook in the study, with no missing data.

Among the 482 approved surveys, 61.00% were from privately owned businesses, with the remaining 39.00% from other types of entities. Businesses established for under a decade, between 10 and 50 years, and over half a century comprised 13.49%, 40.25%, and 46.27% respectively. Enterprises with less than 300 workers, 300 to 600 employees, and over 600 staff members constituted 25.93%, 30.50%, and 43.57% respectively.

Throughout the procedure of participant recruitment and sample collection, ethical guidelines and applicable regulations have been closely followed by this research, ensuring the protection of respondents' rights and interests. Special emphasis has been placed on safeguarding the confidentiality of participants' personal information throughout the survey process. Moreover, prior to their participation in the study, informed consent was secured from each participant, with the assurance that their responses would remain confidential and anonymous, and with the clarification that the research was conducted solely for the purposes of this study.

4 Empirical Analysis

4.1 Common Method Bias Test

Using the Harman one-factor test method in statistical analysis, this research employs SPSS software to perform exploratory factor analysis on all measurement items. The findings from dimension reduction reveal that the first unrotated factor explains 37.187% of the variance, falling short of the 50% threshold as suggested by Podsakoff et al. (2003) and Harman (1976). As a result, it can be inferred that this study is not affected by common method bias.

4.2 Exploratory Factor Analysis

Utilizing SPSS 27.0 software, we performed an exploratory factor analysis on the scale employed in this research. The method employed was principal component analysis. The Kaiser–Meyer–Olkin (KMO) value for the factor analysis stands at 0.884, surpassing the threshold of 0.7. Furthermore, the Bartlett Test of Sphericity yielded a result of 3627.777, with a degrees of freedom (df) value of 105 and a p-value of 0.000, indicating statistical significance below the conventional threshold of 0.05. These outcomes affirm the appropriateness of the data for exploratory factor analysis.

Table 1 Exploratory factor analysis

Variables	Items	Loading	Eigenvalues	Explain the variation amount /%	Explain the cumulative variation amount /%	Cronbach's α
Regulatory Pressure (RP)	RP1	0.827	2.371	15.808	15.808	0.855
	RP2	0.808				
	RP3	0.783				
Food Packaging Sustainable Innovation (FPSI)	FPSI1	0.806	2.328	15.520	31.328	0.914
	FPSI2	0.773				
	FPSI3	0.815				
Competition Pressure (CP)	CP1	0.810	2.221	14.806	46.134	0.783
	CP2	0.762				
	CP3	0.788				
Knowledge Acquisition (KA)	KA1	0.805	2.165	14.433	60.567	0.794
	KA2	0.804				
	KA3	0.769				
Market Demand Pressure (MDP)	MDP1	0.796	2.158	14.389	74.955	0.775
	MDP2	0.820				
	MDP3	0.815				

Source author's analysis

Based on the variables identified by the gravel diagram, a rotation maximizing variance was utilized, setting a threshold for factors to possess eigenvalues exceeding 1. Following the removal of measurement items exhibiting significant cross-loading, five factors were identified, each grouping distinct measurement indicators. Table 1 illustrates the outcomes of the exploratory factor analysis, showcasing factor loadings surpassing 0.600 and Cronbach's α values exceeding 0.700. These findings suggest that the exploratory factor analysis conducted in this research yielded favorable analytic outcomes.

4.3 Confirmatory Factor Analysis

Analyzing the research sample, AMOS 23.0 software was utilized to construct five first-order structural models. A comparison of these models was conducted to determine the optimal factor structure. As evident in Table 2, the data best fit the five-factor model ($\chi 2/df = 1.230$, TLI = 0.993, IFI = 0.995, CFI = 0.995, SRMR = 0.026, RMSEA = 0.022). The validation of the five-factor model indicates strong discriminant validity among the five distinct first-order factors examined in this study,

Table 2 Confirmatory factor analysis

Fit indicators	c2	df	c2/df	RMSEA	CFI	IFI	TLI	GFI	AGFI	SRMR
Five-factor model	98.379	80	1.230	0.022	0.995	0.995	0.993	0.974	0.961	0.026
Four-factor model	661.566	84	7.876	0.120	0.838	0.839	0.798	0.813	0.733	0.1719
Three-factor model	832.442	87	9.568	0.133	0.791	0.792	0.748	0.767	0.678	0.126
Two-factor model	1261.181	89	14.171	0.165	0.671	0.673	0.612	0.654	0.533	0.142
One-factor model	1402.669	90	15.585	0.174	0.632	0.634	0.571	0.648	0.531	0.144

Source author's analysis
Note Five-factor model: RP, MDP, CP, KA, FPSI;
Four-factor model: RP, MDP + CP, KA, FPSI;
Three-factor model: RP, MDP + CP + KA, FPSI;
Two-factor model: RP + DP + CP + KA, FPSI;
One-factor model: RP + DP + CP + KA + FPSI;

each representing unique constructs. Thus, the adherence of the five-factor model to research criteria is confirmed in this investigation.

4.4 Model Fit Test

The results of the model fit testing show that the CMID/DF is 2.994 (Malhotra et al. 2014), the GFI is 0.946 (Bagozzi and Yi 1988), the AGFI is 0.922 (Malhotra et al. 2014), the CFI is 0.954 (Bentler and Bonett 1980; Hu and Bentler 1999), the TLI is 0.941 (Bentler and Bonett 1980; Hu and Bentler 1999), the RMSEA is 0.064 (Browne and Cudeck 1992; Hu and Bentler 1999), and the SRMR is 0.047 (Browne and Cudeck 1992; Hu and Bentler 1999). All fit indices meet the corresponding threshold criteria, and several indicators even surpass the excellent threshold. As a result, it is reasonable to determine that the model employed in this research exhibits a strong fit.

4.5 Measurement Model Analysis

Table 3 presents the outcomes of the assessment on the reliability and validity of the research scales. Each dimension's Cronbach's alpha surpasses 0.7, signifying adequate reliability in the scales (Hair et al. 2019). Moreover, every dimension exhibits a composite reliability (CR) exceeding the 0.7 benchmark (Hair Jr et al.

2021), and an average variance extracted (AVE) above the 0.5 threshold (Fornell and Larcker 1981). Consequently, the research scales demonstrate sound validity. Furthermore, the square root of the AVE for each variable surpasses the correlation coefficients between the variables (refer to Table 4), implying strong discriminant validity in the scales (Zhang et al. 2021).

Table 3 Scale reliability and validity test

Variables	Items	Unstd	S.E	Z	P	Std	Cronbach's α	CR	AVE
Regulatory Pressure (RP)	RP1	1.000				0.795	0.855	0.855	0.662
	RP2	1.050	0.056	18.758	***	0.821			
	RP3	1.097	0.058	18.870	***	0.825			
Market Demand Pressure (MDP)	MDP1	1.000				0.708	0.775	0.777	0.538
	MDP2	1.163	0.089	13.052	***	0.736			
	MDP3	1.254	0.095	13.208	***	0.755			
Competition Pressure (CP)	CP1	1.000				0.749	0.783	0.783	0.547
	CP2	0.932	0.068	13.712	***	0.722			
	CP3	0.972	0.069	14.015	***	0.747			
Knowledge Acquisition (KA)	KA1	1.000				0.738	0.794	0.795	0.565
	KA2	0.938	0.066	14.258	***	0.754			
	KA3	1.062	0.074	14.349	***	0.762			
Food Packaging Sustainable Innovation (FPSI)	FPSI1	1.000				0.882	0.914	0.914	0.780
	FPSI2	1.040	0.039	26.588	***	0.882			
	FPSI3	1.060	0.040	26.723	***	0.885			

Source author's analysis

Table 4 Discriminant validity test

	FPSI	KA	CP	MDP	RP
FPSI	**0.883**				
KA	0.598	**0.751**			
CP	0.637	0.481	**0.739**		
MDP	0.226	0.378	0.150	**0.733**	
RP	0.673	0.375	0.520	−0.277	**0.814**

Source author's analysis

Table 5 Direct effect test

Hypothesis	Path	Unstd. (β)	S.E	Z	P	Std	Test results
H1	RP → FPSI	0.645	0.055	11.832	***	0.577	Confirmed
H2	MDP → FPSI	0.354	0.065	5.464	***	0.271	Confirmed
H3	CP → FPSI	0.338	0.051	6.610	***	0.293	Confirmed
H4	RP → KA	0.350	0.050	6.986	***	0.363	Confirmed
H5	MDP → KA	0.493	0.066	7.493	***	0.438	Confirmed
H6	CP → KA	0.275	0.053	5.202	***	0.276	Confirmed
H7	KA → FPSI	0.202	0.065	3.105	***	0.174	Confirmed

Source author's analysis
Note * $p < 0.10$; ** $p < 0.05$; *** $p < 0.01$

4.6 Structural Model Analysis

Table 5 displays the findings of the research hypotheses and path analysis. The results demonstrate that there are significant positive correlations for all the proposed direct impact hypotheses in this study. Therefore, all the research hypotheses regarding direct impact have been confirmed.

4.7 Mediation Effect Test

This research utilized Hayes' (2009) methodology to examine mediation effects by employing the Bootstrap confidence interval technique. Initially, a 95% confidence level was established, and the Bootstrap sampling was performed 1000 times. Using AMOS software for analysis, if the Z value exceeds 1.96 and the Bias-Corrected and Percentile confidence interval excludes zero, a mediation effect is present. Conversely, if the interval includes zero, no mediation effect is observed.

Based on the findings presented in Table 6, the indirect effect range of KA in the impact of RP on FPSI not only excludes zero in the Bias-Corrected 95% CI (0.124, 0.240), but also in the Percentile 95% CI (0.125, 0.241). Furthermore, the Z score is 6.103, surpassing 1.96. Consequently, the mediating role of KA in the impact of RP on FPSI is statistically significant.

Moreover, in the pathway of MDP influencing FPSI, the indirect effect range of KA excludes zero in both the Bias-Corrected 95% CI and the Percentile 95% CI, with a Z value exceeding 1.96. Hence, the mediation impact of KA in the MDP-FPSI relationship is significant. Similarly, employing a comparable approach, it was evident that the mediating effect of KA holds significance in the CP-FPSI pathway as well.

Table 6 Mediation effect test

Type	Effect paths	Point Estimate	Product of Coefficient		Bootstrapping			
					Bias-Corrected 95% CI		Percentile 95% CI	
			SE	Z	Lower	Upper	Lower	Upper
RP → KA → FPSI								
Indirect effect	RP → KA → FPSI	0.177	0.029	6.103	0.124	0.240	0.125	0.241
Direct effect	RP → FPSI	0.610	0.053	11.509	0.511	0.721	0.510	0.720
Total effect	RP → FPSI	0.787	0.055	14.309	0.678	0.896	0.685	0.901
MDP → KA → FPSI								
Indirect effect	MDP → KA → FPSI	0.309	0.060	5.150	0.200	0.432	0.200	0.434
Direct effect	MDP → FPSI	0.302	0.070	4.314	0.139	0.457	0.143	0.451
Total effect	MDP → FPSI	0.307	0.080	3.838	0.160	0.473	0.157	0.468
CP → KA → FPSI								
Indirect effect	CP → KA → FPSI	0.221	0.040	5.525	0.221	0.150	0.150	0.307
Direct effect	CP → FPSI	0.557	0.074	7.527	0.557	0.414	0.415	0.700
Total effect	CP → FPSI	0.778	0.070	11.114	0.778	0.646	0.646	0.919

Source author's analysis

Furthermore, the findings from the analysis show that the direct effect and total effect of RP on FPSI are significant, as are the direct effect and total effect of MDP on FPSI and CP on FPSI, respectively.

4.8 Multigroup Constancy Test

In order to assess the invariance of the research model across different groups, a multi-group comparison was conducted in this study to examine the model (Kline 2023). The structural model's consistency was assessed by comparing the equality of form, factor loadings, intercepts, factor variances and covariances, paths, structural residuals, and measurement residuals (Arbuckle and Wothke 1999). The outcomes of the examination for the gender variable can be found in Table 7. Equivalent constraints

Table 7 Model constancy test for different groups of respondents by gender

Model	CMIN	DF	CMIN/DF	Δ CMIN	Δ DF	P	TLI	ΔTLI	CFI	ΔCFI
M1	324.534	166	1.955	–	–	–	0.944		0.956	
M2	332.314	176	1.888	7.780	10	0.650	0.948	−0.004	0.956	0.000
M3	346.440	183	1.893	14.126	7	0.049	0.947	0.001	0.954	0.002
M4	350.290	186	1.883	3.850	3	0.278	0.948	−0.001	0.954	0.000
M5	350.334	188	1.863	0.044	2	0.978	0.949	−0.001	0.954	0.000
M6	370.627	203	1.826	20.293	15	0.161	0.951	−0.002	0.953	0.001

Source author's analysis
Note M1: Morphological constancy;
M2: Factor loading constancy;
M3: Path constancy;
M4: Factor variance/covariance constancy;
M5: Structural residual constancy;
M6: Measurement residual constancy

were applied in each evaluation. In accordance with scholarly norms, if the p-value exceeds 0.05, it signifies model consistency (Vandenberg and Lance 2000).

Based on the consistent criteria stated, it is evident that the model's p-values, all exceeding 0.05, indicate no notable distinctions between male and female respondent groups within the model. Thus, one can infer that the research model displayed cross-group invariance in this study.

4.9 Model Stability Test (Cross- Valiadation)

Utilizing Cudeck and Browne's (1983) suggestion for ensuring the validity of the findings, this research employed Amos's functionality for comparing groups to randomly separate the participants into two distinct groups. Subsequently, cross-validation was performed.

The findings, presented in Table 8, demonstrate that all assumptions of equality can be accepted (since all p-values exceed 0.05), aligning with the guidelines set forth by Collier (2020). Thus, the model in this research conforms to the requirements for cross-validation, showcasing its stability.

Table 8 Cross- valiadation test

Model	ΔCMIN	ΔDF	P	ΔNFI	ΔIFI	ΔRFI	ΔTLI	ΔCFI
Measurement weights	6.557	10	0.767	0.002	0.002	−0.004	−0.004	−0.001
Structural weights	6.345	7	0.500	0.002	0.002	−0.002	−0.002	0.000
Structural covariances	2.727	3	0.436	0.001	0.001	−0.001	−0.001	0.000
Structural residuals	0.053	2	0.974	0.000	0.000	−0.001	−0.001	−0.001
Measurement residuals	9.667	15	0.840	0.003	0.003	−0.005	−0.005	−0.001

Source author's analysis

5 Conclusion and Discussion

5.1 Research Finding

In this research, an examination was conducted on the information gathered from a survey conducted among 482 companies in China that specialize in packaging agricultural products. The focus was on exploring the various factors and processes that drive sustainable innovation in the field of food packaging. The results of the study indicate that pressures from regulatory, market demands, and competition play a significant role in not only directly supporting sustainable innovation in food packaging but also indirectly influencing the innovation process by facilitating the acquisition of knowledge. This implies that companies are able to enhance their innovative efforts by obtaining a greater amount of knowledge and information related to sustainable innovation in response to regulatory, market, and competitive pressures.

Moreover, the findings also suggest that pressure from regulatory, market demand, and competition can all drive companies to enhance their knowledge acquisition. This acquisition of knowledge, in return, supports sustainable innovation in the field of food packaging. The interdependent nature of this relationship implies that firms pursuing sustainable innovation must constantly adjust to external pressures, promptly acquire and assimilate new knowledge and competencies (Santos et al. 2021; Sahoo et al. 2023).

These discoveries coincide with previous research findings and echo the sentiments shared by other academics. Earlier investigations have explicitly highlighted that regulatory, market demand, and competition serve as key catalysts for organizational creativity (Chen et al. 2018; Ng et al. 2022). At the same time, distinguished researchers have underscored the crucial importance of acquiring knowledge to facilitate sustainable innovation within businesses. They argue that through continuous education and accumulation of fresh insights can enterprises uphold a competitive advantage amidst intense market rivalry (Sahoo et al. 2023; Bloodgood 2019; Li et al. 2019).

Based on the aforementioned results, it is crucial for businesses to focus on adapting and responding to regulatory, market demand, and competitive pressures during the sustainable innovation of food packaging. Additionally, enhancing the focus on acquiring knowledge is essential for achieving innovation and ongoing growth within the realm of sustainable packaging.

5.2 *Theoretical and Academic Contribution*

The research's theoretical contribution and academic value are evident in various ways. Firstly, there is a lack of research on the factors and mechanisms that impact sustainable innovation in food packaging. This study addresses this gap by examining survey data from 482 Chinese companies that specialize in agricultural product packaging. It offers crucial empirical evidence and theoretical backing for research in this particular area. Secondly, through a thorough examination of the effects of regulatory pressure, market demand pressure, and competitive pressure on sustainable innovation in food packaging, this study contributes to the existing literature. By investigating the influence of these pressures on firms' knowledge acquisition and subsequent impact on sustainable innovation initiatives, this research enhances comprehension of organizational behavior and the innovation process.

This study, carried out in the setting of China's institutional environment, is intricately linked to institutional theory. Through investigating the impact of regulatory pressure on fostering sustainable innovation in firms and the intermediary function of knowledge acquisition in this progression, the study offers fresh empirical validation for institutional theory, enhancing comprehension of the ways in which institutions shape corporate conduct. On the other hand, given the unique institutional framework of China under which the research was conducted, it demonstrates a level of originality. Contrary to other nations, China's distinct government regulations, market conditions, and competitive scenario present a fresh outlook and knowledge on the operations of Chinese businesses within a specific institutional environment. The findings of this study also serve as valuable resources for forthcoming research in similar domains. Delving deeper into the ways in which regulatory, market needs, and competitive forces influence the sustainable innovation of corporations not only introduces novel research angles and approaches but also advances the scholarly progress of relevant fields.

5.3 Management and Practice Contribution

The results of this research have important implications for governmental organizations. To begin with, pressure from regulations directly promotes sustainable innovation in the field of food packaging. This allows government entities to better refine their strategies in environmental protection regulations and standards, as well as incentivize businesses to utilize eco-friendly and sustainable packaging materials and technologies. Additionally, the research indicates that sustainable innovation in food packaging is propelled by regulatory pressure, market demand pressure, and competitive pressure, which stimulate companies to acquire knowledge. Hence, governmental support for research and innovation in the industry can be enhanced through the provision of policy support and incentives, prompting enterprises to boost their investments in sustainable innovation.

The results of the study provide valuable management recommendations for businesses. Companies may enhance their commitment to sustainable innovation by proactively addressing regulatory mandates, market requirements, and competitive forces, as well as recognizing the catalytic effect of knowledge attainment in fostering innovation. Businesses have the opportunity to foster greater partnerships with regulatory bodies and academic organizations to obtain cutting-edge environmental technologies and policy updates, in addition to bolstering market analysis to gain thorough insight into consumer desires and inclinations. Furthermore, companies should improve internal knowledge acquisition and innovation management through education and knowledge management, encouraging employees' innovative abilities, and enhancing the organization's sustainable innovation capacities.

According to the research results, the government has the opportunity to produce specific policies that promote and direct businesses to enhance technological innovation and advancements in sustainable packaging. Furthermore, the government can provide benefits like tax exemptions, grants, or recognitions to motivate businesses to allocate resources to research and innovation for sustainable packaging. This fosters a policy setting that supports sustainable creativity and propels the sector towards greener pathways.

Businesses can adapt to the changing regulatory and market landscapes through the creation of specialized teams focused on innovation, improved collaboration with regulatory bodies and trade organizations, heightened awareness of market dynamics, and the expansion of employee training initiatives. Moreover, companies are encouraged to engage with suppliers to collectively advance sustainable innovation within their respective industries. By attending workshops, conferences, or forming partnerships with fellow businesses, organizations can acquire valuable insights and technologies pertaining to eco-friendly packaging, ultimately strengthening their capacity for innovation. Additionally, governmental entities play a critical role in facilitating knowledge exchange among industry players to foster widespread industry progress.

Simultaneously, it is crucial for companies to prioritize the training of their workforce to guarantee that employees possess the most up-to-date environmental protection technologies and knowledge. Companies have the option to conduct internal training programs, bring in specialists for seminars, or promote employees to engage in outside educational and networking events. By fostering a culture of innovation, businesses can inspire their employees to suggest creative concepts and offer incentives for effective innovation initiatives. This approach aids in nurturing employees' innovative mindset and skills, ultimately propelling the sustainable innovation and growth of the company.

Companies can also develop a system for assessing and reporting environmental performance, where environmental indicators are incorporated into the company's performance evaluation structure. Through continuous assessment of the company's environmental performance, companies can gain a deeper insight into their advancement in sustainable packaging and modify their operational tactics accordingly. Furthermore, companies can periodically release environmental reports showcasing their dedication to sustainable packaging to the public and stakeholders. This not only contributes to strengthening the company's commitment to social responsibility but also enhances its corporate image.

Overall, by implementing these particular strategies and suggestions, both government entities and businesses can collaboratively advance sustainable innovation in food packaging. This involves conducting research, developing, and utilizing eco-friendly food packaging materials to elevate food safety protocols, diminish reliance on natural resources, and foster a balanced relationship between humans and the environment. These actions not only bolster the sustainable edge of enterprises but also establish a robust groundwork for the overall growth of the sector, leading to a prospective society that is more sustainable, secure, and healthy.

5.4 Research Limitation and Future Research

This research has discovered key insights into revealing the driving forces and mechanisms of eco-friendly innovation in packaging for food products. Nevertheless, it is imperative to address the study's constraints and propose areas for further investigation. Initially, a limitation in geographic scope exists, given the diverse nature of the Chinese economy across different regions. The participants in this study were solely from specific regions, hence future studies could enhance the sample size to encompass a wider range of regions, enabling a more thorough exploration of the variations and developments in eco-friendly innovation in food packaging among diverse regions. Secondly, the research is dependent on self-reported information, which could potentially introduce common method bias. However, this study did not show any signs of this issue. In order to improve the trustworthiness of the study, upcoming research endeavors could utilize a panel data methodology, which would mitigate these biases and yield more dependable outcomes.

Thirdly, the study directs attention to the institutional setting in China, suggesting potential for investigating the correlation between regulatory, market demand, and competitive pressures in relation to knowledge acquisition and sustainable innovation in food packaging across varying institutional landscapes. By delving into these connections, scholars can further deepen their comprehension of behavioral patterns within diverse institutional frameworks. Fourthly, the utilization of cross-sectional data in this research does not allow for the capture of the dynamic connections between variables. As a result, upcoming studies may choose to implement a longitudinal research framework in order to monitor the evolution and modifications in sustainable innovation carried out by businesses, thus gaining a deeper comprehension of the temporal correlation between influential factors and outcome mechanisms.

Fifth, the focus of this study is on the agricultural industry in China related to food product packaging. Future research may broaden its scope to encompass sustainable packaging innovation in various sectors. This expansion would allow for a comparison of industry-specific differences and similarities, ultimately yielding tailored management suggestions. Sixthly, while the current study primarily examines China, there is potential for future research to expand its focus to encompass sustainable food packaging innovation in various countries and cultural settings. By doing so, researchers can analyze how different nations and cultures influence corporate innovation practices, thereby enhancing the external validity of both theoretical frameworks and practical findings.

Appendix 1: Measurement items

Variables	Items	References
Regulatory Pressure (RP)	Compliance with strict government environmental protection regulations is very important for our company	Ye et al. (2013), Liao (2018), Yalabik and Fairchild (2011), Kanashiro and Rivera (2019)
	The relevant policies have strengthened our company's willingness to implement sustainable innovation in food packaging	
	Implementing sustainable innovation in food packaging is beneficial for our company to receive preferential treatment from local governments	

(continued)

(continued)

Variables	Items	References
Market Demand Pressure (MDP)	Our customers require that our products comply with environmental regulations	Li (2014), Mowery and Rosenberg (1979), Demirel and Kesidou (2019)
	Our customers place a high value on the green concept embodied in our products	
	Our customers prefer products that have green attributes	
Competition Pressure (CP)	Establishing a company's environmental protection image through green concepts can provide a competitive advantage	Li (2014), Yalabik and Fairchild (2011), Vives (2008)
	Green concepts can help increase a company's market share	
	Green concepts can enhance a company's competitive advantage relative to its competitors	
Knowledge Acquisition (KA)	Our company maintains contact with professionals and expert technical personnel	Pérez López et al. (2005), Liao (2018), Pattinson and Preece (2014), Dahiyat (2015)
	Company employees regularly attend trade fairs and exhibitions	
	The company encourages employees to join formal or informal networks composed of external personnel	
Food Packaging Sustainable Innovation (FPSI)	My company is committed to sustainable innovation in food packaging to reduce the negative impact on the environment	Cheng and Shiu (2012), De Pilli et al. (2021), Guillard et al. (2018)
	My company develops more environmentally friendly food packaging through new technologies that easily decompose its materials	

(continued)

(continued)

Variables	Items	References
	My company reduces the use of packaging materials, minimizes resource consumption, and lessens waste production by optimizing packaging design and processes	

Source author's analysis

References

Abdulla SF, Shams R, Dash KK (2024) Edible packaging as sustainable alternative to synthetic plastic: a comprehensive review. Environ Sci Pollut Res, 1–15

Agenda I (2016) The new plastics economy rethinking the future of plastics. In: World economic forum (Vol 36)

Ahmad K, Li Y, Tu C, Guo Y, Yang X, Xia C, Hou H (2024) Nanotechnology in food packaging with implications for sustainable outlook and safety concerns. Food Bioscience, 103625

Al Mahmud MZ, Mobarak MH, Hossain N (2024) Emerging trends in biomaterials for sustainable food packaging: a comprehensive review. Heliyon

Arbuckle JL, Wothke W (1999) AMOS 4.0 users guide. Chicago: Smallwaters Corporation. Google Scholar

Babiak K, Trendafilova S (2011) CSR and environmental responsibility: motives and pressures to adopt green management practices. Corp Soc Responsib Environ Manag 18(1):11–24

Bagozzi RP, Yi Y (1988) On the evaluation of structural equation models. J Acad Mark Sci 16:74–94

Bansal P, Roth K (2000) Why companies go green: a model of ecological responsiveness. Acad Manag J 43(4):717–736

Bentler PM, Bonett DG (1980) Significance tests and goodness of fit in the analysis of covariance structures. Psychol Bull 88(3):588

Berrone P, Fosfuri A, Gelabert L, Gomez-Mejia LR (2013) Necessity as the mother of 'green'inventions: Institutional pressures and environmental innovations. Strateg Manag J 34(8):891–909

Bloodgood JM (2019) Knowledge acquisition and firm competitiveness: the role of complements and knowledge source. J Knowl Manag 23(1):46–66

Browne MW, Cudeck R (1992) Alternative ways of assessing model fit. Sociol Methods Res 21(2):230–258

Brulé JF, Blount A (1989) Knowledge acquisition. McGraw-Hill, Inc.

Campbell JL (2007) Why would corporations behave in socially responsible ways? An institutional theory of corporate social responsibility. Acad Manag Rev 32(3):946–967

Carley K (1986) Knowledge acquisition as a social phenomenon. Instr Sci 14:381–438

Cavusoglu H, Cavusoglu H, Son JY, Benbasat I (2015) Institutional pressures in security management: direct and indirect influences on organizational investment in information security control resources. Inform Manag 52(4):385–400

Chen X, Yi N, Zhang L, Li D (2018) Does institutional pressure foster corporate green innovation? Evidence from China's top 100 companies. J Clean Prod 188:304–311

Cheng CC, Shiu EC (2012) Validation of a proposed instrument for measuring eco-innovation: an implementation perspective. Technovation 32(6):329–344

Choua J, Ng L, Sibilkov V, Wang Q (2011) Product market competition and corporate governance. Rev Dev Financ 1(2):114–130

Collier J (2020) Applied structural equation modeling using AMOS: basic to advanced techniques. Routledge

Croley SP (2008) Regulation and public interests: the possibility of good regulatory government. Princeton University Press

Crossan MM (1996) The knowledge-creating company: how Japanese companies create the dynamics of innovation

Cudeck R, Browne MW (1983) Cross-validation of covariance structures. Multi-Variate Behav Res 18:147–167

Cui R, Wang J (2022) Shaping sustainable development: external environmental pressure, exploratory green learning, and radical green innovation. Corp Soc Responsib Environ Manag 29(3):481–495

Dahiyat SE (2015) An integrated model of knowledge acquisition and innovation: examining the mediation effects of knowledge integration and knowledge application. Int J Learn Chang 8(2):101–135

De Pilli T, Baiano A, Lopriore G, Russo C, Cappelletti GM (2021) Sustainable Innovations in food packaging. Springer International Publishing

Demirel P, Kesidou E (2019) Sustainability-oriented capabilities for eco-innovation: Meeting the regulatory, technology, and market demands. Bus Strateg Environ 28(5):847–857

DiMaggio PJ, Powell WW (1983) The iron cage revisited: institutional isomorphism and collective rationality in organizational fields. Am Sociol Rev 48(2):147–160

Eiadat Y, Kelly A, Roche F, Eyadat H (2008) Green and competitive? An empirical test of the mediating role of environmental innovation strategy. J World Bus 43(2):131–145

Esper TL, Ellinger AE, Stank TP, Flint DJ, Moon M (2010) Demand and supply integration: a conceptual framework of value creation through knowledge management. J Acad Mark Sci 38:5–18

Fornell C, Larcker DF (1981) Evaluating structural equation models with unobservable variables and measurement error. J Mark Res 18(1):39–50

Gaines BR (2013) Knowledge acquisition: Past, present and future. Int J Hum Comput Stud 71(2):135–156

Guillard V, Gaucel S, Fornaciari C, Angellier-Coussy H, Buche P, Gontard N (2018) The next generation of sustainable food packaging to preserve our environment in a circular economy context. Front Nutr 5:121

Hair Jr J, Hair Jr JF, Hult GTM, Ringle CM, Sarstedt M (2021) A primer on partial least squares structural equation modeling (PLS-SEM). Sage Publications

Hair JF, Risher JJ, Sarstedt M, Ringle CM (2019) When to use and how to report the results of PLS-SEM. Eur Bus Rev 31(1):2–24

Han JW, Ruiz-Garcia L, Qian JP, Yang XT (2018) Food packaging: a comprehensive review and future trends. Compreh Rev Food Sci Food Safety 17(4):860–877

Harman HH (1976) Modern factor analysis. University of Chicago Press

Harvey F (2019) Mismanaged waste'kills up to a million people a year globally. The Guardian, Tuesday, 14th May

Hassoun A, Boukid F, Ozogul F, Aït-Kaddour A, Soriano JM, Lorenzo JM, ... Câmara JS (2023) Creating new opportunities for sustainable food packaging through dimensions of industry 4.0: New insights into the food waste perspective. Trends Food Sci Technol, 104238

Hayes AF (2009) Beyond Baron and Kenny: statistical mediation analysis in the new millennium. Commun Monogr 76(4):408–420

Hu LT, Bentler PM (1999) Cutoff criteria for fit indexes in covariance structure analysis: conventional criteria versus new alternatives. Struct Equ Modeling 6(1):1–55

Iqbal R, Shahzad K, Donia MB (2023) Environmentally specific transformational leadership and employee green attitude and behavior: An affective events theory perspective. J Environ Psychol 92:102181

Ismail T (2015) The influence of competitive pressure on innovative creativity. Acad Strateg Manag J 14(2):117

Jambeck JR, Geyer R, Wilcox C, Siegler TR, Perryman M, Andrady A, ... Law KL (2015) Plastic waste inputs from land into the ocean. Science 347(6223):768–771

Jiao XQ, Mongol N, Zhang FS (2018) The transformation of agriculture in China: looking back and looking forward. J Integrat Agric 17(4):755–764

Jurić M, Bandić LM, Carullo D, Jurić S (2024) Technological advancements in edible coatings: Emerging trends and applications in sustainable food preservation. Food Bioscience, 103835

Kalfas D, Kalogiannidis S, Papaevangelou O, Melfou K, Chatzitheodoridis F (2024) Integration of technology in agricultural practices towards agricultural sustainability: a case study of Greece. Sustainability 16(7):2664

Kamalanon P, Chen JS, Le TTY (2022) "Why do we buy green products?" An extended theory of the planned behavior model for green product purchase behavior. Sustainability 14(2):689

Kanashiro P, Rivera J (2019) Do chief sustainability officers make companies greener? The moderating role of regulatory pressures. J Bus Ethics 155:687–701

Keränen O, Komulainen H, Lehtimäki T, Ulkuniemi P (2021) Restructuring existing value networks to diffuse sustainable innovations in food packaging. Ind Mark Manage 93:509–519

Khassawneh O, Elrehail H (2022) The effect of participative leadership style on employees' performance: the contingent role of institutional theory. Administr Sci 12(4):195

Kim K (2018) Proactive versus reactive corporate environmental practices and environmental performance. Sustainability 10(1):97

Kline RB (2023) Principles and practice of structural equation modeling. Guilford Publications

Kuhn D, Garcia-Mila M, Zohar A, Andersen C, White SH, Klahr D, Carver SM (1995) Strategies of knowledge acquisition. Monographs of the society for research in child development, i–157

Kumar N, Garg P, Singh S (2022) Pro-environmental purchase intention towards eco-friendly apparel: augmenting the theory of planned behavior with perceived consumer effectiveness and environmental concern. J Glob Fash Market 13(2):134–150

Lammers JC, Barbour JB (2006) An institutional theory of organizational communication. Commun Theory 16(3):356–377

Lau CC, Wong CW (2024) Achieving sustainable development with sustainable packaging: a natural-resource-based view perspective. Bus Strategy Environ

Li L, Li G, Yang X, Yang Z (2019) Pursuing superior performance of service innovation through improved corporate social responsibility: from a knowledge acquisition perspective. Asia Pac J Mark Logist 31(4):925–943

Li Y (2014) Environmental innovation practices and performance: moderating effect of resource commitment. J Clean Prod 66:450–458

Li Y, Chen Y (2024) The spatiotemporal characteristics and obstacle factors of the coupled and coordinated development of agricultural and rural digitalization and food system sustainability in China. Front Sustain Food Syst 8:1357752

Liao Z (2018) Institutional pressure, knowledge acquisition and a firm's environmental innovation. Bus Strateg Environ 27(7):849–857

Lin B, Qiao Q (2023) Exploring the participation willingness and potential carbon emission reduction of Chinese residential green electricity market. Energy Policy 174:113452

Lindh H, Olsson A, Williams H (2016) Consumer perceptions of food packaging: contributing to or counteracting environmentally sustainable development? Packag Technol Sci 29(1):3–23

Lo AY, Jim CY (2010) Willingness of residents to pay and motives for conservation of urban green spaces in the compact city of Hong Kong. Urban Forestry Urban Greening 9(2):113–120

Madhani PM (2010) Resource based view (RBV) of competitive advantage: an overview. Resource based view: concepts and practices, Pankaj Madhani, ed, 3–22

Malhotra NK, Lopes E, Veiga RT (2014) Structural equation modeling with Lisrel: an initial vision. Brazilian J Market 13(2)

Martínez-Ferrero J, García-Sánchez IM (2017) Coercive, normative and mimetic isomorphism as determinants of the voluntary assurance of sustainability reports. Int Bus Rev 26(1):102–118

Marzucchi A, Montresor S (2017) Forms of knowledge and eco-innovation modes: Evidence from Spanish manufacturing firms. Ecol Econ 131:208–221

Masocha R, Fatoki O (2018) The impact of coercive pressures on sustainability practices of small businesses in South Africa. Sustainability 10(9):3032

Mateus ARS, Pena A, Sendón R, Almeida C, Nieto GA, Khwaldia K, Silva AS (2023) By-products of dates, cherries, plums and artichokes: a source of valuable bioactive compounds. Trends Food Sci Technol 131:220–243

McGraw KL, Harbison-Briggs K (1989) Knowledge acquisition: principles and guidelines. Prentice-Hall, Inc.

Menguc B, Auh S, Ozanne L (2010) The interactive effect of internal and external factors on a proactive environmental strategy and its influence on a firm's performance. J Bus Ethics 94:279–298

Montiel I (2008) Corporate social responsibility and corporate sustainability: separate pasts, common futures. Organ Environ 21(3):245–269

Montiel I, Delgado-Ceballos J (2014) Defining and measuring corporate sustainability: are we there yet? Organ Environ 27(2):113–139

Mowery D, Rosenberg N (1979) The influence of market demand upon innovation: a critical review of some recent empirical studies. Res Policy 8(2):102–153

Ng PM, Wut TM, Lit KK, Cheung CT (2022) Drivers of corporate social responsibility and firm performance for sustainable development—an institutional theory approach. Corp Soc Responsib Environ Manag 29(4):871–886

Oliver C (1991) Strategic responses to institutional processes. Acad Manag Rev 16(1):145–179

Oukil A, Nourani A, Soltani AA, Boulassel MR, Bencheikh A (2024) Improving agricultural sustainability through farm mergers: an energy efficiency perspective. Int J Agric Sustain 22(1):2293598

Pattinson S, Preece D (2014) Communities of practice, knowledge acquisition and innovation: a case study of science-based SMEs. J Knowl Manag 18(1):107–120

Paul D, Powell Walter W (1983) The iron cage revisited: collective rationality and institutional isomorphism in organizational fields. Am Sociol Rev 48(2):147–160

Pérez López S, Manuel Montes Peón J, José Vazquez Ordás C (2005) Organizational learning as a determining factor in business performance. Learn Organ 12(3):227–245

Peters BG (2022) Institutional theory. In: Handbook on theories of governance, pp 323–335. Edward Elgar Publishing

Podsakoff PM, MacKenzie SB, Lee JY, Podsakoff NP (2003) Common method biases in behavioral research: a critical review of the literature and recommended remedies. J Appl Psychol 88(5):879

Prajogo D, Tang AK, Lai KH (2012) Do firms get what they want from ISO 14001 adoption?: an Australian perspective. J Clean Prod 33:117–126

Sahoo S, Kumar A, Upadhyay A (2023) How do green knowledge management and green technology innovation impact corporate environmental performance? Understanding the role of green knowledge acquisition. Bus Strateg Environ 32(1):551–569

Santi R, Garrone P, Iannantuoni M, Del Curto B (2022) Sustainable food packaging: an integrative framework. Sustainability 14(13):8045

Santos GMC, Marques CS, Ratten V, Ferreira JJ (2021) The impact of knowledge creation and acquisition on innovation, coopetition and international opportunity development. Eur J Int Manag 16(3):450–472

Sarkis J, Gonzalez-Torre P, Adenso-Diaz B (2010) Stakeholder pressure and the adoption of environmental practices: the mediating effect of training. J Oper Manag 28(2):163–176

Sin KY, Osman A, Salahuddin SN, Abdullah S, Lim YJ, Sim CL (2016) Relative advantage and competitive pressure towards implementation of e-commerce: overview of small and medium enterprises (SMEs). Procedia Econ Financ 35:434–443

Sousa VD, Rojjanasrirat W (2011) Translation, adaptation and validation of instruments or scales for use in cross-cultural health care research: a clear and user-friendly guideline. J Eval Clin Pract 17(2):268–274

Suchman MC (1995) Managing legitimacy: strategic and institutional approaches. Acad Manag Rev 20(3):571–610
Sundqvist-Andberg H, Åkerman M (2021) Sustainability governance and contested plastic food packaging–an integrative review. J Clean Prod 306:127111
Tao F, Zhou Y, Bian J, Lai KK (2023) Optimal channel structure for a green supply chain with consumer green-awareness demand. Ann Oper Res 324(1):601–628
Trotta F, Da Silva S, Massironi A, Mirpoor SF, Lignou S, Ghawi SK, Charalampopoulos D (2024) Advancing food preservation: sustainable Green-AgNPs bionanocomposites in paper-starch flexible packaging for prolonged shelf life. Polymers 16(7):941
Van Beveren J (2002) A model of knowledge acquisition that refocuses knowledge management. J Knowl Manag 6(1):18–22
Vandenberg RJ, Lance CE (2000) A review and synthesis of the measurement invariance literature: suggestions, practices, and recommendations for organizational research. Organ Res Methods 3(1):4–70
Varghese SA, Pulikkalparambil H, Promhuad K, Srisa A, Laorenza Y, Jarupan L, ... Harnkarnsujarit N (2023) Renovation of Agro-Waste for sustainable food packaging: a Review. Polymers 15(3):648
Vătămănescu EM, Bratianu C, Dabija DC, Popa S (2023) Capitalizing online knowledge networks: from individual knowledge acquisition towards organizational achievements. J Knowl Manag 27(5):1366–1389
Versino F, Ortega F, Monroy Y, Rivero S, López OV, García MA (2023) Sustainable and bio-based food packaging: a review on past and current design innovations. Foods 12(5):1057
Vives X (2008) Innovation and competitive pressure. J Ind Econ 56(3):419–469
Wernerfelt B (1984) A resource-based view of the firm. Strateg Manag J 5(2):171–180
Wu H, Li J, Zhao Y (2023) Foreign demand shocks, product switching, and export product quality: evidence from China. The World Economy 46(1):276–301
Xiang X, Liu C, Yang M (2022) Who is financing corporate green innovation? Int Rev Econ Financ 78:321–337
Xie X, Wang L, Zeng S (2018) Inter-organizational knowledge acquisition and firms' radical innovation: a moderated mediation analysis. J Bus Res 90:295–306
Xu K, Huang KF, Gao S (2012) The effect of institutional ties on knowledge acquisition in uncertain environments. Asia Pacific J Manag 29(2):387–408
Yalabik B, Fairchild RJ (2011) Customer, regulatory, and competitive pressure as drivers of environmental innovation. Int J Prod Econ 131(2):519–527
Ye F, Zhao X, Prahinski C, Li Y (2013) The impact of institutional pressures, top managers' posture and reverse logistics on performance—evidence from China. Int J Prod Econ 143(1):132–143
Yli-Renko H, Autio E, Sapienza HJ (2001) Social capital, knowledge acquisition, and knowledge exploitation in young technology-based firms. Strateg Manag J 22(6–7):587–613
Zhang B, Wang Z, Lai KH (2015) Mediating effect of managers' environmental concern: bridge between external pressures and firms' practices of energy conservation in China. J Environ Psychol 43:203–215
Zhang F, Zhu L (2019) Enhancing corporate sustainable development: Stakeholder pressures, organizational learning, and green innovation. Bus Strateg Environ 28(6):1012–1026
Zhang SN, Li YQ, Liu CH, Ruan WQ (2021) A study on China's time-honored catering brands: achieving new inheritance of traditional brands. J Retail Consum Serv 58:102290
Zhang W, Luo Q, Liu S (2022) Is government regulation a push for corporate environmental performance? evidence from China. Econ Anal Policy 74:105–121
Zhu Q, Geng Y (2013) Drivers and barriers of extended supply chain practices for energy saving and emission reduction among Chinese manufacturers. J Clean Prod 40:6–12
Zhu Q, Sarkis J (2007) The moderating effects of institutional pressures on emergent green supply chain practices and performance. Int J Prod Res 45(18–19):4333–4355

Corporate Social Responsibility Practices Toward Promoting Environmental Sustainability: Case Studies in Vietnam

Phuong Mai Nguyen, Huong-Linh Le, Xuan-Hoa Nghiem, and Luu Ly Tran

Abstract This study aims to discover the corporate social responsibility (CSR) practices implemented by Vietnamese businesses and to shed light on their specific initiatives to promote environmental sustainability. We selected several renowned companies in Vietnam for sustainable development initiatives to study. By applying the content analysis method to sustainability reports of these companies from 2018 to 2022, we categorized and analyzed how these companies have implemented CSR practices to promote environmental sustainability. Our findings confirm a growing recognition among businesses in Vietnam regarding the significance of incorporating CSR into their overarching strategies. Consequently, these entities have engaged in a series of initiatives designed to minimize their environmental impact and contribute to sustainable development. Notable actions include adopting eco-friendly technologies, reducing greenhouse gas emissions, proactive waste management and recycling, and supporting environmental conservation endeavors. The case studies effectively highlight the positive outcomes of these initiatives, such as improved environmental performance, heightened corporate reputation, and increased stakeholder engagement. These case studies in Vietnam serve as compelling evidence that CSR has an indispensable role to play in promoting environmental sustainability. Firms possess huge potential to address environmental challenges by incorporating CSR practices into their operations. This study provides valuable insights and several implications for policymakers, businesses, and other stakeholders committed to leveraging CSR to enhance environmental sustainability.

Keywords Corporate social responsibility · Environmental sustainability · Manufacturing sector · Service sector · Vietnam

P. M. Nguyen · H.-L. Le (✉) · X.-H. Nghiem
International School, Vietnam National University Hanoi, Hanoi, Vietnam
e-mail: linhlh@vnuis.edu.vn

L. L. Tran
Thuyloi University, Hanoi, Vietnam

P. Singh et al. (eds.), *Integrating Environmental Sustainability and Corporate Social Responsibility: A Move Beyond Profit*, World Sustainability Series,
https://doi.org/10.1007/978-3-031-81469-3_14

1 Introduction

In modern society, humanity grapples with a myriad of pressing environmental issues that can profoundly affect our daily lives and the future of our planet. It is widely agreed that the spread of the COVID-19 pandemic, and international political instability (such as the Russia-Ukraine war, the Israel-Hamas conflict) only distracted and diverted attention and resources from environmental challenges, thus making people neglect and/or overlook other issues, such as sustainable development goals (SDGs) (Bakry et al. 2023a, 2023b). Chief among these challenges are climate change and its consequences, such as natural disasters, rising sea levels, as well as the severe impacts they may have on people's health (Nghiem et al. 2023). Climate change, driven primarily by human activities such as the burning of fossil fuels and deforestation, has led to unprecedented shifts in global weather patterns, thus resulting in extreme weather events and natural calamities such as hurricanes, droughts, and wildfires. These natural disasters not only cause immediate devastation to communities and ecosystems but also exacerbate long-term environmental degradation. For example, rising sea levels, a direct consequence of climate change, pose a severe threat to coastal regions worldwide, leading to coastal erosion, inundation of low-lying areas, and the displacement of millions of people. Worse still, the adverse effects of environmental degradation on human health have become increasingly severe and evident, with air and water pollution contributing to respiratory diseases, cardiovascular problems, and other health issues. These dire consequences of climate change and environmental degradation make it imperative to address these interconnected environmental challenges through collaborative and sustainable efforts. Therefore, it is urgently needed to promote environmental sustainability, which may offer substantial benefits for countries such as a healthier, more resilient future for human beings (Nasreen et al. 2023), necessitating collective, decisive, and concrete measures at all levels.

At the country level, numerous strategies have been put forth to attain environmental sustainability, spanning from broad international or national initiatives such as commitments to achieving net-zero emissions, the use of information and communication technology (ICT) (Nghiem et al. 2023), the integration of digital and energy industries (Wang et al. 2023), use of renewable energy (Guo et al. 2023). At the firm level, businesses have also been taking significant steps toward realizing sustainability through initiatives like green investment and the adoption/integration of Environment-Social-Governance (ESG) practices. ESG, in turn, is usually regarded as CSR, as applied in the finance literature, thus highlighting the critical importance of integrating CSR practices in firms' daily activities. There is a growing emphasis on facilitating CSR as a means to achieve environmental goals. CSR presents a multi-faceted approach that can yield benefits on multiple fronts (Garcia et al. 2017). Firstly, it serves as a mechanism to address environmental issues at large and plays a vital role in curbing carbon emissions, thereby contributing to global efforts to combat climate change. Secondly, the adoption of CSR practices has the potential to enhance firms' reputations and overall performance. By demonstrating a commitment to sustainability, companies can cultivate a positive image among stakeholders, attract socially conscious investors, and even gain a competitive edge in the market. Hence, CSR emerges as a compelling and mutually beneficial solution that not only addresses environmental challenges but also fosters corporate success and societal well-being. The quest for greener, more sustainable corporate practices is driven not only by external reasons (customers' pressures, government requirements, and regulations, pressures from partners and stakeholders) but also by internal factors (such as self-motivation, managerial awareness, and attitudes, resource and capability constraints) (Cordeiro et al. 2020).

Among different sectors/industries exerting significant effects on both the economy and the environment, the financial industry and the dairy industry stand out thanks to their utmost importance. For banks, they may have an impact on the environment through their lending and daily operations. The crucial role played by banks and the financial sector is well-established in the literature. The banking industry is also believed to play a critical role in the green transition. For milk companies, the production process can have a significant, negative impact on the environment and exacerbate environmental degradation. The financial and dairy industries exemplify sectors that warrant focused sustainability efforts, given their substantial scale and influence within the global economy. Research shows over $3.8 trillion in annual banking activities still enable carbon-intensive industries, contrary to limiting warming to 1.5 °C (Ma et al. 2021). Beyond this direct impact, the daily operations of financial services also considerably contribute to emissions. Bank branch networks numbering over 200,000 globally consume vast energy for facilities and IT systems (Wang et al. 2022). Meanwhile, an estimated 284 million tons of CO_2 result yearly from nearly 10 million bank employee flights for business travel (Wolkomir 2022).

Likewise, as the largest agricultural commodity by value of $600 billion, dairy production relies on immense land and water resources globally (Food and Agriculture Organization-FAO 2020). Home to over 1.5 billion cattle, this sector accounts for a sizable portion of anthropogenic methane due to cow enteric fermentation and manure management (FAO 2020). Given methane's high heat-trapping potential, rising emissions from expanding herds seriously threaten climate stabilization goals (Intergovernmental Panel on Climate Change 2021). Additionally, nutrient runoff from dairy farms remains a primary source of water pollution, impairing freshwater quality (Graham and VanDerZanden 2016). Where cropland conversion supplies cattle feed, dairy-driven deforestation occurs (Ritchie and Roser 2013).

Being a developing, fast-growing Southeast Asian nation, Vietnam is highly vulnerable to the impacts and negative consequences of climate change. According to the United States Agency for International Development (USAID 2022), Vietnam is one of the top five countries most vulnerable to climate change impact.[1] On the one hand, being an agriculture-dependent country with long coastlines, Vietnam is confronted with severe threats from rising sea levels, increased flooding, droughts,

[1] https://www.usaid.gov/climate/country-profiles/vietnam#:~:text=Vietnam%20is%20one%20of%20the,long%2C%20densely%20populated%2C%20coast.

saltwater intrusion, and storms that can negatively impact food production, infrastructure, and livelihoods (World Bank 2021). On the other hand, Vietnam has registered robust economic growth and urbanization, which has led to increasing greenhouse gas emissions. These facts emphasize the need for Vietnam to strike a balance between growth and sustainability. In other words, it is critical for Vietnam to take urgent climate actions (at the national- and firm-levels) to build resilience and facilitate the transition towards a more sustainable development pathway.

The financial and dairy sectors[2] in particular play important role in Vietnam's economy and have significant potential to contribute to environmental and social solutions through climate-friendly practices. However, despite the strong benefits and risks associated with these industries in regard to climate change and sustainability, there has been little research investigating CSR case studies in Vietnam. This represents a gap in understanding best practices and how Vietnamese corporations are addressing environmental and social challenges.

This chapter aims to fill this void by closely examining CSR practices in two leading Vietnamese corporations—Vietnam Dairy Products Joint Stock Company (Vinamilk), representing the dairy industry, and Bao Viet Group (Bao Viet), representing the financial industry. As major players in Vietnam's financial and food sectors respectively, the CSR approaches taken by Vinamilk and Bao Viet have important implications for sustainable development across the country. Case studies of their initiatives will provide valuable insights into how Vietnamese businesses are beginning to integrate environmental responsibility into their operations and promote low-carbon growth. Such analysis can also help identify gaps and opportunities to scale up enterprise contributions to Vietnam's climate change responses.

2 Materials and Methods

We start by collecting sustainability reports of two large corporations in Vietnam from 2018 to 2022 to analyze CSR activities towards environmental sustainability that these companies have implemented. We chose the Vietnam Dairy Products Joint Stock Company (Vinamilk) and Bao Viet Group—two of the largest corporations in Vietnam. In the next step, we employ content analysis techniques to perform the analysis as follows.

Firstly, we scanned the sustainability reports to detect CSR activities. We applied the stakeholder theory approach to select some keywords for the scanning process. Secondly, we picked up phrases that contain the keywords to sort CSR activities, which are related to environmental sustainability into specific categories. Finally, we extracted the data and analyzed sentences to evaluate how the chosen companies have implemented their CSR activities toward environmental sustainability.

[2] While the terms "industry" and "sector" may have different meanings, in this stuyd, they are assumed to have the same implications and therefore, will be used interchangeably.

3 CSR Practices Towards Environmental Sustainability: Case Studies in Vietnam

3.1 The Case Study of Vinamilk

Vinamilk, founded in 1976 in Vinh Long Province, is Vietnam's leading dairy company and one of Southeast Asia's largest food and beverage producers. Initially formed by the Vietnamese government with 30 employees to boost local milk production, Vinamilk has grown into a national powerhouse. Over the past four decades, it has expanded through large-scale investments, strategic partnerships, and acquisitions. Currently, Vinamilk operates over 20 world-class manufacturing facilities across Vietnam, utilizes advanced technologies and processes from dairy farming production of raw materials, manufacturing and processing beverage and dairy products, distribution, and trading to healthcare and nutrition (see Fig. 1).

The company's robust supply chain sources fresh milk from over 30 dairy cooperatives and a network of 29,000 cattle farms, ensuring strict quality controls at every stage. This has established Vinamilk as the leading supplier of dairy products in Vietnam and abroad. Its diverse product lineup includes household brands like Truc Bach (Trúc Bạch) and Gia vi phung (Gia vị phụng), offering fresh milk, yogurt, soy milk, nutritional products, and infant formula. Vinamilk exports to over 40 countries across Asia, Africa, Oceania, and the Middle East. Its environmental and social initiatives have garnered numerous awards.

Based on data and information from the sustainability reports of Vinamilk from 2018 to 2022, we found some types of CSR activities that Vinamilk has conducted to protect the environment and reduce its negative impact on the environment.

3.1.1 Energy Consumption

Vinamilk has been actively and heavily engaging in transitioning its energy consumption and sustainability practices. This section focuses on the company's energy use effectiveness and its sustainability development from 2018 to 2022. The energy consumption data for Vinamilk covers various energy sources, including gasoline, gas, electricity, biomass, and CNG (compressed natural gas). Vinamilk has proactively applied CSR practices in energy consumption, such as integrating renewable energy sources, using cleaner alternative energy sources, reducing traditional energy

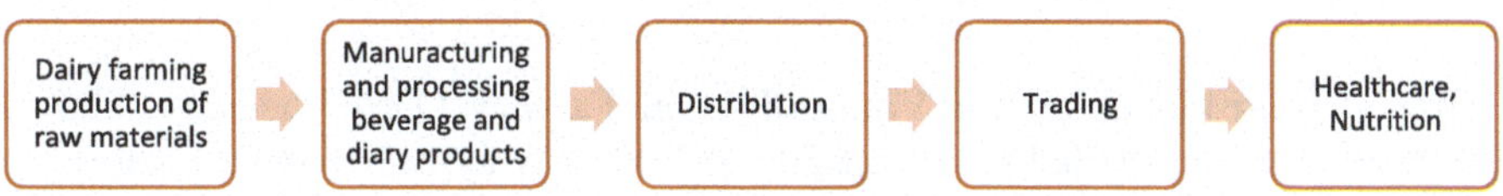

Fig. 1 Vinamilk value chain. (*Source* Compiled from information in Vinamilk Sustainable Development Report 2019)

use, improving energy saving, and developing initiatives in energy-use efficient technologies. These practices have contributed significantly to reducing the company's carbon footprint. The results of these initiatives are summarized in Table 1 below.

Reduction of Conventional Energy and Improvement in Energy-Saving

In 2022, Vinamilk showcased its unwavering commitment to sustainability through a diverse combination of solutions. These included promoting green energy sources from biomass and biogas, investing in and expanding solar energy systems, improving technology, and optimizing transportation routes. This comprehensive approach played a crucial role in reducing the use of traditional energy sources like gasoline and gas, leading to a considerable decrease in greenhouse gas emissions. These efforts are a testament to Vinamilk's ongoing commitment to sustainable energy practices and its vision for future sustainability.

Gasoline consumption decreased from 241,302 kg in 2018 to 195,800 kg and 179,782 kg in 2022. The reason for the reduction in gasoline consumption was partly impacted by social distancing during the COVID-19 pandemic in 2021. However, the average gasoline consumption per ton of products decreased significantly, from 0.48 kg/ton in 2018 to 0.16 kg/ton in 2022 (see Table 1). This substantial reduction in gasoline consumption is a clear indicator of improved energy efficiency, likely due to better fuel management and more efficient logistics, and has led to a significant decrease in Vinamilk's carbon footprint.

Gas consumption per ton of products fluctuated, peaking at 0.26 kg/ton in 2020 and decreasing to 0.21 kg/ton in 2022. Total gas consumption increased from 289,329 kg in 2018 to 323,122 kg in 2020, then decreased to 244,022 kg in 2022 (See Table 1). Despite the fluctuations, the overall trend shows efforts to stabilize and reduce gas consumption relative to production output, underscoring Vinamilk's commitment to energy efficiency and sustainability.

Integration and Extension of Renewable Energy Sources (Solar Energy, Biomass)

Vinamilk began using ***solar energy*** for its farms in 2016 and expanded this model in 2022 to include its factories and central office. Currently, 11 out of 16 factories (14 in Vietnam and two abroad) and 13 in 15 farms use solar energy, equivalent to 11 million kWh/year (Vinamilk 2022 Sustainable Report). Solar power usage increased significantly from 2,637,718 kWh in 2021 to 15,471,029 kWh in 2022. Solar power has helped reduce grid electricity consumption from 203,881,784 kWh in 2020 to 187,298,535 kWh in 2022 (See Table 1). Solar energy accounted for 7.6% of electricity consumption in 2022, highlighting a shift towards renewable energy, reducing reliance on grid electricity, and improving sustainability.

Box 1: Solar Power Used in Vinamilk Mega Factory One of the latest solar power systems installed has a capacity of 3,394 kW at the Vinamilk Factory. This system uses 7,542 high-efficiency solar panels and 27 inverters (110 kW each) with a lifespan of over 25 years. The Vinamilk Factory, also known as

Table 1 Energy consumption from 2018–2022

Indicator	2018	2019	2020	2021	2022
Gasoline					
Consumption(kg)	241,302	214,640	202,419	195,800	179,782
Average (kg)/ton of products	0.48	0.48	0.16	0.16	0.16
Gas					
Consumption(kg)	289,329	312,581	323,122	312,038	244,022
Average (kg)/ton of products	0.13	0.14	0.26	0.25	0.21
Electricity					
Grid Electricity Consumption (kWh)	N/A	N/A	203,881,784	205,731,848	187,298,535
%	N/A	N/A	N/A	98.7%	92.4%
Solar power Consumption (kWh)	N/A	N/A	N/A	2,637,718	15,471,029
%	N/A	N/A	N/A	1.3%	7.6%
Total Consumption (kWh)	180,093,719	196,234,594	203,881,784	208,369,566	202,769,565
Average (kWh)/ ton of products	109.38	116.75	165.80	169.21	177.46
Biomass					
Consumption (ton of steam)	233,381	254,053	249,417	243,224	235,846
Average (ton of steam)/ton of products	0.20	0.21	0.20	0.20	0.21
CNG					
Consumption(mm BTU)	247,180	281,096	254,097	230,590	217,962
Average (mm BTU)/ton of products	0.21	0.23	0.21	0.19	0.19
Total energy consumption (MJ)	2,031,050,994	2,189,265,519	2,258,405,520	2,239,377,766	2,163,324,010
Average (MJ)/ ton of products	1,233	1,303	1,837	1,819	1,893

(*Source* Compiled from Vinamilk's Sustainability Reports During 2018–2022 Period)

the "Mega Factory," has a high level of automation and a production capacity of 800 million liters per year (approximately 2.2 million liters per day), which will be increased to over one billion liters per year shortly. The clean energy from this solar power system is used for production activities at the factory, saving a significant amount of traditional energy and contributing positively to the transition to sustainable energy, thereby reducing the carbon footprint in production activities.

(*Source* Vinamilk Sustainability Report 2022)

Use of Biomass

Vinamilk has implemented sustainable energy use through comprehensive measures and large-scale projects. Vinamilk has used ***biomass*** energy in production since 2010. Biomass is only used in the milk production sector at Vinamilk. The farms continuously research and apply improvement initiatives to maximize the use of available biogas. This biogas is used as fuel, saving substantial costs. Vinamilk has equipped its operations with saturated steam using fluidized bed boilers fueled by biomass instead of traditional energy sources. In 2021, biomass energy accounted for 65.58% of the energy used at Vinamilk (Vinamilk 2021 Sustainable Report). Biomass consumption remained relatively stable, slightly increasing from 233,381 tons of steam in 2018 to 235,846 tons in 2022. The average biomass consumption per ton of products has been steady at around 0.20–0.21 tons of steam per ton of products (See Table 1). Stable biomass consumption indicates consistent use of renewable energy.

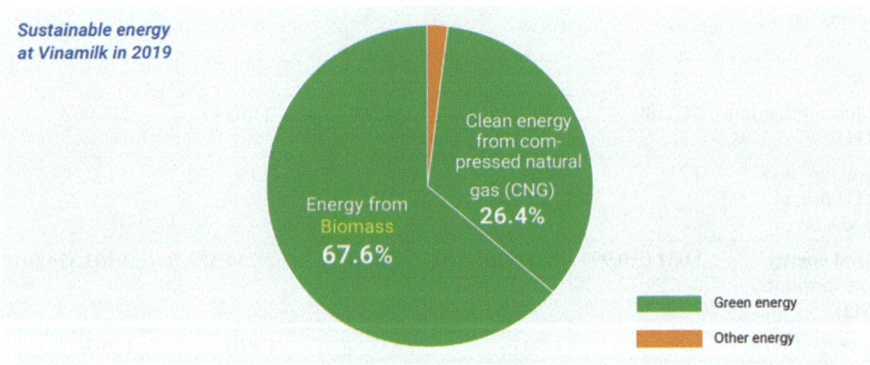

Fig. 2 Sustainable energy at Vinamilk in 2019 (*Source* Vinamilk Sustainability report 2019)

Using Cleaner Alternative Energy Sources (CNG Instead of DO/FO)

Vinamilk is using CNG to replace DO/FO oil in the boilers across its factories. This transition to using CNG, a cleaner alternative to fossil fuels like DO and FO, significantly contributes to the company's goal of reducing CO_2 emissions (Fig. 2). CNG produces less carbon dioxide per unit of energy than coal or oil, making it a relatively cleaner fossil fuel in the short term.

From 2018 to 2022, Vinamilk's CNG consumption decreased from 247,180 mm BTU to 217,962 mm BTU. The average CNG consumption per ton of products decreased from 0.21 mm BTU/ton in 2017 to 0.19 mm BTU/ton in 2022 (see Table 1). While CNG is considered a cleaner fossil fuel, it is not entirely a green energy source. Therefore, the reduction in CNG consumption while increasing solar energy usage reflects improved efficiency and Vinamilk's ongoing transition towards more sustainable energy sources in the long run.

Development of Initiatives in Energy-Use Efficient Technologies

Vinamilk has invested in modern, energy-efficient machinery and equipment. For example, by using biogas-powered dryers for towels and work clothes, Vinamilk cut electricity consumption by over 140,000 kWh/year, saving 274 million VND/year (See Box 2). The company also implemented several initiatives in production processes, such as optimizing the distribution of operating time for equipment, reducing electricity consumption during peak hours, and maintaining machines according to planned schedules. These initiatives led to significant energy savings, including 538,871 kWh/year of electricity, 100,979 kg/year of oil, and 1,323.7 mm BTU/year of CNG in 2020 (See Table 1).

Box 2: Innovative Use of Biogas The company has continuously researched many solutions to make the most of energy from the Biogas system, including successfully manufacturing a grass dryer operated from Biogas. Before this improvement, Vinamilk manually processed dry grass by drying it directly under the sun. Therefore, the output was very little, and the quality was uneven. This innovation has helped the company solve the above problem effectively. The dryer can produce more than 2,100 tons of fresh grass/year, equivalent to 462 tons of dry grass/year, ensuring a stable and continuous supply of food sources to fully meet the health and nutritional needs of cows and calves. This reduces costs for imported grass, ensures effective use of available resources and green energy, and limits greenhouse gas emissions. Seeing the economic value and environmental protection benefits from this initiative, Vinamilk plans to invest in increasing dryer capacity and deploying it across all farms in their system.

(Source Vinamilk Sustainability Report 2021, 2022)

Vinamilk's efforts to improve energy use effectiveness and sustainability development have resulted in notable achievements, particularly in reducing gasoline and CNG consumption, increasing solar power usage, maintaining stable biomass consumption, using cleaner alternative energy sources, and developing initiatives in energy-use efficient technologies. Despite the increased energy demand to serve the living needs of employees during the "3 on-site" process, the energy index in 2021 recorded positive results with a slight decrease compared to 2020. However, the average energy consumption per ton of products increased from 1,233 MJ/ton in 2017 to 1,893 MJ/ton in 2022, indicating a need for further efficiency improvements (see Table 1). The company's ongoing commitment to sustainability through technological upgrades, process optimization, and renewable energy integration sets a strong example for the industry.

3.1.2 Water Sustainability Measures at Vinamilk

Vinamilk has implemented various measures to manage and conserve water resources sustainably. This analysis examines water consumption data from 2018 to 2022 and incorporates additional information on recent business expansions and their potential impact on water use.

Water Reuse in Production

Vinamilk's efforts to reuse water in production demonstrate a solid commitment to sustainability. In 2020, the company reused 237,309 m^3 of water, which increased to 246,608 m^3 in 2021 and then decreased to 216,872 m^3 in 2022. The percentage of water reused was 5.15% in 2020, 5.40% in 2021, and 4.99% in 2022 (see Table 2). Although there was a slight decline in reuse in 2022, the overall trend indicates significant water conservation. These practices help reduce the demand for fresh water and minimize environmental impact.

Table 2 Vinamilk water consumption from 2018–2022

Indicator	2018	2019	2020	2021	2022
Water reused in production (m^3)	N/A	N/A	237,309	246,608	216,872
Water reused in production (%)	N/A	N/A	5.15	5.40	4.99
Groundwater consumption (m^3)	2,063,301	1,722,245	3,055,910	3,301,649	2,910,214
Surface water consumption (m^3)	N/A	N/A	79,601	65,696	916,823
Tap water consumption (m^3)	2,366,864	3,022,895	3,163,150	3,228,854	3,301,727
Average (m^3)/ton of products	N/A	N/A	5.12	5.84	6.24

(*Source* Compiled from Vinamilk's Sustainability Reports During 2018–2022 Period)

Groundwater and Surface Water Consumption

The fluctuations in Vinamilk's groundwater consumption reflect production and operational needs changes. Groundwater consumption peaked at 3,301,649 m^3 in 2021 before reducing to 2,910,214 m^3 in 2022. Surface water consumption substantially increased from 79,601 m^3 in 2020 to 916,823 m^3 in 2022 (see Table 2). This significant rise in surface water use suggests a strategic shift to balance water sources, likely aimed at reducing pressure on groundwater resources. Sustainable sourcing practices are essential to ensure long-term water availability and to mitigate the environmental impact.

Tap Water Consumption and Water Use Efficiency

The data indicates a steady increase in tap water consumption, rising from 2,366,864 m^3 in 2018 to 3,301,727 m^3 in 2022. Similarly, the average water consumption per ton of products rose from 5.12 m^3/ton in 2020 to 6.24 m^3/ton in 2022 (see Table 2). This trend reflects increased production demands, likely driven by Vinamilk's expansions and new product offerings. To address this, Vinamilk needs to implement more efficient water use practices and technologies to mitigate the growing water demand.

Recent business expansions have contributed to the increased water use observed in the data. In 2018, Vinamilk invested in holding 51% of shares in Lao-Jagro, a significant dairy farming project. By 2022, Vinamilk had increased its ownership ratio to 87.3%. The Lao-Jagro facilities include a large-scale dairy farm with a capacity of over 24,000 cows and an extensive 5,000-ha scale in the first phase. These investments likely increased water consumption due to expanded operations and production needs at Lao-Jagro facilities (see Fig. 3).

In 2019, Vinamilk acquired 75% of the shares in GTNFoods JSC,[3] thereby participating in managing Moc Chau Milk JSC, which has a herd size of 27,500 cows.[4] Managing such a large herd would significantly increase water use for livestock maintenance, cleaning, and milk production. This acquisition undoubtedly contributed to the higher water consumption observed in subsequent years.

In 2021, Vinamilk established a joint venture with Del Monte Philippines, Inc., contributing 50% of its capital (Vinamilk Sustainable Development Report 2022). This venture allowed Vinamilk to distribute products to consumers in the Philippines from Q4/2021. Establishing and operating this joint venture increased water use to support production and distribution in a new market, further driving up overall water consumption. In 2022, Vinamilk exchanged all GTNFoods shares for Vietnam Livestock Corporation (Vilico) shares after merging GTNFoods into Vilico.[5] This merger likely required additional water resources to support integrated operations and increased production capacity, contributing to the rise in water consumption. Besides, due to the influence of climate change and the COVID-19 pandemic, the demand for water at farms to serve the cattle and the living needs of employees

[3] https://vietnamnews.vn/economy/521009/vinamilk-ups-stake-to-383-in-gtnfoods.html and https://www.marketscreener.com/quote/stock/GTNFOODS-JSC-53979443/company/

[4] https://finance.vietstock.vn/GTN-ctcp-gtnfoods.htm?languageid=2.

[5] https://vietnaminsider.vn/vinamilk-acquires-majority-stake-in-competitor/

Fig. 3 Miles stones of Vinamilk business development 2016–2022 (*Source* Vinamilk's Sustainability report 2023)

during the "3 on-site" process at the units has been increasing. Additionally, in 2021, Vinamilk included the data from the Quang Ngai farm in the reporting scope; this new farm had just come into operation, so there was no milk production in the initial period. These are the reasons for the increased water use index per ton of product in 2021 (Vinamilk Sustainable Development Report 2021).

Vinamilk's water sustainability measures reflect a solid commitment to efficient water management, with significant efforts in water reuse, balanced sourcing, and technological upgrades. However, recent business expansions have increased water use, including investments in Lao-Jagro and GTNFoods JSC and the joint venture with Del Monte Philippines. This underscores the need for ongoing improvements in water efficiency. Continuous investment in water-saving technologies, optimized production processes, and sustainable sourcing strategies will be crucial for Vinamilk to achieve its water sustainability goals while supporting its growth and expansion initiatives.

3.1.3 Vinamilk's Efforts to Reduce Emissions

Vinamilk, Vietnam's leading dairy company, has made substantial efforts to manage and reduce emissions as part of its sustainability initiatives. These efforts, as evidenced by the provided emissions data from 2018 to 2022 and additional information from Vinamilk's sustainability reports, have led to significant strides in emissions reduction, instilling hope and optimism for a greener future. Table 3 below demonstrates the results of these efforts.

Hazardous Waste Management

Vinamilk has shown a steadfast commitment to managing hazardous waste effectively. The data indicates that hazardous waste increased slightly from 159,090 kg in 2018 to 167,504 kg in 2022, with the average hazardous waste per ton of products rising from 0.10 kg to 0.15 kg over the same period (see Table 3). However, to mitigate this, Vinamilk has implemented advanced waste management systems, ensuring that all hazardous waste is collected, stored, and treated by licensed contractors. This commitment to compliance with environmental regulations provides reassurance about Vinamilk's responsible waste management practices.

Table 3 Vinamilk Emissions from 2018–2022

Indicator	2018	2019	2020	2021	2022
Emissions					
Hazadous waste (kg)	159,090	163,611	168,571	179,701	167,504
Average (kg)/ton of products	0.10	0.11	0.14	0.15	0.15
Scrap (kg)	10,728,628	10,052,809	9,577,945	9,354,444	8,839,282
Average (kg)/ton of products	7.05	6.44	7.79	7.60	7.74
Domestic waste (kg)	944,048	1,095,750	1,201,407	1,658,329	1,632,141
Average (kg)/ton of products	0.58	0.65	0.98	1.35	1.43
Total amount of waste resued, recycled and by other recovery method	N/A	N/A	9,577,945	9,354,444	8,839,282
Total amount of waste treated by methods such as incineration landfill (kg)	N/A	N/A	1,369,978	1,838,030	1,799,645
Total wastewater output (m^3)	3,395,142	3,538,777	3,535,194	3,560,789	3,396,444
Total water consumption (m^3)	994,912	1,082,947	2,763,468	3,626,674	3,732,321
Carbon Footprint					
Total emissions (kg)	192,331,768	206,481,184	253,476,971	296,391,550	266,161,432
Average (kg)/ton of products	14.55	122.93	206.14	240.69	232.95

(*Source* Compiled from Vinamilk's Sustainability Reports During 2018–2022 Period)

Domestic Waste Reduction

Vinamilk has also made significant strides in reducing domestic waste. The data indicates a notable increase in domestic waste from 944,048 kg in 2018 to 1,632,141 kg in 2022, with an average increase from 0.58 kg to 1.43 kg per ton of products (see Table 3). However, Vinamilk has effectively tackled this issue by enhancing waste segregation at the source and promoting recycling among employees. They have implemented successful waste reduction programs, such as the use of eco-friendly packaging materials and the encouragement of reducing single-use plastics within their operations. In 2022, the company introduced 44 initiatives to reduce plastic use, save energy, and conserve resources, resulting in significant savings of up to 4.3 billion VND (Vinamilk Sustainable Report 2022).

Scrap and Waste Treatment

Vinamilk's scrap generated decreased from 10,728,628 kg in 2018 to 8,839,282 kg in 2022 (see Table 3). The company has established comprehensive recycling programs to manage this effectively. The corporation has set up systems to recycle packaging materials and other production waste, minimizing the amount sent to landfills (Vinamilk Sustainable Report 2021). The total amount of waste treated by incineration and landfills increased from 1,369,978 kg in 2020 to 1,799,645 kg in 2022, indicating an ongoing need to balance waste reduction and effective treatment.

Wastewater Treatment

Vinamilk's wastewater output remained relatively stable from 3,395,142 m^3 in 2018 to 3,396,444 m^3 in 2022, while total water consumption increased from 994,912 m^3 in 2018 to 3,732,321 m^3 in 2022 (See Table 3). The company has adopted water-saving technologies and practices, such as closed-loop cooling systems and water recycling. All production wastewater is treated to meet environmental standards before discharge (Vinamilk Sustainable Report 2022).

Carbon Footprint and Energy Use

Vinamilk's total emissions increased from 192,331,768 kg in 2018 to 266,161,432 kg in 2022. The average emissions per ton of products rose from 14.55 kg to 232.95 kg in the same period (See Table 3). This unexpected increase can be attributed to the fact that, in 2021, the company adjusted its greenhouse gas emission calculation formula to reflect the updated emission factor of the Vietnamese power grid, leading to an increase in the emission index per ton of product. Despite the higher demand for energy due to the "3 on-site" process during the COVID-19 pandemic, Vinamilk achieved a slight reduction in energy consumption compared to 2020 by implementing effective energy management and using renewable energy sources like solar power.

Box 3: The Circular Economy at Vinamilk At Vinamilk, the circular economy aligns with environmental protection and sustainable development goals, primarily through the efficient and economical use of resources, reducing and ultimately eliminating environmentally harmful waste. Vinamilk transforms waste into resources. At dairy farms, strict waste management ensures environmental protection. Modern systems convert collected manure into organic fertilizer and biogas. This organic fertilizer replaces chemical fertilizers, improving soil structure and fertility while reducing methane and nitrous oxide emissions from waste decomposition. Renewable energy from biogas is used for farm operations, including pasteurizing milk for calves, heating water, drying towels and clothes, and drying hay.

(Source Vinamilk's Sustainable Report 2022)

Vinamilk's efforts to reduce emissions are reflected in its various sustainability initiatives, including modernizing production facilities, integrating renewable energy, effective waste management, and adopting efficient water use practices. Continuous improvements and investments in sustainable technologies will be crucial to achieving its long-term emissions reduction goals. The coporation's proactive approach to waste management, energy efficiency, and resource conservation demonstrates its commitment to sustainable development and environmental stewardship.

Concrete Agenda Towards Sustainable Development

Vinamilk has developed a comprehensive Corporate Social Responsibility (CSR) strategy that emphasizes environmental sustainability and improved working conditions. As a pathway to dairy Net Zero 2050, Vinamilk is committed to achieving ***net-zero emissions*** by 2050. This plan includes reducing greenhouse gas emissions through a variety of initiatives, such as planting over 100,000 trees to regenerate 25 hectares of mangrove forest in Ca Mau National Park and installing solar power systems across their production facilities.[6] In 2023, Vinamilk achieved ***carbon–neutral certification*** for one of its factories and one farm, marking a significant milestone in their sustainability efforts.[7] The company has set goals to save at least 3% energy over five years, reduce CO2 emissions, and apply ISO 50001 standards across all its factories for energy management. This involves using clean and renewable energy sources and energy-saving equipment.[8] Vinamilk's dairy farms adhere

---

[6] https://www.vinamilk.com.vn/en/sustainable-development/

[7] https://vir.com.vn/vinamilk-dedicated-to-long-term-roadmap-in-aid-of-net-zero-goals-107232.html.

[8] https://www.vinamilk.com.vn/en/sustainable-development/

to Global G.A.P standards, focusing on sustainable farming practices. Their production facilities are certified with ISO 9001 and ISO 22000, ensuring high standards in quality management and food safety.[9]

3.2 The Case Study of Bao Viet Group

Bao Viet Group is a leading financial services conglomerate based in Vietnam. Established in 1988, it has grown to include various subsidiaries operating across the banking, insurance, and investment sectors. Bao Viet Group is a commercial banking arm with over 260 branches and transaction points nationwide. It provides various financial products and services for retail and corporate customers. Its insurance portfolio comprises three leading companies—Bao Viet Insurance Corporation, Bao Viet Life Insurance, and Bao Viet General Insurance. These entities have a substantial market share in Vietnam's fast-growing insurance industry. Bao Viet is also interested in securities, asset management, and private equity firms that handle investments, trading, and asset management. This includes subsidiaries like Bao Viet Securities and Bao Viet Fund Management Company. Headquartered in Hanoi, Vietnam, Bao Viet employs over 8,000 people and has total assets exceeding $3 billion, making it a prominent player in the country's financial landscape. The conglomerate aims to further expand regionally across Southeast Asia. In summary, Bao Viet is one of Vietnam's largest and most diverse financial services corporations, playing an influential role through its network of banking, insurance, and investment subsidiaries. Some CSR initiatives that Bao Viet has been taking towards environmental sustainability are anlyzes in our study as follows.

3.2.1 Energy Saving

Electricity Consumption

Bao Viet has firmly committed to reducing electricity consumption, significantly dropping 948 thousand kWh in 2022 from 2,537 thousand kWh in 2019 (see Table 4). In 2022, Bao Viet leveraged its experience in adapting and applying 100% online system management technologies during the COVID-19 lockdown to continue improving and implementing flexible working regulations across the entire system. This approach created a modern, dynamic working environment suited to the young workforce.

Recognizing the impacts of climate change and human activities on the environment, Bao Viet has strengthened internal communication to raise employees' awareness of environmental protection through energy conservation at headquarters

[9] https://www.vinamilk.com.vn/en/sustainable-development/

Table 4 Bao Viet's Environmental Sustainable Indicators 2018–2022

Indicator	2018	2019	2020	2021	2022
Gasoline consumption	20,030	18,029	12,384	11,769	28,931
Electricity consumption (Thousand kWh)	1,900	2,537	2,445	2,407	948
Water consumption (m3)	10,665	4,355	4,439	3,711	8,229
Wastewater (m3)	N/A	12,639	8,652	6,898	8,229
Greenhouse gas (GHG)	N/A	N/A	1,125	1,104	761
Scope 1	N/A	N/A	122.18	120.97	318.26
Scope 2	N/A	N/A	989.61	970.10	400.62
Scope 3	N/A	N/A	13.56	12.87	41.71
Greenhouse gas emission intensity at Bao Viet (tons of CO2/1 billion VND of revenue)	N/A	N/A	0.02	0.02	0.01

(*Source* Compiled from Bao Viet Sustainable Development Reports 2018–2022)

and branches. The group has invested in information technology to digitize transactions, reduce paper waste, and adapt to the complex COVID-19 situation, improving efficiency and minimizing its environmental footprint.

Bao Viet buildings utilize and maximize the efficiency of the Building Management System (BMS). This integrated system controls and manages all technical systems within the building, such as electrical systems, water supply, ventilation, environmental warning, security, and fire-fighting systems. The BMS ensures equipment's precise and timely operation, leading to significant energy savings and operational efficiency. It helps monitor and adjust energy use in real-time, prevent waste, and optimize resource use.

Bao Viet's offices are designed with the ability to open windows for ventilation, natural lighting, and sun shading to reduce heat when necessary, limiting the impact of solar radiation and preventing heat loss when air conditioning is used. When renovating office areas, Bao Viet requires contractors to provide solutions to limit indoor temperature increases caused by the greenhouse effect from large glass panels or primarily glass structures. Recognizing that many office areas may have increased overall costs if electrical devices are used when unoccupied, Bao Viet has installed sensors in areas such as hallways and restrooms at the headquarters. These sensors handle power and water shut-off when unnecessary, ensuring energy is used only when needed.

The main energy-consuming components in buildings include air conditioning systems, auxiliary equipment (water pumps, fans), lighting systems, office equipment, and elevators. Optimizing energy use and operating costs in buildings is projected to save approximately 10–40%. Bao Viet focuses on maximizing efficiency in these areas to achieve significant cost savings and reduce environmental impact.

The air conditioning system is turned off after 5:30 PM to save energy. Low-efficiency lighting systems are replaced with energy-saving devices like compact fluorescent bulbs and electronic ballasts. Outdoor lighting is controlled automatically based on time or ambient light levels, reducing unnecessary energy consumption. These measures ensure optimal energy use, reduce operating costs, and contribute to Bao Viet's sustainability goals.. As a result, the electricity saved in 2022 amounted to 1,458,339 kWh (5,250 GJ), equivalent to a cost saving of 4.0 million VND (Bao Viet Sustainable Development Report 2022). By continuously seeking innovative solutions and engaging employees in energy conservation, Bao Viet sets a benchmark for sustainability practices in the industry, showcasing how businesses can effectively reduce their environmental impact while maintaining operational efficiency.

Gasoline Consumption

To better manage energy consumption during operations, Bao Viet has implemented monthly monitoring of electricity and gasoline usage while also promoting energy-saving awareness among employees throughout the organization. Gasoline consumption has generally decreased, with significant drops in 2020 and 2021, followed by an increase in 2022 (see Table 4), reflecting ongoing adjustments in operational efficiencies.

From 2020 to 2021, the COVID-19 pandemic severely impacted the business operations of many companies, including Bao Viet. During this period, both travel activities and the gasoline demand decreased significantly due to lockdowns and restricted movement. In 2022, businesses, including Bao Viet, began to recover and rebuild their supply chains and operations. As a result, Bao Viet's gasoline consumption increased by 146% compared to 2021 and 134% compared to 2020 (see Table 4). This rise was primarily due to the higher number of vehicles and the resumption of frequent inter-provincial business trips aimed at promoting new business ventures and implementing social welfare activities.

The increase in gasoline consumption and travel distance in 2022 can also be attributed to Bao Viet expanding its data audit scope from one office group (the parent company) to three office groups. This broader audit provided a more comprehensive view of the organization's gasoline usage, contributing to the apparent increase. Despite these increases, Bao Viet's gasoline consumption in 2022 remained comparable to the levels seen in 2018 and 2019 before the pandemic (see Table 4). The organization ensures that energy consumption targets are met through regular internal audits conducted by Bao Viet Group's internal audit department. These audits help identify areas for improvement and ensure that energy efficiency remains a priority.

Bao Viet balances operational efficiency with sustainable energy practices by closely monitoring and auditing energy use. The company's ongoing efforts to promote energy-saving awareness and adjust operational practices highlight its commitment to environmental responsibility while supporting business growth and recovery post-pandemic.

Bao Viet started tracking water consumption and wastewater treatment in 2019. They have managed to decrease water usage from 12,639 m^3 in 2019 to 8,229 m^3 in 2022. The corresponding wastewater treatment figures reflect similar reductions, indicating improved water management practices (see Table 4).

3.2.2 Water Consumption

In 2022, water consumption was 8,229 m3, which increased by 122% compared to the COVID-19 pandemic in 2021 (3711 m3), when offices and buildings temporarily halted operations for social distancing measures. Compared to 2020 and 2019, water consumption increased by 85% and 89%, respectively. However, in 2022, during normal business operations, water consumption decreased by 23% compared to 2018 (See Table 4).

The company policies now include water-saving measures to promote water conservation. Building water management teams are committed to continuing education on water-saving practices and reducing water waste to protect the environment.

Monitoring and Controlling Wastewater

As a company engaged in business activities related to financial services, Bao Viet's operations do not have significant environmental impacts. However, for its office headquarters and buildings managed by Bao Viet, controlling the amount of waste and wastewater from these buildings is the environmental protection measure that Bao Viet is implementing. The corresponding wastewater treatment figures reflect a similar reduction in 2021 (6,898 m3) due to COVID-19 social distancing compared to 2018 and 2020. Following the recovery in 2022, water treatment increased to 8229 m3, indicating improved water management practices (see Table 4).

Bao Viet has two types of wastewater: rainwater runoff and building wastewater. Rainwater runoff is considered clean, but it can collect dirt and debris when it flows through the premises. It requires only primary treatment before being discharged. Rainwater from roofs is collected via gutters and pipes directly to storm drains, avoiding surface runoff and regularly clearing sediment from catch basins. Grates are also placed at drainage channels to filter larger debris. Building wastewater is mainly used for restrooms, floor cleaning, and equipment cooling. This wastewater is channeled through the treatment system before discharge. After categorization, it continues through the treatment process, meeting QCVN 14:2008/BTNMT standards before entering the area's sewer system.

As a financial services company, Bảo Việt's operations have minimal environmental impact. Therefore, controlling wastewater discharge is a crucial environmental protection measure for Bảo Việt's buildings. With a wastewater treatment system meeting QCVN14:2008/BTNMT standards and strict adherence to environmental and water resource laws, wastewater is treated, disinfected, and de-CO2ed before being released into the environment. The wastewater reception points function well, even during heavy rains.

3.2.3 Greenhouse Gas Control

The company reported its greenhouse gas (GHG) emissions starting in 2020, decreasing from 1,125 tons in 2020 to 761 tons in 2022. The GHG emission intensity, measured in tons of CO2 per billion VND of revenue, decreased from 0.02 in 2020 to 0.01 in 2022, indicating better emission control relative to their revenue (see Table 5).

Monitoring and Controlling Greenhouse Gas Emissions

Bao Viet's operations as a financial services company inherently produce fewer greenhouse gas emissions compared to manufacturing sectors. However, the company is committed to managing its environmental impact by focusing on emissions from its office buildings. Bao Viet is developing regulations based on the Greenhouse Gas (GHG) Protocol to manage these emissions effectively. The GHG Protocol categorizes emissions into three scopes: direct emissions from company-controlled sources (Scope 1), indirect emissions from purchased electricity (Scope 2), and other indirect emissions from sources not owned or controlled by the company (Scope 3) (see Table 5).

From 2020 to 2022, Bao Viet saw a ***significant increase in Scope 1 emissions***, which include direct emissions from fuel burning for system operations, vehicle operations, and air conditioning. Specifically, emissions from burning fuel to operate the system rose from 28.04 in 2021 to 68.92 in 2022. Similarly, emissions from fuel used for enterprise vehicles increased from 41.13 in 2021 to 98.56 in 2022. The most

Table 5 Green House Gas (GHG) Protocol in Bao Viet's buildings

Scope of Operations		2020	2021	2022
Scope 1	Burning fuel to operate the system	29.50	28.04	68.92
	Burning fuel to operate the enterprise's vehicles	41.71	41.13	98.56
	Emissions from air conditioner	50.96	51.80	150.78
	Scope 1—Total	***122.18***	***120.97***	***318.26***
Scope 2	Electricity consumed for power generation system	962.26	947.36	390.43
	Electricity consumed for hot and cold water system	27.35	22.73	10.20
	Scope 2—Total	***989.61***	***970.10***	***400.62***
Scope 3	Use of transportation means work	13.53	12.85	30.34
	Use vehicles to transport products	0.03	0.02	0.02
	Use vehicles to transport employees	0.00	0.00	11.35
	Scope 3- Total	***13.56***	***12.87***	***41.71***
Total		**1125.35**	**1103.94**	**760.6**

(*Source* Bao Viet Sustainability Report 2022)

notable rise was seen in emissions from air conditioners, which jumped from 51.80 in 2021 to 150.78 in 2022. Consequently, total Scope 1 emissions grew significantly from 120.97 in 2021 to 318.26 in 2022. This trend likely reflects the increased operational activities as Bao Viet resumed normal operations post-pandemic.

In contrast to the rise in Scope 1 emissions, Bao Viet ***substantially reduced Scope 2 emissions***, which cover indirect emissions from purchased electricity used for power generation and water heating/cooling systems. The electricity consumed for the power generation system decreased dramatically from 947.36 in 2021 to 390.43 in 2022. Additionally, the electricity consumed for hot and cold water systems dropped from 22.73 in 2021 to 10.20 in 2022. As a result, total Scope 2 emissions fell from 970.10 in 2021 to 400.62 in 2022. This significant reduction suggests the implementation of successful energy-saving measures or a transition to more efficient energy sources.

Bao Viet also experienced an ***increase in Scope 3 emissions***, which include other indirect emissions such as those from work transportation, product transport, and employee transport. The use of transportation for work rose from 12.85 in 2021 to 30.34 in 2022. Moreover, the emissions from vehicles used to transport employees increased to 11.35 in 2022, up from zero in previous years, indicating a rise in employee travel. Consequently, total Scope 3 emissions increased from 12.87 in 2021 to 41.71 in 2022. This trend reflects the higher indirect emissions associated with increased transportation activities as business operations resumed.

Despite the increases in Scope 1 and Scope 3 emissions, Bao Viet achieved ***an overall decrease in total GHG emissions***, primarily due to the significant reduction in Scope 2 emissions. The total GHG emissions decreased from 1103.94 in 2021 to 760.60 in 2022. This overall decline indicates that the company's efforts in energy conservation and efficiency improvements in electricity use have been practical, even as other areas saw increased emissions. Bao Viet's overall strategy seems to be reducing its carbon footprint while managing the fluctuating demands of its operational activities.

Box 4: Changes in Bao Viet to Reduce GHG Bao Viet integrates ***green spaces*** within its buildings to create a healthier work environment and reduce CO2 emissions. These green areas, such as the 50 square meters at the IT Center at 71 Ngo Si Lien, 140 square meters at the headquarters at 8 Le Thai To, and 300 square meters at the Financial Center at 233 Dong Khoi, act as natural CO2 filters. Bao Viet enhances its commitment to sustainability and environmental responsibility by incorporating these green spaces.

In response to the need for greater sustainability, Bao Viet has implemented various ***operational changes to reduce its carbon footprint.*** These include limiting travel, hosting external events, and promoting online business activities. Such measures have led to a significant reduction in fuel costs and CO2 emissions. In 2022, Bao Viet emitted 0.013 tons of CO2 for every 1 billion VND in revenue, achieving a 38% reduction in greenhouse gas emission intensity compared to 2021. This effort underscores Bao Viet's dedication to operational efficiency and environmental sustainability.

Given its extensive network across 63 provinces and cities, air travel is essential for Bao Viet. 2022 the company's business flights covered 157,492 km, with 111 flights to 31 domestic and international destinations. To ***mitigate the environmental impact of air travel,*** Bao Viet has optimized flight routes, chosen airlines with high-efficiency aircraft, and increased the use of virtual meetings. Despite a recorded 18.11 tons of CO2 emissions from air travel in 2022, these measures demonstrate Bao Viet's commitment to reducing its environmental footprint. The company aims to reduce flights by 5% in the coming years and continues exploring new sustainable business travel solutions.

Bao Viet also prioritizes ***air quality control within its offices***. The company conducts quarterly assessments of airborne disease transmission risks, maintains an environmental sanitation system, ensures proper air circulation, and uses Demand-Controlled Ventilation (DCV) based on occupancy and temperature. By maintaining central air filtration systems at full capacity, Bao Viet safeguards the health and well-being of its employees while upholding its environmental responsibility.

(Source Bao Viet Sustainability Report 2022)

Emissions Management

These initiatives highlight Bao Viet's comprehensive approach to sustainability, balancing operational efficiency with environmental stewardship.

Office building operations inevitably produce various emissions, solid waste, and wastewater. Bao Viet has mandated that the units operating its office buildings fully comply with legal waste management regulations and submit annual environmental protection reports to manage these waste types tightly and minimize environmental impact. Emissions primarily come from Bao Viet's vehicles, visitor vehicles, and traffic around the buildings. Without mitigation measures, dust and emissions from these vehicles can contribute to the greenhouse effect and increase air pollution, posing health risks to the community.

Box 5: Technical Measures for Emission Reduction in Bao Viet To address these challenges, Bao Viet has implemented several technical measures. The company has arranged numerous green areas to purify the air, required vehicles to turn off their engines when entering the building premises, and cleaned the lobby and parking garage daily to reduce pollution. Daily checks of the parking garage ventilation systems ensure they are always operational, effectively managing emissions from vehicles entering the underground levels. High-emission equipment such as power generators, ventilation systems, and water pumps are placed in designated areas with separate air treatment systems.

Bao Viet also focuses on community-based solutions to further reduce emissions. The company organizes collective activities and employee initiatives to plant trees, thereby reducing emissions in localities where Bao Viet has branches and transaction offices. An annual 5S evaluation program is implemented to build a green, clean, and scientifically organized working environment. Bao Viet encourages each workstation to have a green plant and ensures each office has at least two large potted plants to purify the air. Additionally, specific areas are designated for office equipment like printers and fax machines, which are equipped with separate air treatment systems.

(Source Bao Viet Sustainability Report During 2018–2022 Period)

By integrating these measures, Bao Viet aims to manage emissions effectively, reduce environmental impact, and contribute to a healthier, more sustainable community.

3.2.4 Waste Management

Domestic Waste

Domestic waste at Bao Viet primarily comes from office activities of staff and customers, as well as cleaning operations. This includes damaged paper, used pens, single-use packaging, restroom waste, leftover food, plastic bags, cans, and bottles. Bao Viet has implemented several technical measures to manage this waste. Large composite bins and small trash cans are placed at the end of staircases and in each office to collect waste. Cleaning staff are assigned to clean offices, hallways, and surrounding areas daily and to collect and move waste to a centralized location for daily disposal by a sanitation company. Community solutions include raising awareness among staff to reduce paper use, reusing office supplies, and regularly maintaining equipment to prevent environmental pollution. These efforts have led to collecting and recycling 1,800 kg of domestic waste, reducing CO_2 emissions by 4.8 tons annually.

Hazardous Waste

Hazardous waste in Bao Viet's office buildings mainly consists of used fluorescent bulbs, batteries, ink cartridges, and cleaning rags with chemicals. These wastes are dangerous because they can potentially cause fires, explosions, poisoning, corrosion, or infections. Bao Viet stores hazardous waste in separate containers for no longer than one year, labels them according to regulations, and gathers them in designated storage areas. The company submits reports on hazardous waste collection and disposal to the relevant authorities. Community solutions involve educating staff on hazardous waste sorting and employing various methods for waste disposal, such as biological treatment, incineration, landfill, and recycling. As a result, 146 kg of bulbs and 31 kg of batteries were collected and recycled, reducing CO2 emissions by 0.5 tons annually.

Wastewater

Wastewater from Bao Viet's office buildings includes effluent from restrooms and rainwater drainage. If not properly treated, wastewater can cause leaks, environmental pollution, and affect the overall drainage system, posing health risks. Technical measures include treating wastewater to meet QCVN 14:2008/BTNMT standards before discharge, contracting specialists for quarterly monitoring, and submitting annual reports on wastewater management. Additionally, the company regularly adds biological agents to enhance septic tank efficiency and disinfects wastewater before release. Regular inspections prevent blockages and overflows, even in adverse weather conditions. Community solutions include placing signs in restrooms to guide water use, promoting awareness among staff to maintain cleanliness and avoid blockages, and organizing annual training to improve wastewater management skills. These measures resulted in 6,898 m^3 of treated wastewater, reducing CO2 emissions by 6 tons annually.

4 Lessons Learned and Areas for Improvement

This study explored how Vinamilk and Bao Viet utilized CSR activities over five years to enhance environmental sustainability performance. The findings demonstrate that with committed multi-stakeholder engagement, priority issue identification, target-setting, systematic tracking, supply chain integration, innovation, transparency, and continuous improvement, CSR practices can meaningfully support organizations' environmental stewardship journeys. Overall, experiences of Vinamilk and Bao Viet offer valuable insights for maximizing nature-positive impact through responsive and responsible actions.

Vinamilk and BaoViet strategically sought to address their environmental responsibilities by integrating their CSR practices into their core business operations. Their key CSR practices towards environmental sustainability include (1) Energy saving, (2) Emissions control; (3) Waste and water management. However, due to the differences in the nature of business, the two companies are slightly different in their priority on these CSR practices. As a manufacturing company, Vinamilk has put more effort into waste management and water reservation. Meanwhile, Bao Viet has achieved significant results in reducing GHG emissions thanks to their protocol and strict application of measures. Our findings are supported by several studies, such as Nguyen et al. (2022), Tran (2020) and Le et al. (2021).

Moreover, Vinamilk and BaoViet leveraged CSR strategically to make meaningful strides in combatting various pressing sustainability issues. Through materiality assessments, both companies prioritized climate change mitigation and established reduction targets and reporting mechanisms to manage and shrink their greenhouse gas emissions systematically. The two case studies show that efforts to boost resource efficiency involve on-site initiatives like renewable energy adoption and waste minimization. Our finding is in line with the studies of Pham et al. (2021) and Nguyen et al. (2022). Additionally, Vinamilk aided biodiversity protection through community environmental projects (Vinamilk 2020). Regulatory compliance was strengthened by transparent target-setting and performance disclosure (DOST 2018). Overall, their CSR application constructively addressed multifaceted sustainability challenges.

From the two case studies, we also found some areas for improvement for not only Vinamilk and Bao Viet, but also other companies. We suggest that companies can greatly improve environmental sustainability by implementing a comprehensive strategy. First and foremost, it is advisable for them to allocate resources towards the adoption of sustainable energy sources such as solar, wind, and hydroelectric power to mitigate their carbon emissions. By incorporating energy-efficient technologies and practices, such as LED lighting and smart thermostats, it is possible to further reduce energy use. Additionally, it is imperative for enterprises to give precedence to sustainable supply chain management by procuring raw materials from environmentally conscious suppliers and advocating for the utilization of recycled or biodegradable materials in their goods. In addition, implementing extensive recycling programs and adopting circular economy ideas, such as product take-back schemes and remanufacturing, can significantly reduce the amount of garbage sent to landfills. Finally, cultivating a culture of sustainability within the firm by providing employee education and promoting environmentally friendly activities, such as telecommuting and digital offices, can guarantee a lasting dedication to environmental responsibility. By incorporating these tactics, firms may safeguard the environment and reap economic advantages by reducing expenses and improving their brand image.

5 Conclusions, Limitations, and Future Research Directions

There are a few limitations to consider in this research on CSR practices toward promoting environmental sustainability at Vinamilk and BaoViet. First, the analysis relied solely on data and information disclosed publicly in sustainability reports by the companies over 2018–2022. There is no ability to verify the accuracy or completeness of implemented initiatives independently and reported performance. Second, the short five-year study period precludes drawing definitive conclusions about long-term impacts—sustainability progress requires enduring commitment.

Third, contextual factors like evolving national regulations and socioeconomic conditions were not fully accounted for. Finally, the perspectives of external stakeholders were not incorporated due to the secondary nature of the research design (Nguyen et al. 2022). Future studies could address such constraints through primary data collection involving company interviews and surveys of community members (Vo and Nguyen 2021). Despite boundaries, valuable insights were still gleaned from this initial examination of how prominent Vietnamese organizations strategically leverage CSR. Future research could involve supplemental interviews, pre-post assessments of progress towards targets, and comparative analyses across sectors and global peers to define best practices (Nguyen 2020). As sustainability challenges persist globally, continued examination of how strategic CSR application benefits communities and ecosystems can guide other businesses seeking to operationalize responsibility.

This preliminary exploration provided valuable baselines, but numerous opportunities exist to build on its foundations moving forward. Primary data collection methods could capture in-depth perspectives from companies and stakeholders regarding best practices, challenges, and opportunities for enhancing environmental programs (Le et al. 2022). Longitudinal analyses evaluating targets and progress over extended time horizons would strengthen understanding CSR long-term sustainability impacts (Nguyen et al. 2022). Comparative case studies across geographic regions and economic sectors may highlight transferable solutions to common issues like climate change (Dang et al. 2021). A deeper examination of shifting national policy contexts could offer guidance on practical advocacy approaches. As corporate responsibility remains dynamic, continued multidisciplinary research exploring emerging issues and cross-cutting themes holds the potential to meaningfully support organizations and communities navigating sustainability (Tran et al. 2020). Overall, furthering this work with applied, mixed-methods investigations provides valuable insights for both practice and theory development surrounding strategic CSR application.

References

Bakry W, Mallik G, Nghiem X-H, Sinha A, Vo XV (2023a) Is green finance really "green"? Examining the long-run relationship between green finance, renewable energy and environmental performance in developing countries. Renew Energy 208:341–355

Bakry W, Nghiem X-H, Farouk S, Vo XV (2023b) Does it hurt or help? Revisiting the effects of ICT on economic growth and energy consumption: a nonlinear panel ARDL approach. Econ Anal Policy 78:597–617

BaoViet Group (2018) Sustainability report. Available at http://www.baoviet.com.vn

BaoViet Group (2019) Sustainability report. Available at http://www.baoviet.com.vn

BaoViet Group (2020) Sustainability report. Available at http://www.baoviet.com.vn

BaoViet Group (2021) Sustainability report. Available at http://www.baoviet.com.vn

BaoViet Group (2022) Sustainability report. Available at http://www.baoviet.com.vn

Cordeiro JJ, Profumo G, Tutore I (2020) Board gender diversity and corporate environmental performance: the moderating role of family and dual-class majority Nguyen VT, Tran HQ, Dang

CN (2022) Greening agricultural supply chains in Vietnam. Sustainability 14(6):3511. https://doi.org/10.3390/su14063511
Dang TTL, Vu LV, Nguyen HB (2021) Corporate social responsibility in emerging economies: a systematic literature review. Sustainability 13(10):5715. https://doi.org/10.3390/su13105715
DOST (2018) Vietnam state of the environment report. Available at https://en.tiki.vn/
FAO (2020) Greenhouse gas emissions from the dairy sector: a life cycle assessment. Food and agriculture organization of the united nations. Available at http://www.fao.org/3/cb2863en/cb2863en.pdf
Forbes (2020) Data analytics: the future of sustainability reporting. Available at https://www.forbes.com/
Forbes (2020) Empowering smallholder farmers through data. Available at https://www.forbes.com
Graham JP, VanDerZanden AM (2016) Environmental health indicators for sustainable food systems. Environ Health Perspect 124(12):1820–1827. https://doi.org/10.1289/EHP117
Guo B, Guo Q, Amri F, Abbas S, Nghiem XH (2023) Powering environmental sustainability through renewable energy and natural resources: a Dynamic ARDL simulation approach. Environ Sci Pollut Res 30:90906–90923
Intergovernmental Panel on Climate Change (2021) Climate change 2021: the physical science basis. Cambridge University Press. https://www.ipcc.ch/report/ar6/wg1/
Le TTH, Blome C, Thuy PNB (2021) Climate change strategies in supply chains: the context of sustainable procurement in Vietnam. Bus Strateg Environ 30(4):2054–2068
Le TTH, Nguyen PT, Pham TN (2022) Decarbonizing dairy through 4R nutrient stewardship in Vietnam. Sustainability 14(3):1583. https://doi.org/10.3390/su14031583
Ma X, Rao X, Li M (2021) Global warming impact of China's commercial banks' project finance. Nat Commun 12(1):1–10. https://doi.org/10.1038/s41467-021-23031-6
Nguyen VT, Tran HQ, Dang CN (2022) Greening agricultural supply chains in Vietnam. Sustainability 14(6):3511. https://doi.org/10.3390/su14063511
Nghiem X-H, Bakry W, Al-Malkawi H-AN, Farouk S (2023) Does technological progress make OECD countries greener? New evidence from panel CS-ARDL. Manag Environ Qual Int J 34(6):1555–1579
Nasreen S, Khan FI, Nghiem X-H (2023) The effects of financial development and technological progress on environmental sustainability: novel evidence from Asian countries. Environ Sci Pollut Res 30:53712–53724
Ownership structures. Bus Strategy Environ 29:1127–1144
Pham TV, Nguyen TTH, Duong HL (2021) Energy retrofits for buildings in Vietnam. Renew Sustain Energy Rev 144:111068
Ritchie H, Roser M (2013) Emissions by sector. Our World in Data. https://ourworldindata.org/emissions-by-sector
The Economist (2021) Supply-chain scrutiny is driving sustainability in business. https://www.economist.com/
Tran BX (2020) Barriers to green procurement in emerging economies. Bus Strateg Environ 29(3):1370–1382
Tran BX, Bui TD, Nguyen TT (2020) Strategic integration of corporate sustainability in emerging markets. Bus Strateg Environ 29(3):1338–1351. https://doi.org/10.1002/bse.2464
Vinamilk (2018) Sustainability report. Available at https://www.vinamilk.com.vn/
Vinamilk (2019) Sustainability report. Available at https://www.vinamilk.com.vn/
Vinamilk (2020). Sustainability report. Available at https://www.vinamilk.com.vn/
Vinamilk (2021) Sustainability report. Available at https://www.vinamilk.com.vn/
Vinamilk (2022) Sustainability report. Available at https://www.vinamilk.com.vn/
Vo TT, Nguyen HT (2021) Stakeholder engagement for sustainability in emerging economies. Corp Soc Responsib Environ Manag 28(3):1164–1175. https://doi.org/10.1002/csr.2092
Wang Z, Palacios M, Hambrick G, Rajkumar R, Yu H (2022) Understanding data center energy consumption characteristics. ACM Trans Cyber-Phys Syst 6(2):1–24. https://doi.org/10.1145/3504812

Wang J, Nghiem X-H, Jabeen F, Luqman A, Song M (2023) Integrated development of digital and energy industries: paving the way for carbon emission reduction. Technol Forecast Soc Chang 187:122236
World Bank (2021) Climate risk country profile: Vietnam. The world bank group. Washington DC, USA. Available at https://www.worldbank.org/
Wolkomir R (2022) Little-noticed emissions from business travel could offset climate progress. Eos 103. https://doi.org/10.1029/2022EO220021

Correction to: Towards Sustainability: Exploring Sustainable Innovation in Chinese Agricultural Food Packaging

Zhibin Tao

Correction to:
Chapter 13 in:
P. Singh et al. (eds.), *Integrating Environmental Sustainability and Corporate Social Responsibility: A Move Beyond Profit*, World Sustainability Series, https://doi.org/10.1007/978-3-031-81469-3_13

In the original version of the book, Fig. 1 has been updated. The correction chapter and the book have been updated with the change.

The updated version of this chapter can be found at
https://doi.org/10.1007/978-3-031-81469-3_13

P. Singh et al. (eds.), *Integrating Environmental Sustainability and Corporate Social Responsibility: A Move Beyond Profit*, World Sustainability Series,
https://doi.org/10.1007/978-3-031-81469-3_15

The manufacturer's authorised representative in the EU is Springer Nature Customer Service Centre GmbH, Europaplatz 3, 69115 Heidelberg, Germany. If you have any concerns regarding our products, please contact ProductSafety@springernature.com

Printed and bound by CPI Group (UK) Ltd, Croydon, CR0 4YY
30/09/2026
02225602-0011